World Economy in Transition

Other titles of interest

AMARA, R. and LIPINSKI, A. J.
Business Planning for an Uncertain Future

GIARINI, O.
Cycles, Value and Employment: Responses to the Economic Crisis

GIARINI, O.
Dialogue on Wealth and Welfare

GIARINI, O. and LOUBERGE, H.
The Diminishing Returns of Technology

LASZLO, E.
The Inner Limits of Mankind

PECCEI, A.
The Human Quality

A related Journal

ECONOMIC BULLETIN FOR EUROPE*
The Journal of the United Nations Economic Commission for Europe
Edited by United Nations Economic Commission for Europe, Palais des Nations, Geneva 10, Switzerland

The Bulletin is one of the foremost European economic publications. It publishes studies on contemporary economic problems affecting Europe and North America in the context of the world economy. It reflects the full range of topics addressed by the UNECE in its economic research programmes and is particularly noted for its studies of East-West economic relations.

*Free specimen copy available on request.

World Economy in Transition

Essays presented to Surendra Patel on his sixtieth birthday

Edited by Krishna Ahooja-Patel

Anne Gordon Drabek

and Marc Nerfin

PERGAMON PRESS

OXFORD · NEW YORK · TORONTO · SYDNEY
FRANKFURT · TOKYO · SAO PAULO · BEIJING

U.K.	Pergamon Press, Headington Hill Hall, Oxford OX3 0BW, England
U.S.A.	Pergamon Press Inc., Maxwell House, Fairview Park, Elmsford, New York 10523, U.S.A.
CANADA	Pergamon Press Canada, Suite 104, 150 Consumers Road, Willowdale, Ontario M2J 1P9, Canada
AUSTRALIA	Pergamon Press (Aust.) Pty. Ltd., P.O. Box 544, Potts Point, N.S.W. 2011, Australia
FEDERAL REPUBLIC OF GERMANY	Pergamon Press GmbH, Hammerweg 6, D-6242 Kronberg, Federal Republic of Germany
JAPAN	Pergamon Press, 8th Floor, Matsuoka Central Building, 1-7-1 Nishishinjuku, Shinjuku-ku, Tokyo 160, Japan
BRAZIL	Pergamon Editora Ltda., Rua Eça de Queiros, 346, CEP 04011, São Paulo, Brazil
PEOPLE'S REPUBLIC OF CHINA	Pergamon Press, Qianmen Hotel, Beijing, People's Republic of China

First edition 1986

Library of Congress Cataloging in Publication Data
Main entry under title:
World economy in transition.
1. Economic development—Addresses, essays, lectures. 2. Technology transfer—Addresses, essays, lectures. 3. Patel, Surendra, J.—Addresses, essays, lectures. I. Ahooja-Patel, Krishna. II. Drabek, Anne Gordon. III. Nerfin, Marc. IV. Patel, Surendra J. HD75.W67 1986 338.9 85-21721

British Library Cataloguing in Publication Data
World economy in transition
1. Developing countries—Economic conditions
I. Ahooja-Patel, Krishna II. Drabek, Anne Gordon
III. Nerfin, Marc IV. Patel, Surendra
330.9172′4 HC59.7

ISBN 0-08-031285-3

Printed in Great Britain by A. Wheaton & Co Ltd, Exeter

Contents

A Personal Prologue ix
KRISHNA AHOOJA-PATEL

Surendra Patel: The Man and the Economist xi
ARTHUR F. EWING

1. Wherein Lies the Crisis?

The Dynamic Role of the Periphery 3
RAUL PREBISCH

Development Economics: What Have We Learnt? 11
JAGDISH N. BHAGWATI

The Crisis: Some Reflections 29
GAMANI COREA

The Winners and the Losers: The Rise of the South-East, the Decline of the North-West 37
JOHAN GALTUNG

Neither Prince nor Merchant: Citizen—An Introduction to the Third System 47
MARC NERFIN

2. Some Dimensions of the Development Debate

Economic Distance and the Living Standard 63
AMARTYA K. SEN

Energy and International Security in the 1980s: Realities or Misperceptions? 75
MIGUEL S. WIONCZEK

The North-South Negotiating Process 85
MUCHKUND DUBEY

A View from the South: Growth of World Trade in Manufactures 105
VIJAY L. KELKAR

India: Development of the Idea of Development 123
ALICE THORNER

India: What's Missing? 133
NIKHIL CHAKRAVARTHY

Transnational Conglomerates: Reflections on Global Power 139
FREDERICK F. CLAIRMONTE

3. Perceptions of the South

Is an Endogenous Development Strategy Possible in Africa? 159
SAMIR AMIN

Africa in the 1980s: Some Key Political Economic Issues, Problems and Prospects 173
REGINALD HERBOLD GREEN

The Cultural Dimension in Asian Development 195
GODFREY GUNATILLEKE

Reflections on Development Perspectives since the 1950s 213
NURUL ISLAM

Some Considerations on the Transfer of Technology: A Latin American Perspective 219
ENRIQUE IGLESIAS

4. Technological Transformation of the World

Transfer of Technology: Theory and Policy 229
TOM GANIATSOS

Science, Society and Endogenous Development 253
GEORGE SKOROV

ANNEX I 267
Bibliography of Surendra Patel's Personal and Principal Official Publications, 1945 to 1983

ANNEX II 271
Governmental Meetings in UNCTAD on Technology Issues, 1971 to 1984.

Annex III 275
Resolutions, Decisions and Agreed Conclusions on UNCTAD's Work in Technology, 1968 to 1984

Annex IV 281
A Bibliography of UNCTAD's Major Studies and Reports on the Subject of Technology

List of Contributors 287

Index 289

A Personal Prologue

A volume on the complex web of the international economy is more than the sum total of articles arranged together under a neat cover. It is, in fact, a reflection of personal and professional initiatives and efforts of a select group of people. Out of words and phrases, each has carved a concrete structure of thoughts encompassing different historical and geographical perceptions. Ideas, unlike people and goods, cross national frontiers without a visa. At what point they are transformed into "issues" remains hidden from the human eye until they are brought out in reports, studies, documents and resolutions onto negotiating tables.

With a series of technological revolutions, we are rapidly moving towards the 21st century without really touching the ordinary lives of the majority of the world's population. Several major questions remain unsolved which will be shifted onto the shoulders of the next generation: When will the existing rules of the game between the North and the South be changed in favor of those who make up the outsize basket labeled the "Third World"? How can the number of advocates who take up the cause of this part of humanity be multiplied? What is the special mechanism by which the bargaining power of this large majority could be enhanced and strengthened?

This volume is about conflict and consensus, confrontation and cooperation in handling or managing the changing economic and political relationships at the world level. It is devoted to those who will pick up the fine threads of the international argument on how to survive and share equally the resources, assets, income and the common cultural heritage of humankind. This tremendous task falls heavily on the new generation who will bring urgency to the agenda items of the 21st century to accord a legitimate place to the majority in the international scene. Technically, practically and actually, it is now conceivable and possible to have the voice of this majority heard through sound-proof window panes. The authors who are Surendra's personal and professional friends, and who have worked with him for a generation within and outside the UN system wished to make a collective contribution—an expression of their participation in his efforts towards bringing technology from complete anarchy to the rules of the market-place. The Code of Conduct for the transfer of technology incorporating new legal and economic norms now being negotiated in UNCTAD, when adopted could represent a breakthrough for the technological transformation of the world.

A Personal Prologue

The list of those favorite people who made this volume possible is long. First and foremost is Arthur F. Ewing who supported various efforts and gave them a form and a symmetry; he gave full encouragement to the initiative, lending his experience and advice on all aspects of the project. From the beginning and throughout the processes of collecting articles and binding the book the contribution of Marc Nerfin cannot be described in words alone. He handled communication and coordination among authors who are spread across several continents. Anne Gordon Drabek showed such enthusiasm through the editorial working sessions that it was difficult not to become infected by her warmth and generosity. She provided editorial insight and was instrumental in arranging publication of this volume. Above all, behind the scenes, the team which held this project together were all of Surendra's colleagues in the UNCTAD Division of Technology* who saw in its realization a culminating point of their interest in the subject (see the article by Tom Ganiatsos in this volume).

This work is dedicated to the children of the Third World whose dreams of an equitable world must now become a reality. Our three sons, Mahesh, Shail and Samir, join in the struggle to move on . . .

Krishna Ahooja-Patel†
Geneva

* These members of the Technology Division included: Mr. R. Andreasson, Mr. J. Barut, Ms. D. Blow, Ms. V. Clément, Mr. S. Dalawelle, Mr. R. Dhanjee, Ms. E. Erb-Paniagua, Mr. T. Ganiatsos, Mr. O. Hassan, Mr. E. Herbolzheimer, Ms. G-V. Koch, Mr. A. Kouznetsov, Ms. K. Makhetha, Mr. M. Nagao, Ms. S. Navarro, Ms. A. Neale, Mr. J. D'Oliveira e Sousa, Mr. A. Omer, Mr. H. Ouane, Mr. P. Roffe, Mr. Y. Soubra, Ms. I. Tan-Be, Ms. A. von Wartensleben, Ms. H. Wicks, and Mr. A. Yusuf.

† Editor in her personal capacity.

INTRODUCTION

Surendra Patel: The Man and the Economist

ARTHUR F. EWING

My first professional contact with Surendra Patel was in Geneva in 1956, soon after we had come to the UN Economic Commission for Europe (ECE). About this time, perhaps a few months earlier, we must have met at children's parties since we lived in the same street. The annual session of the ECE, earlier in the year, had requested a study on the economic implications of automation. The responsibility, so far as the Secretariat was concerned, fell to my division; but we were supposed to be "technicians", and we needed proper economic guidance from the elite other world of the Research and Planning Division. I had, of course, already taken part in such joint exercises which were fruitful, if uncomfortable, for those of us who were lesser mortals. But Surendra was something very different! The study was prepared by three rapporteurs, one a Scotsman and the others from the USSR, with our assistance; the work was most agreeable. The conclusions are of little interest in the light of another quarter of a century of vast technological change, but they made good sense at the time.[1] I have no doubt Surendra wrote much of the text, since some of his work as the creator and leader of UNCTAD's Technology Division is foreshadowed.

Surendra Patel was born in India in 1923; he was the product of a rural and provincial world deep in a colonized country. The early formative influences were family agriculture and teaching—and the passionate belief of educated Indians in the struggle for independence. He was educated in India until he was ready for the university. He holds his B.Com. from the H.L. College of Commerce (Ahmedabad), and an M.B.A. degree and doctoral degree from the Wharton School of Business of the University of Pennsylvania in the United States.

His early intellectual development was influenced by Marx and Keynes, and one of his first published works was a solid contribution on the relationship of their respective thinking.[2] But just as important—and more directly so—was the influence of Professor Simon Kuznets and Professor Daniel Thorner while he was at the University of Pennsylvania.

He joined the United Nations Secretariat in 1950 when the headquarters were at Lake Success and was a member of the division headed by Professor M. Kalecki, who was to be another significant influence on his thinking. Also influential were two other colleagues at the UN Headquarters—Hans Singer and Sidney Dell. Listing those who have contributed to an economist's thinking can be invidious, possibly arid; having started, however, it might also be mentioned that others who contributed during Surendra's early years were Raul Prebisch, Gunnar Myrdal, K. N. Raj, Amartya Sen, Ashok Rudra, Osvaldo Sunkel and Dudley Seers.

Surendra soon developed his own characteristic style and approach. First, there was the profound hatred of oppression and exploitation; and the equally deep sense of history and especially technical change therein. A lifelong socialist, he has been and continues to be one who believes in freedom, of people and of thought; there is little sympathy with the doctrinaire left. A real believer in "the future" as something man himself makes, he has made this explicit by exceptional statistical ability, together with, what is rarer, statistical imagination. At a later period, when he was in Africa, there was a request from a government—to the UN Economic Commission for Africa (ECA)—for assistance in drawing up a development plan. One of his colleagues was to be sent there and sought guidance, expressing despair at the almost total absence of statistics. Surendra asked him what was the principal export and on being told "bananas," led him gently yet relentlessly on to estimates of foreign trade, GDP, its structure and hypotheses about how this might be changed over a period of time. Yet Surendra has never been a futurist, or any sort of Utopian. He is fully conversant with contemporary economic theory, yet has a limited interest in theory for its own sake; for him, this is useful simply as a tool for empirical, policy-oriented research. He has long been sardonic about the "gloomy science" and is fond of citing the growing list of doom-laden prophesies, even from the greatest of economists. Sharing *filets de perche* with him on the lakeside of Geneva just after the publication of Myrdal's *Asian Drama,* he was ironical about my enthusiasm and pointed to the subtitle, with its reference to the poverty of nations, contrasting it with Adam Smith's challenging belief in the wealth of nations.

Surendra Patel's early work on his own account was mainly about India. His first book is a thorough study of rural poverty in the Indian subcontinent, and his grasp of both historical and statistical analysis is immediately apparent.[3] He shows that in the closing decades of the 19th century landless laborers accounted for no more than 13% of the total agricultural population and that the proportion had tripled 50 years later. It was in this period that the great polarization of Indian society arose, with at the apex 4 million absentee landlords living on rents and at the base 42 million landless laborers, 27 million tenants-at-will and sharecroppers and 10 million more tenants cultivating about five acres but with no secure rights; more than 70% of the agricultural population had no rights in land. Hence the poverty, the low productivity of land and the shortage of food. Furthermore, he showed that the increase in landlessness was unevenly spread over the subcontinent, with three broad regions, in each of which a distinct system of land settlement had been introduced under British rule. Over half the essays reprinted in a collection of his papers published in the mid-1960s are about Indian problems, with a recurring theme of contrasts between India and China—and one to which he returned in 1980.[4]

Surendra has been back to India almost every year—invariably finding time to spend with his parents at the family farm, which has become a technologically advanced "capitalist enterprise"; but his professional life has been with the United Nations, other than two brief spells in his home country. The first was enforced. In 1951, while there on leave, he was refused a visa to return to his UN post in New York. This was the McCarthy era. The matter was raised in the Indian Parliament by Krishna Menon, at the time leader of the Indian Delegation to the UN and a close collaborator of Prime Minister Nehru. It was six months before Surendra got back to his job. Not surprisingly, he became a member of the Staff Council soon after. The United States Government, no doubt unintentionally, had started a new fashion. Most governments now consider nationals from their countries as "theirs," and the posts occupied by them as "theirs," in direct conflict with Articles 100 and 101 of the UN Charter. Much later when Surendra became the director of the Technology Division and faced a refusal by the Administration to appoint a competent person he wanted, invariably on "geographical distribution" grounds, he left the post open until he got something like his way. There was a second period in India in 1961, on secondment to the Indian Planning Commission. One's impression is that he considered this a positive experience. The ideas he would have liked to put into practice were set out in a book published a few years later.[5]

It is time to resume the chronology of this story. With Surendra Patel's transfer to the Economic Commission for Europe in the mid-1950s, there began to unfold his most significant contribution to economics. This was set out in concise form in an article published in the *Economic Journal* in 1964 with a title which is explicit: "Economic Distance between Nations: Its Origin, Measurement and Outlook"; but his ideas had been developed, discussed and employed in both UN and personal publications over a period of several years.[6] The essence of the thesis, derived from delving into the facts about past patterns of growth, is the short span of time during which the now industrialized countries had transformed their economies, how it had been done and how the Third World could do likewise, without impossibly high rates of growth and within half a century, given the right policies. Surendra's approach has remained consistent. He does not project or predict, but argues that a particular outcome is perfectly feasible within a relatively short space of time on the basis of past experience and with the production factors and technological possibilities available, provided appropriate policies are pursued and suitable institutions established.

The first time the thesis was applied, and in some ways most thoroughly and effectively, was in a study published by the ECE entitled "Development Problems in Southern Europe and Ireland."[7] When Surendra arrived on the scene Gunnar Myrdal was Executive Secretary and his favorite child—the Research Division—was a remarkable team, the members of which were given the freedom to carry out their research. Surendra, though a young newcomer, was no exception; his exceptional talents were quickly recognized. The study was from "the Secretariat," but throughout, in the organization of the data and its presentation, it bears the mark of his pen. From then on, the analysis by ECE Secretariat of the European economy has regularly comprised not only the Western and Eastern countries, but also the "developing" countries in the South; and this had made possible some of its most original research.[8]

At the end of 1961, I moved to the UN Economic Commission for Africa (ECA), which was still in its infancy, with a relatively free hand to build up a division (concerned with industry, transport and natural resources); it was logical to try and raid the ECE as much as possible, through loan or transfer. Surendra came first on loan. Godfrey Lardner, at the time Permanent Secretary of the Nigerian Ministry of Economic Affairs and Nigerian delegate to the Commission (he was later to become a member of my team and subsequently a Division Director), had thrown down a challenge to the almost non-existent industry group to "produce something." Surendra had a very large hand in our reply, in a study which set out a bold plan for the industrialization of Africa in half a century.[9] The growth rates assumed were little higher than those adopted in the first United Nations Development Decade. A conscious policy of industrial development with emphasis on intermediate and capital goods was sketched out; and the need and scope for subregional cooperation was also demonstrated. Much of what was advocated has happened in developing countries in other parts of the world, if not yet, alas, very much in Africa. But sooner or later the programs and policies that were the ECA's in its early days will have to be implemented.

Later Surendra came back to Addis Ababa, on transfer, as part of the ECA Economic Survey Division under Dudley Seers, which had been started by Stein Rossen and Jacques Royer, both of whom moved later to Geneva; there was already the nucleus of a lively team which included Bella Abbai and Erling Nypan. My own somewhat amorphous group expanded and included at different times Duccio Turin, Marc Nerfin, Pierre Sevette (on regular loan), Krishna Ahooja, Howard Goldsmith, Louis Gelineau and Gerrald Dekker. The lines between divisions were wavy, and we worked together on projects. We became a sort of gang. But the arteries of the Commission Secretariat hardened too soon. It was inevitable and right that Africans should fill a very high proportion of the posts as quickly as possible—this was indeed a principal preoccupation of all of us. But it was also urgent for the sake of African development to build on the substance of what had been done, and this required freedom for bright people, who also had knowledge and experience, to work out their ideas. Surendra was not the only one to become somewhat frustrated as the Commission "settled down" under its second Executive Secretary. Moreover, at that stage of his career, he had a limited capacity to suffer fools gladly—this was also true of some of the others. The gang was dispersed, and he was transferred to the Economic Commission for Asia and the Far East (ECAFE, now ESCAP). This brought to an end our second period of direct association.

Surendra Patel's period in Bangkok seems to have been something of an interlude. There was a close association and friendship with the Director of the Research Division there, Dr. Hirushi Kitamura, and also time to write. The "distance" principle was applied to education—foreshadowing the later work on technological development.[10] It was also applied to Asia.[11] When I moved to Cambodia in 1968, he sent me two papers written while with ECAFE: one on the principal problems of the Cambodian economy; and the other the first piece of economic analysis I saw of the Lower Mekong Project (a UN-sponsored effort to harness for power, irrigation and flood control the river and its tributaries in the four riparian countries—Thailand, Cambodia, Laos and the Republic of Vietnam—which somehow managed to keep going during the war, but is now virtually dead). Both these papers were of immense help, whether in attempts to assist the development of Cambodia, or further the Mekong Project objectives, since economic insights—as distinct from high-class technical expertise—were scarce. Surendra has always been fascinated by the possibilities of multipurpose river development. While with the ECA, he was requested by the Government of Zaire to prepare an outline of a development plan; and he sketched out dramatically but also realistically the possibilities of the fabulous Inga project near the mouth of the Congo River.[12]

Surendra Patel was transferred to UNCTAD in 1967, at the request of Dr. Prebisch. At first there was a period of marking time in the quieter corners of UNCTAD.[13] Contributions to learned journals continued, including some good sense on commodities.[14] There also appeared at this time the culmination of the distance thesis, in a contribution to Maurice Dobb's festschrift.[15] But the UN career seemed to be checked. Promotion had always been slow, a consequence of the McCarthy days, and perhaps also the difficulties non-conformists generally experience in climbing the greasy pole in the UN Secretariat. I recall a conversation with him in Geneva in the late 1960s. He was perfectly happy. He was writing articles and had planned a major book. But he did not think he would get much further in the UN.

At the beginning of the 1970s, there was a sea change. In UNCTAD Dr. Wladek Malinowski, who was the wisest of the Secretary-General's advisers—and before that of the Under Secretary-General of the UN in charge of economic and social affairs on regional commission matters (and for that matter of the Executive Secretaries of these bodies)—was responsible for what were called "invisibles", which included technology, until then almost completely invisible both in economic theory and development strategy. He urged the setting up of a Technology Division—at first concerned mainly with transfer of technology—under the leadership of Surendra Patel. Within a matter of months, Surendra became not only the leader of a growing research team, one he was was able to shape himself, in a new and vitally important field; but he also became in time a major international negotiator on practical problems of the transfer and development of technology, above all the down-to-earth details of devising suitable national policies and institutions within the developing countries, and obtaining assistance and training for this purpose. Some years earlier, in a foreword I contributed to a collection of his essays, I suggested tentatively that he had "perhaps not yet sufficiently examined the immense practical problems that have to be solved if the perfectly realistic perspectives he has opened up are to be realized—the formulation of coherent plans, their execution, the coordination of plans, particularly among small countries, patterns of foreign aid, technical assistance, management, and above all the massive training effort required at all levels".[16] Any doubts on this score were to be totally and triumphantly dissipated.

There is no need to spend much time in this note on the next decade and more, the climax of Surendra Patel's career with the United Nations. Tom Ganiatsos's paper in this volume puts it into perspective.[17] UNCTAD's new Technology Division started up on a minute scale, and most of the pioneers are still there. In 1975, after I had retired from the UN Secretariat, he asked me to help on a part-time basis with the technical assistance and training side of his work. By then the whole operation was in full swing; the Division was steadily expanding and—contrary to Parkinson's law—the volume of work still faster. I was there for the next 5 years and feel justified in attempting to assess the achievements of Surendra and his team during more than a decade. First—the starting point—was a detailed study of the industrial property system, as a basis for a revision of the Paris Convention (negotiations to this end were, of course, carried out under the aegis of the World Intellectual Property Organization but the "needling" and the preparation of the developing countries' position was the result of UNCTAD's studies).[18]

The second is the introduction at an early stage of a draft of an international code of conduct on the transfer of technology. Negotiations on this code have been going on for nearly a decade, but there is now every prospect of early agreement—to coincide approximately with Surendra Patel's retirement from the United Nations. The code is perhaps the major achievement of his group. Although it will not be mandatory, it has already changed irrevocably the whole climate of commercial exchange of technology in all its forms; and it has been, through the process of learning via immensely thorough legal and economic studies and patient negotiation, the principal means of strengthening the policies, institutions and legislation of developing countries in the field of technology.[19]

In the context of the code negotiations—and indeed all those on technology carried out under UNCTAD auspices—the contribution of Surendra Patel has been decisive; it has also been of far wider significance. There is room for difference of opinion on the negotiating policy of the Group of 77 as it has evolved, particularly within the UN system. Furthermore, there can also be more than one view of the UNCTAD Secretariat's somewhat ambiguous dual role—on the one hand as the secretariat of a world organization with universal membership and on the other as the principal means of assisting the Group of 77 in the negotiating process. But the importance of collective self-reliance among the developing countries can hardly be doubted. And the developing countries have no comprehensive permanent secretariat of their own, as is the case of Western and Eastern countries in, respectively, the OECD and COMECON. Surendra was one of the first to set out effectively the case for collective self-reliance and the need for a permanent Third World secretariat to implement such a policy.[20]

Third, Surendra Patel and his group, in close association with a number of university economists—notably Amartya Sen, Charles Cooper, Frances Stewart and K. K. Subrahmanian—were the first to analyze the way in which transfer of technology to developing countries actually takes place, and the predominant part played by multinational corporations, in a path-breaking study entitled *Guidelines for the Study of the Transfer of Technology to Developing Countries.*[21] The title is significant. There was no pretense that all the answers were known.

Fourth, the need for developing countries to work out their own policies and set up their own institutions, not only for the transfer—perhaps a better term is acquisition—of technology, but also the adaptation and development thereof, was fully established. This need crystallized around the idea of a national technology center which was the subject of a resolution at UNCTAD IV in 1976. Now such centers have been set up in one form or another in most developing countries.

Fifth, the need to assist the developing countries in implementing the recommendations made and unanimously endorsed through *suitable* kinds of technical assistance and training was also established. The setting up of an Advisory Service on the Transfer and Development of Technology, within UNCTAD's Technology Division, was authorized. Subsequently, there has been a long struggle to obtain the resources to staff this service, but this has not prevented the carrying out of a comprehensive program of advisory missions and training courses.

Sixth, the need to build up certain industrial sectors in developing countries, particularly capital goods, energy and essential consumer goods—food, pharmaceuticals—has been demonstrated and worked out. This is becoming increasingly common ground in development thinking. But it is largely wishful thinking without consciously building in the ability to strengthen domestic technological capacity; and this is the particular contribution of the group.

Surendra Patel's subject is, of course, political economy. His last major paper for UNCTAD, his Division's contribution to UNCTAD VI in 1983, has the challenging title: *A Strategy for the Technological Transformation of Developing Countries.*[22] His most recent personal paper is called *Age of the Third World.*[23] Some may think that his reporting of the demise of the First World, like Mark Twain's, is exaggerated; but in these two papers there is, undiminished, the profound, yet realistic conviction as to what developing countries have done and can do through their own efforts. Paradoxical though it may seem, in view of his socialist convictions and reservations about free international trade, Surendra seems to me to be more akin to Adam Smith, in outlook and style, than to any other classical economist. And the style has always been vivid, picturesque, and sharply ironical, whether in speaking or writing, yet without ever losing sight of the most austere classical principles of the unfolding of a logical argument. Under his leadership, the Division of Technology in UNCTAD became more of an academic institution than a bureaucratic department within the United Nations.

Surendra Patel is determined, sometimes single-minded, in pursuit of his political objectives. But he is also the simplest and gentlest of men, with a capacity for unswerving friendship. Despite a vast capacity for work, he has always been able, in his limited spare time, to relax completely, to enjoy good food and drink, to work in the garden tending his roses, and to talk and laugh about almost everything. His interest in history and imaginative writing is apparent from his own works. He is genuinely and completely at home with the young—from children of all ages to his own junior colleagues. But until the last decade—as pointed out already—he had little capacity for tolerance, especially of the pretentious and the pompous. He has now demonstrated a real talent to get on with all ranks and races, like the legendary English dukes and dustmen, because his interest in other peoples' ideas and problems is not a bit of political apparatus but perfectly genuine.

No one is indispensable, but quite evidently, finding a successor as leader of UNCTAD's Technology Division is going to be a hard task. The gain will be the release of Surendra's energy elsewhere. He has still much to teach and write. And it is to be hoped he will spare himself a little, for the first time. All his friends have admired profoundly the stoical courage with which he has lived with some muscular disability in the last few years. If they can have but one wish it is that he will allow himself to relax enough for that to be overcome.

Notes

1. *The Economic Implications of Automation*, United Nations Publication, Sales No. 59.II.E/Mim. 8.
2. "Marxism and recent economic thought", *Science and Society*, Vol. **XI,** No. 1 (New York: 1947).
3. *Agricultural Labourers in Modern India and Pakistan* (Bombay: Current Book House, 1952).
4. *Essays on Economic Transition* (Bombay: Asia Publishing House, 1965), Parts One, Two and Three; and "Planned development in India," *Mainstream* Part I (Republic Day Number, 1980) and Part II (16 February 1980).
5. *The India We Want. Its Economic Transition* (Bombay: Manaktalasi, 1966).
6. The main articles are reprinted in *Essays on Economic Transition, op. cit.*, the other two being

entitled "Rates of industrial growth in the last century, 1860–1958" and "Main features of economic growth over the last century".

7. *Economic Survey of Europe in 1959*, United Nations Publication, Sales No. 60.II.E.I, Part IV.
8. Recent examples are two studies in the ECE, *Economic Bulletin for Europe:* one entitled "Changing comparative advantages in manufactured exports from southern Europe, 1965–1978" (Vol. **33**) and the other, "The relative performance of south European exports of manufactures to OECD countries in the 1970s: An analysis of demand factors and competitiveness" (Vol. **34**). Original studies from this source are now less frequent; there is much more of the kind of surveying that Surendra Patel describes as "This went up and that went down".
9. *Industrial Growth in Africa,* United Nations Publication, Sales No. 62.II.K.3. Surendra Patel published an outline of an overall development strategy for Africa entitled "Economic transition in Africa", in the *Journal of Modern Africa Studies,* Vol. **II**, No. 3 (1964).
10. "Educational distance between nations", *The Indian Economic Journal,* Vol. **XIII**, No. 1 (July–September, 1965).
11. "Asia in the changing world economy", *The Journal of Development Studies,* Vol. **3**, No. 2 (January 1967).
12. *Données Générales pour le Premier Plan Quinquennal de la République du Congo, 1965–1969* (Léopoldville: Ministere du Plan et du Développement Industriel, February 1964).
13. For example, working on problems of trade of developing countries with socialist countries.
14. See in particular "Depressed exporters: A special case", *Foreign Trade Review* (April–June 1969).
15. "World economy in transition, 1850–2060", in C. H. Feinstein (Ed.), *Socialism, Capitalism and Economic Growth* (Cambridge University Press, 1967).
16. *Essays on Economic Transition, op. cit.*, p. xi.
17. I have also written about it elsewhere; see in particular *Journal of World Trade Law,* Vol. **14**, No. 2 (1980); and Vol. **16**, No. 1 (1982).
18. There are several major UNCTAD reports. Here it must suffice to refer to two of S. J. Patel's signed articles: "The patent system and the Third World", and "Trademarks and the Third World", respectively, in *World Development,* Vol. **2**, No. 9 (September 1974), and Vol. **7**, No. 7 (July 1979).
19. For an assessment of the code as an instrument helpful to all countries see Dennis Thompson, "The UNCTAD Code on Transfer of Technology", *Journal of World Trade Law,* Vol. **16**, No. 4 (July–August 1982).
20. "Collective self-reliance of developing countries", *The Journal of Modern African Studies,* Vol. **13**, No. 4 (1975), pp. 569 ff.
21. UN Sales Publication No. E.72.II.D.19.
22. TD/277.
23. First published in a French version as "L'Age du tiers monde est proche" in *Le Temps Stratégique,* No. 2 (Geneva: Autumn 1982).

1

WHEREIN LIES THE CRISIS?

The Dynamic Role of the Periphery

RAUL PREBISCH

When I visited UNCTAD in Geneva recently, Surendra Patel, my old friend and respected former colleague, explained to me his views regarding the historical expansion of capitalism; since then, I also read some of his notes that added precision and perspective to his previous remarks.

Capitalism had its beginning in the United Kingdom; later, new centers emerged, particularly during a wave of depression of the former. This process was not so much a spontaneous expansion, but rather a deliberate effort of the new centers to industrialize through protection, taking advantage of successive technological innovations. And now, Patel says, the trend has reached some of the more advanced countries of the periphery.

This thesis is fascinating, and I fully agree with Patel's view of the role of the periphery in this context. I would like to refer to the Latin American countries, where the process of industrialization begun during the First World War gathered strength during the Great Depression. This was not the result of a doctrinary preference but rather an imposition of facts. In my country, Argentina, we had to embark resolutely on import substitution, supported by protection, to obtain the goods that we could no longer import due to the drastic fall in exports and the sharp deterioration of the terms of trade (this does not mean, of course, that protection was not excessive and abusive, as indeed was generally the case).

Now a new stage can be visualized. Those countries that have advanced further in their process of industrialization—in Latin America as well as in other regions of the periphery—may be in a position to play a dynamic role of paramount importance in the development of the centers. Until a few years ago, they had been but passive elements within the global process of development; but now, for the first time in the history of capitalism, the periphery is entering a new dynamic stage. It was not so before, when those countries exported only primary goods, nor when their development was buttressed by import substitution. But at the present stage, they have learned to export manufactures and—as it is well known—this triggered an adverse reaction in the centers resulting in considerable obstacles to the import of manufactures from the periphery, not only during the present recession but also during the long years of prosperity.

The Secretariat of UNCTAD has clearly explained on several occasions that while—as a consequence of the Kennedy and Tokyo Rounds—there was an outstanding liberalization of trade between developed countries, the periphery as a whole did not benefit from this except in a very marginal way. The reason is that the nature of the interchange between the centers has been increasingly prompted by their technological innovations. There was a growing demand, but nevertheless the periphery was left behind, mainly—as I will explain later—because its exports generally corresponded to types of manufactures where the growth of demand was relatively slow.

It cannot be doubted that the periphery has an enormous potential demand for those manufactures produced by the centers in response to their incessant technological innovations. Nevertheless, to make this potential effective, the peripheral countries need to have the means to pay for such purchases through their own exports of manufactures.

This interchange of technologically advanced goods coming from the centers for the less technologically advanced manufactures from the periphery is a clear example of comparative advantage in international trade. These advantages could improve the rates of productivity on both sides; unfortunately, the facts have proven that the problem of the centers is not so easy to solve, especially now that the adverse effects of the recession are being so strongly felt.

When—years ago—I was at UNCTAD, I put this question to myself: Why, if development in the centers has been characterized by the persistent internal transfer of manpower from those activities where demand tends to grow at a relatively slow pace to undustries or other activities where demand tends to grow at a relatively high rate, are the centers so reluctant to accept the growing competition of manufactured goods from the periphery and the corresponding transfer of manpower?

I think that the clue to this puzzle lies in the different nature of these two phenomena. Let me explain: the continuous increase in demand for technologically advanced goods and services arises from the increase in productivity and income and the larger profits afforded by that increase; in other words, changes in demand lead to a larger capacity to accumulate through profits. Competition from the periphery—achieved through the lowering of prices—does not increase profits but reduces them and, consequently, it does not improve the potential for capital accumulation either, to the detriment of employment.

Let us be frank on this score: this is a problem that in my view cannot be corrected by the free play of the market forces or by unilateral trade restrictions, be those "voluntary" export limitations or any of the many other forms that the UNCTAD Secretariat has so thoroughly elucidated. New formulas must be found that allow for the progressive participation of manufactures from the periphery in the growth of consumption in the centers. This would minimize the disruptive effects of competition through lower prices, which are harmful both to the centers and to the periphery, since the latter transfers part of its productivity through a lowering of prices.

But that is not all there is to the solution of the problem. Far from it. The problem has a dynamic solution, and it is here that we can find the new role to be played by the periphery, through its great and growing demand for technologically advanced manufactures from the centers. The more this potential demand is translated into an effective one, the better the possibilities for a gradual transfer of manpower from industries affected by the above-mentioned competition to higher categories of industry.

It is in this field that the World Bank and the regional banks could make concrete contributions. At the UNCTAD VI session in Belgrade, Mr. Clausen, President of the World Bank, stressed his determination to augment substantially the operations of the Bank in developing countries, and a few days later, in a statement made to the *Washington Post,* he recognized the paramount importance that these lending operations have for increasing the exports of the developing countries, to the reciprocal advantage of those countries and of the center nations. However, there are clear limitations to these expansionary operations referred to by Mr. Clausen, resulting from the restrictive monetary policies prevailing in the centers and the extraordinarily high rates of interest.

In my address at UNCTAD VI, I mentioned another possibility for strengthening the development operations of the World Bank and the regional banks. What I have in mind is the transfer of the proceeds of Special Drawing Rights issues—in so far as these are needed for better world liquidity—to the aforementioned financial institutions. As the cost of the Special Drawing Rights could be nil, this transfer would result in a lower average rate of interest for their lending operations.

Obviously this runs counter to the fact that the creation of world liquidity has been left entirely in the hands of the United States, which for a long period has been taking advantage of the extraordinary privilege of being able to use the dollars it issues to pay for a sizeable portion of the imports coming from the rest of the world. Let me say in passing that I do not conceive of any international monetary reform that does not place in the hands of an international authority the creation of new world money.

Notwithstanding the great importance of trade between the centers and the periphery, I do not mean to say that this could be the main element for tackling the fall of productivity prevailing in the centers for some years. On the contrary, I believe that this fall in productivity is the result of a fundamental structural imbalance.

I do not wish to be accused of putting forward a simplistic interpretation, but I must say that I am more and more inclined to detect a tendency in the centers for the rate of increase in expenditures—both private and public—to surpass the rate of increase of reproductive capital accumulation: that is, capital needed for the multiplication of employment and productivity. This imbalance can only be corrected through a proper readjustment of the rate of expenditure to make possible an increase in the rate of reproductive capital accumulation. I do not think that this is the right place for the deployment of theoretical arguments, but I do indeed wish to underline the importance of this imbalance and the need to face it squarely in order to give impetus to the dynamics of development. I firmly believe that the fall in productivity is a reflection of a lack of adaptability of theories and forms of action to the great structural changes that have accompanied the development of new technologies and the corresponding changes in power relations.

The substantial deficit experienced in the United States is a clear expression of the disequilibrium to which I am referring. It is due to the considerable growth of social as well as military expenditure, especially in the high-income groups. There has been no rational distribution of the global income resulting from the growth of productivity between the different types of expenditure and the rate of reproductive capital accumulation.

How is this fiscal deficit being handled? The idea prevailing some time ago—and still advocated in some circles—was to reduce income taxes so that the high-income groups could increase their savings and, in this fashion, lever up the growth rate of the economy and thus contribute to the elimination of the deficit. Independently of the social implications of this measure, it is quite inflationary since even in the best of cases, its immediate impact would be to aggravate the deficit while waiting for the positive effects of the operation.

Without abandoning this idea, the United States Administration is now following another path, that is the use of public savings to cover the deficit and reduce inflation. The tool used to achieve this end has been the extraordinary raising of interest rates, something that had never taken place before—at any rate in such a persistent fashion—in the whole history of capitalism. And this has been accompanied by credit restrictions that have been most detrimental to economic activity and employment. It is true that inflation has been reduced to a very low figure, but at a tremendously high economic and social cost.

The beginning of a moderate expansion observed in the United States to counteract the forces of recession is, therefore, quite understandable.

I emphasize all this because I am far from believing that what is being done represents a fundamental solution of the problem; and the finding of a solution is of paramount importance, not only for that country but for the whole world economy—particularly the periphery, where the adverse effects of this policy have been very serious. In this last respect, it should be remembered that in a recent OECD study it is said that the cost of reducing inflation has fallen on the prices of primary products (*The Economist*, 11 June 1983).

In an interview published in *The Economist* in 1983, Martin Feldstein, then chief of President Reagan's economic advisers, gave a very interesting insight into this problem. European leaders have insistently recommended proper coordination of expansionary policies between the main centers. However, Mr. Feldstein warned that stepping-up the expansion in the United States could bring about new inflationary pressures, thus defeating the positive results already obtained. It is very significant that he added that it is necessary to have recourse to fiscal tools to tackle the deficit and reduce interest rates without bringing about inflationary consequences. And what is still more significant is that the President's Chief Advisor mentioned, in a rather subtle way, the need for a far-reaching fiscal reform. In his view, in order to readjust the rate of expenditure so as to improve the rate of capital accumulation, it would be necessary to tax what the corporations distribute, leaving untouched what they save and accumulate. But that is not all. He suggested complementing this measure with a progressive tax on individual expenditure instead of a progressive income tax, also leaving untouched what is saved and invested.

I recall that a few years ago, when William Simon was Secretary of the Treasury under President Nixon, he formed a group of experts to deal with fiscal reform. This group submitted two proposals: one for an improvement of the present system, and the other for its radical change in the form of the progressive taxing of expenditures, as suggested by Mr. Feldstein, exempting from taxation what is devoted to savings (this report has not been printed but is available).

I consider this extremely meaningful as it is a clear proof that even such highly qualified experts as Mr. Feldstein—not to mention others—see ways, other than those put into effect by the US Administration, to solve the imbalance between accumulation and expenditure rates. I would not say that a reform of this type can take care of the whole problem of accumulation and income distribution, but it could be a very important element for an income policy intended to eradicate the structural sources of inflation in the centers as well as in the periphery. A very difficult political problem indeed.

The Swedes are being very active in trying to find solutions to this structural imbalance. The Social Democratic Party, as well as the labor unions, have recognized that due to the social struggle for distributing the fruits of productivity, the rate of capital accumulation has been seriously flagging to the detriment of productivity, employment and the competitiveness of Swedish exports in world markets. They recommend another way of dealing with the problem: that is, to transfer part of the profits of enterprises (above a certain level) to form an accumulation fund to be managed by the workers. It can be easily understood why this proposal has met with considerable resistance. For my part, I do not think that the solution lies either in the free play of the market forces or in an unregulated distributive struggle. It is necessary to introduce some form of rational regulation. The market is very important indeed, both from the economic and the political point of view, but it is not everything. It is necessary to deal with the structural factors lying behind the market. Far from suffering, the functioning of the market would thus be considerably improved.

In this, as well as in other regards, I am an admirer of Japanese development. First of all, their capitalism is austere, just as much in the high-income groups as in the rest of society; and this austerity has allowed Japan—as is well known—to reach a high rate of reproductive capital accumulation. Although this rate has slackened, as it has everywhere else, this has not been due to internal factors, but mainly to the detrimental consequences of the high interest rates prevailing in the United States. Furthermore, the Japanese have been wise enough to avoid falling into the temptation of improving social welfare through increases in wages and salaries that correspondingly augment production costs and are transferred to prices in an inflationary manner; they have tackled the problem by granting bonuses to the workers twice a year, in line with the financial results of the companies. Obviously, this has had no inflationary effects, and in boom years the bonuses have been several times the amount of a monthly salary. I am not advocating a mere imitation of Japan, being quite aware of the cultural background and characteristics that differentiate it from other capitalist countries, including peripheral ones; but it is important to ponder the ways in which they are dealing with problems that in other countries are quite far from being solved.

I have referred above to the need to give impetus to trade between the centers and the periphery in order to help attain equilibrium and so counteract the structural factors responsible for the imbalance. But I do not think this is enough, because of the serious disparities in demand between the centers and the periphery caused by the historical lag in the industrialization of the latter. So, even in the best of cases, it would be necessary to have recourse to import substitution. It should be recalled that during the long years of prosperity of the centers, ending at the end of the first half of the 1970s, there was a strong reaction against import substitution. Indeed, the centers strongly criticized this policy, recommending again an open-market policy; and, what is worse, in some peripheral countries—mainly Latin American—this idea of an open-market policy gathered force and was translated into the dismantling of protection, with very serious consequences. I am afraid that some circles were dazzled by that period of prosperity and believed that the opening of the center would mean an indefinite enlargement of the market for the exports of their manufactured products. The advocates of this mistaken policy did not consider that even during those years of great prosperity, the centers did not liberalize substantially their restrictions to imports of manufactured products from the periphery. Now this situation has been aggravated by the adverse effects of the recession.

At present, there is recognition that import substitution is being dictated by the facts rather than by a mere doctrinaire preference. Also, it is more and more acknowledged that this substitution must be made in markets that are wider than the national ones. Here, too, the belief in the open world market theory for peripheral manufactured products, together with other factors, weakened the tendency of previous years towards the enlargement of national markets through broader trade agreements.

This is the only constructive way to tackle seriously the structural trade imbalance, in conjunction with persistent efforts to increase exports to the centers. This enlargement of national markets will gradually correct the economic fragmentation of the territory that is another adverse result of the asymmetry applied by the centers, which before the First World War and the Great Depression gave the peripheral countries the appendicular role of supplying primary products, keeping them on the sidelines of the industrialization process.

Clearly, this was in the interests of the centers at the time, as well as in the interests of the dominating groups of the periphery, and the old concept of the international division of labor suited both interest groups very well. It is paradoxical that this theory is still recommended in some quarters for the periphery, when the centers ignore it by imposing all kinds of restrictions not only on imports of some primary products from the periphery, but also on imports of the latter's manufactured goods.

Such, in brief, are the comments prompted in me by the original and always incisive thinking of Surendra Patel.

Development Economics: What Have We Learnt?*

JAGDISH N. BHAGWATI

The wealth and poverty of nations and their peoples constitute the domain of development economics. The complexity of the problems it studies is thus immense: and the difficulties facing its practitioners, either the Suslovs or the *aparatchiks*, are heroic.

Few foresaw, at the turn of the century, the dramatic rise of modern Japan or, for that matter, at the end of the Second World War, the later emergence of the Far Eastern economies of Taiwan, South Korea, Singapore and Hong Kong as the more successful success stories in development. Those who enjoy Japanese literature as I do may be interested to know that Natsume Sōseki, the great Japanese novelist, is reported to have wanted to become an architect instead when he entered college in 1884, but was advised that "there was no glory in being an architect in such a poor country as Japan, where there would never arise the opportunity of building a great monument of the order of St. Paul's"![1]

It is perhaps true that the only valid generalization in development economics is that no generalization is possible. And that, as Joan Robinson once said of India, whatever you say about it, the opposite is equally true. Yet, one need not despair. For, as one scrutinizes the evolution of developmental experience, and the concomitant rise and fall of ideas in development economics, one cannot but learn some lessons even if the learning implies unlearning and learning by undoing dominates learning by doing. I must warn you that the following essay will be subjective and reflect my own development as a development economist over the last 25 years. Indeed, in a field so vast and

*This text is based on a lecture delivered as part of the Distinguished Speakers Series Program of the Asian Development Bank, Manila, on 6th October 1983. The central ideas developed here were originally presented in my Keynote lecture at the tenth anniversary celebration of the Center for Development Studies at the University of Antwerp in Belgium. Thanks are due to Irma Adelman, Hollis Chenery, Ian Little, Theodore Schultz and Paul Streeten for helpful comments on the Antwerp version. This paper appears in the *Asian Development Review*, published by the Asian Development Bank (March 1984), with minor editorial changes. I am happy to offer it as a token of my great esteem for Surendra Patel whose scholarship and dedicated efforts at UNCTAD in the cause of development have made him a legend among his countless admirers. I hope that this long retrospective, which matches his historical and intellectual interests as well, serves adequately as my tribute to a gifted scholar and a wonderful friend.

complex, and so central to our concerns, one can only speak to advantage if one speaks from the immediacy of one's own experience.

The remarkable aspect of development economics at its creation in the early postwar years was the spirit of optimism that characterized it. The discipline grew directly out of policy concerns. This is true of Paul Rosenstein-Rodan's famous 1943 *Economic Journal* article and equally so of Ragnar Nurkse's classic 1953 work on *Problems of Capital Formation in Underdeveloped Countries*, for example.[2] I woud say that the optimism, in the sense that development of the underdeveloped countries and the resulting removal of poverty would be a relatively easy task, that we would be able to collapse what had been a long, drawn-out process of improvement in the standards of living of the masses in many of today's developed countries into a fairly quick phenomenon in the lives of the present-day underdeveloped areas, came from two fundamental factors.

First, there was simply the fact that the major figures in this nascent discipline were looking for simple, almost uni-causal, theories of development, thus building implicit or explicit models that carried within them what appeared to be immediate lessons for policymakers set upon the cause of developing their economies. Now, I certainly do not wish to argue that such theorizing is inappropriate. As the philosopher Henri Bergsen once said, the great virtue of time is that it prevents everything from happening at once. And equally, the chief advantage of economic theory is that it prevents us from considering everything at the same time. But the simple economic theories and attendant policy prescriptions did generate the optimism that I have noted.

There were really two classes of such theories at the time, both focusing on capital formation as the key to growth. I shall return later to the question: why growth and not poverty or income distribution? Indeed I shall argue that growth was regarded by many of us then, *not* as an objective in itself, but as a way of making a sustained assault on poverty and that it is no more than ignorance to suggest otherwise when we address these issues today. But let me turn presently to the question of different developmental theories.

(1) Among the two classes of such theories, there was one set that focused on how to get investment going. These theories identified the problem of development as simply one where investment had failed to materialize and the task of the planner or policymaker was to identify the constraint that led to this unfortunate outcome and to break it. Paul Rosenstein-Rodan's thesis was that this constraint came from what we would call today a Prisoners' Dilemma problem and what he characterized as the problem of decentralized entrepreneurs failing to invest since none of them in isolation had the assurance that others would invest simultaneously and create demand for one's output. Thus came Rosenstein-Rodan's celebrated prescription for a coordinated plan of investments by the government, a "balanced growth" scenario: and with it, of course, the intellectual rationale also for planning in the developmental literature.

Curiously, Albert Hirschman's 1958 critique of the Rosenstein-Rodan approach, in *The Strategy of Economic Development*, shared the same diagnosis of the basic developmental problem.[3] It viewed the developmental task as one of creating the inducement to invest. However, it fell into the anti-planning mode of thinking, suggesting that the necessary inducement to invest was created readily by cutting off imports and dangling the resulting, assured markets before domestic entrepreneurs. I have called this, following the agricultural strategy of "slash and burn", the import-substitution developmental strategy of "slash and grow", though I fear the resulting growth is likely to be so inefficient that it may well turn eventually into a self-defeating strategy of "slash and stagnate".

(2) But these theories of development were complemented soon by theories that focused, not on how to get investment going, which might be termed the Primitive Problem of Accumulation, but on the subsequent, separate but interlinked, problems of: how much to invest (the Optimal Savings Problem); where to invest (the Investment Allocation Problem); and what techniques to employ (the Choice of Techniques Problem).

The main thrust in policymaking was provided, however, by certain simple growth models, among which the Harrod-Domar model takes the pride of place. As you doubtless know, the Harrod-Domar model is a simple construct: the growth rate is given simply by the ratio s/v where s is the average savings ratio and v is the marginal capital-output ratio. From initial income Yo, the savings are yielded by the savings ratio s as sYo; when next turned into investment, these savings increase income by $\{\frac{s}{v}Yo\}$ since the capital-output ratio is v. With fixed savings and capital-output ratios, therefore, the economy is characterized by an exponential growth at rate s/v: much as if you were telling someone to take the first left, than the next right, and then to keep repeating that sequence forever!

The implications of this popular model were dramatic and reassuring. It suggested that the central developmental problem was simply to increase resources devoted to investment. Two policy implications immediately followed: one for domestic savings and the other for foreign aid. If governments were to achieve, say, 5% growth rates, which were both respectable and could be flexible, and if the marginal capital-output ratio was 3:1 on the average, the net investment rate would have to be targetted at 15%. Therefore, if an underdeveloped country was starting from low savings rates around 5 to 7%, it was in no position to achieve the target investment rates right away. First, therefore, a manageable policy target had to be set for raising the *domestic savings* rate gradually by tax effort until, at the end of the specified period (say, three Five-Year Plans), the country had reached a net savings rate of 15%. The rate at which the domestic savings rate could be so raised would have to reflect the usual domestic political considerations that afflict Finance Ministers. Second, while the savings rate was being forced upwards towards the "self-sufficiency" target of 15% the gap between required domestic investment and possible domestic savings would be met by *foreign aid*. As the domestic savings ratio rose, however, through time, the need for foreign aid would diminish until it self-destructed when the target savings ratio of 15% was achieved.

This was then the essential, key policy-mix that underlay many developmental plans during the 1950s and 1960s. The formula seemed a sure thing. It also underlay Walt Rostow's famous 1960 work on *The Stages of Economic Growth* where, interestingly, the last stage (reached on raising the domestic savings rate to its self-sustaining target level) was the celebrated take-off and the program or prognosis did not seem to admit the pessimistic possibility of an extra stage called crash-landing![4]

This sense of optimism about the possibility of accelerated development also stemmed from the prevalent notion that there were "hidden resources" waiting to be tapped in the underdeveloped countries. Ragnar Nurkse, more than anyone else, discussed at length in his 1953 classic, how workers could be taken from the countryside and put to work on investment projects; if their consumption was not eaten away by those left behind, and if agricultural output was sustained (as could be the case if there was surplus labor or "disguised unemployment"), then there was here a distinct possibility of "disguised investment" potential! One remembers well how fetching were these notions of "disguised unemployment" and "disguised investment potential": and, in ways that I have no time to pursue here, they recurred in different forms and degrees in the independent writings of Paul Rosenstein-Rodan and W. Arthur Lewis,[5] to cite only a few major economists.

In a very different sense, and from a unique Toynbeesque perspective (which he has come to reject later in favor of the thesis that alternative responses such as Exit, Voice and Loyalty can occur), Albert Hirschman also fuelled this optimism concerning hidden resources. The process of development was not constrained by resources. As strains and stresses were generated in the economy, creative disequilibria would pose challenges and generate responses: latent entrepreneurship would emerge, technological breakthroughs would occur, and so on. Whereas Nurkse and others saw their hidden resources as tappable through what can only imply a planning approach, Hirschman's hidden resources would emerge into view as an uncoordinated, "anti-planning" developmental process generating the necessary disequilibria. In this sense, the Nurkse conception of hidden resources is closer to the optimism underlying the occasional Marxist notion that the Revolution would usher in the New Man who could be mobilized to perform other-directed, heroic tasks.

The optimism extended not just to growth, but also to the critical question of its effects in ultimately ameliorating poverty. Not everyone then discussed explicitly the link between growth and poverty. It is astonishing how many development economists during the 1950s, for instance, analyzed problems such as the choice of techniques without explicitly bringing into view the possible effects on poverty or on income distribution. I am not surprised therefore that some of them have turned belatedly to questions of poverty with an air of discovery. But it is no more than vulgar nonsense to allege that therefore growth was regarded universally as an end in itself and that poverty was not in the picture. If I may turn to my own writings, on hearing the claims that poverty had only recently been discovered and elevated as a target of development, I fully expected to find that Chapter 1 of my 1966 volume on *The Economics of Underdeveloped Countries* would be titled "Growth"; behold my surprise when it turned out to be "Poverty and Income Distribution"![6] Mind you, this was no accident, a chance stumbling into what would become the new orthodoxy many years later. For, when I was working in the early 1960s with the great Indian planner, Pitambar Pant, who propagated the phrase "minimum incomes" and stressed that the goal of planning should be meeting basic needs by providing minimum incomes to all, we were looking at income distribution, functional and general, in different countries to see if we could learn anything about the income generation and distributive process that would suggest policy measures to make a sustained impact on (absolute) poverty.[7] The strategy of rapid growth was decided upon, in light of the resulting reflections, as providing the only sure way of making a dent in poverty: by providing jobs, we would provide incomes and pull more into gainful employment and above the poverty line. While this strategy would provide the *central* thrust of the assault on poverty, it was to be combined with land reforms that would include politically feasible redistribution and with increasing levels of public distribution of social consumption such as clean water, sanitation, health services and education.

This is the "pull up" strategy for attacking poverty, and it seemed to make eminent sense in India where even the immediate impact on poverty from simply redistributive measures could not be expected to make a significant impact. As Michal Kalecki, the distinguished Polish economist, remarked to me in India at the time, "The trouble with India is that there are too few exploiters and too many exploited!" The "pull up" strategy is, of course, what has come to be described in Western discussions as the "trickle down" strategy. In combining it with an assault on the Welfare State, in the world's most advanced economy, President Reagan has made this a strategy of opprobrium. But put into the context of underdeveloped countries, and combined with feasible redistributive reforms and some provision of social consumption, it can hardly be considered less than progressive. And indeed our optimism extended to this game plan: that indeed growth would be rapid and sustained enough to pull up an increasing number of people above the poverty line and that the dent made in poverty at the end of a generation of such planning efforts would be very substantial.

As I look back at the early era of optimism in developmental economics, I must qualify for two important strands of pessimism: one disappeared soon but the other was to prove central to much developmental thought. The pessimism that vanished related to the notion that the underdeveloped areas were inhabited by a species that belonged doubtless to *homo sapiens*, but which had no claim to the virtues with which the Economic Man was endowed. Thus, if price incentives were offered, farmers would not respond with more production; given higher wages, the natives supplied less labor, not more; custom and values prevailed over greed and prices. To many, it seemed as if such behavior was "irrational", though sophisticated economists have always shied away from that epithet since behavior that reflects one's preference is only irrationally described as such. The sense that the peoples in the underdeveloped countries were radically different in their economic motivations had a long intellectual history, of course. You may recall the chronicles of the Spanish conquests of the Americas in the 16th century that record remarkable thoughts on the native Indians. Oviedo, the official chronicler of the Indians, declared the natives to be generally "lazy and vicious, melancholic and cowardly . . . a lying and shiftless people," whereas the Dominican friars, anxious to see a spiritual role for themselves, saw in them a lovable nature and, in the view of Las Casas, the fulfillment of "all the requirements laid down by Aristotle for the Good Life"![8]

It did not take long for economists to discover that the received orthodoxy was extreme, and asserted a false dichotomy in consequence. Empirical studies through the early postwar years increasingly showed, for instance, that the farmers in the poor countries used techniques that minimized costs, and that they responded fairly substantially to price incentives. The important work of Theodore Schultz on *Transforming Traditional Agriculture* in 1964 was the tract that did most to change our traditional pessimism on this score and properly won him the 1979 Nobel Prize.[9] After this work, the question was changed from "Would farmers respond to prices?" to "At *what* price would *this* set of farmers respond?" This reminds me that the Russians, when looking over a prospective son-in-law, used to ask not whether he drank, but how he behaved when he was drunk! In the same vein, when reassuring my Western friends that we in India are really not so vastly different from them in our economic responses, I tell them that while we believe in an infinity of lifetimes, we maximize in the present one much like everyone else!

But if the pessimism about the poor countries being handicapped by the absence of Economic Men disappeared very quickly from the scene, the other source of pessimism lingered. It was at the heart of important doctrines of development, implicitly or explicitly, and seemed self-evident to many for a long time. This pessimism related to the *external* environment in which the underdeveloped countries would operate in the postwar period. Astonishingly, many major development economists were pessimistic about foreign trade opportunities. Either they argued, like Ragnar Nurkse, that the era of export-led growth was over and that trade could not be expected to be an "engine of growth"; or they argued, in varying ways, that trade opportunities were so restricted now that development strategies would have to turn increasingly inward-looking, import-substituting and the like.

This "export-pessimism" is explicit in Ragnar Nurkse's 1953 work, as also in his 1959 Wicksell Lectures where he deduces from it his prescription that the underdeveloped countries must pursue balanced growth reflecting internal demand configuration. It is explicit also in Raul Prebisch's writings that advocated import-substitution in Latin America because he feared that the terms of trade were steadily worsening for primary exports. It is implicit in both Rosenstein-Rodan and Hirschman because the inducement to invest is hardly likely to be a difficult problem if potential entrepreneurs face fixed world prices at which they can sell with assurance.

In the end, therefore, the development economists were optimistic about the possibility of development in the sense that rapid growth was possible and would impact speedily on poverty, reflecting a naiveté of sorts about the difficulties that such *internal* transformations could pose. Balancing and contrasting with this optimism was the pessimism regarding the *external* conditions within which the internal developmental process would have to work.

As it happened, the developments during the 1950s and 1960s turned out to show that the *internal* problems were more complex, whereas the *external* pessimism had been unwarranted. That the few countries, especially the Far Eastern Four that did not share this pessimism and quickly turned to the EP (Exporting Promoting) strategy, profitted immensely is now beyond doubt except for those who would substitute preconceptions for empirical evidence or those who hold the EP strategy in contempt because they equate the pursuit of an EP strategy with reactionary regimes. Both reasons for rejecting the EP strategy particularly afflict the extreme fringe of radical economists and their friends.

Their refusal to face up to the evidence, produced now in depth by many distinguished economists and research projects at the World Bank (Balassa), the OECD (Little, Scitovsky and Scott), the NBER (Bhagwati and Kruger) and the Kiel Institute (Donges), is reminiscent of the continued faith in the immiserization of the proletariat that one can still encounter. It also reminds me of a story told to me by an eminent American political scientist who had met a Ceylonese member of what is probably the last surviving organized Trotskyite party today. The Trotskyite claimed that the Americans were devious, to which my friend replied with humor: yes, I know that, but why do *you* think so? Ah, the Trotskyite replied, have you seen all these binoculars, cameras and the splendid products of advanced engineering that come into Colombo? They all carry names such as Matsushita, Sanyo and Sony. But then everyone knows that the Japanese cannot possibly make these things: they are good only at "one-dollar blouses". Evidently, the Americans are masquerading under Japanese names to get into Asian markets!

As for the equation of the EP strategy with Reaction, I cannot accept that this is a necessary connection. Admittedly, the Far Eastern Four have had varying degrees to which the Iron Fist, rather than the Invisible Hand, has ruled. Besides, reflection suggests that regimes that assume political control through *direct*, authoritarian means are less likely to be bothered by the fact that external orientation of industries generally implies less political power and patronage than inward-looking, homemarket-dependent import-substituting industries that can be controlled through licensing and quantitative allocations and whose control then yields rents and hence political power to the parties that can appropriate these rents. The authoritarian regimes may thus be much more willing to shift to the EP strategy, whereas the pluralistic ones are not. But, this is to say that, paradoxically and unhappily, the authoritarian regimes may be more likely to adopt the superior EP strategy; it is simply a *non sequitur* to count this as a shining point in favor of the IS (Import Substitution) strategy and as a dark blot in the EP book!

Perhaps the chief lesson of the success story of the EP-strategy countries is not in the demonstration of the success of this strategy, *per se*. Rather, it may be in the demonstration that economic success comes from taking risks, from recognizing and seizing opportunities. The risk-averse export pessimists saw the postwar trade opportunities pass them by; the Schumpeterian risk-takers of the Far Eastern Four, and those such as Brazil who followed quickly in their footsteps, seized them and prospered. The economic miracles were made, here and now, not in heaven. I am inclined to assert this in the teeth of the witticism in the Latin American story which runs in the form of a dialogue: I do not think there are any economic miracles; but then what about the German miracle; ah, that can be explained by German industriousness; but what about the Japanese miracle; ah, that can be explained by Japanese inventiveness; then what about the Brazilian and Mexican miracles; ah, those *are* miracles! Incidentally, I wonder if Lévi-Strauss would make something of the fact that, in keeping with Carlyle's description of Economics as the "dismal science", we invariably react to superlative economic performance by describing it as a miracle!

But if the external pessimism of the early developmental years was reversed by the glorious trade expansion in the world economy during the golden age of the 1950s and 1960s, the internal optimism of the developmental economists has yielded to somewhat sobering reality in the opposite direction.

I must hasten to say that the growth rates in most of the developing countries in the postwar period have been higher than the historic growth rates of the older, developed countries at their peak performance in the 19th century. They have also exceeded the growth rates of the developing countries themselves in the pre-Second-World War period, generally prior to their independence. Moreover, in many (though not most, if we weight India appropriately) developing countries, the growth rates projected or planned for them by development theorists or practitioners such as Paul Rosenstein-Rodan and Hollis Chenery in the 1950s and 1960s, respectively, were exceeded by the outcomes. David Landes has described the epoch introduced by the Industrial Revolution as *Prometheus Unbound*; the fetching phrase is applicable, only with mild exaggeration, to the average growth performance of the developing countries.[10] Assembled and analyzed by my former MIT student David Morawetz, in his splendid 1977 analysis of *Twenty-five Years of Economic Development 1950–1975*, these facts and propositions are only a fitting testimony to the view that the early growth-optimism was not altogether misplaced.[11]. But I must proceed to accent the "altogether" here.

Investment rates did go up generally *à la* Harrod-Domar-Rostow wherever growth rates moved up. Indeed, historically, I would speculate—and speculation and hypothesis are all that is possible since facts tend to disappear as we move back in time: it has been wittily remarked, in riposte to the rigorous cliometricians, that "true" history begins where facts end—that the history of the last three centuries may well be divided into three phases: the pre-Industrial-Revolution period where growth rates were characteristically negligible, since both technical change and rates of accumulation were small; the Industrial Revolution era that ushered in higher growth rates, still in the range however of 1–2% annually, reflecting an unprecedented acceleration of technological know-how but not a matching, planned or unplanned, dramatic and sustained rise in investment rates; and then our modern post-Second-World-War period where the developing countries registered substantially greater growth rates, generally in excess of 4%, reflecting *both* high rates of technological absorption and steadily rising investment rates supported by planned increases in domestic savings and infusion of foreign assistance. But if high growth rates in the developing countries were associated with high investment rates, high investment rates did not necessarily lead to high growth rates. The relationship, contrary to the early notions, turned out to be a necessary one, not one of sufficiency.

Moreover, and more fundamentally, the early predictions were often off the mark on how the countries would rank-order on their growth rates. Morawetz records wryly that Burma, India and Argentina were on the list of four highly successful countries in Rosenstein-Rodan's predictions. Raul Prebisch, who comes from Argentina, likes to regale his friends by saying that there are four kinds of countries in the world: developed, developing, Japan and Argentina! But many of us could tell this story as if it applied to our own country.

That is precisely where, I think, we have learnt the virtue of scholarly hesitation. We have learnt that sustained growth often can be a difficult process. The behavioral relationships we had worked with in our simple models, whether explicit or implicit, were really far too simplistic: things could go wrong in unforeseen ways. I shall illustrate with some telling examples, relating to foreign aid. You will recall how the Harrod-Domar model was used in the 1950s to calculate aid as a supplement to domestic savings. A typical assumption in these exercises was that domestic savings was behaviorally independent of the aid inflow, except in the sense that the tax-policy efforts were to be geared to raising the domestic savings rate. In practice, however, it seems as if aid may have often led instead to a disincentive effect on the domestic savings effort. While this evidence, which is persistent to an excess, is based on conceptually weak foundations, there is some ground for our worrying now that the aid process may, in certain cases, harm rather than help the domestic effort at raising the savings ratio. Our early models and argumentation simply did not allow for this complexity that afflicts the real-world recipients of aid! Again, when food aid was given, it was assumed that it would assist the recipient's development programs by increasing its real resource availability. But then, as Theodore Schultz was to remind us, it may have created a disincentive effect on the recipient's agricultural production. But, if aid was to prove counterproductive in unforeseen ways in these instances, the politics of the aid process were also seen to be occasionally disconcerting. Much as some may scoff at the notion of dependence, it must be recognized that the interruption of aid funds to a recipient can create a "resource crunch" that can be the source of intended destabilization: a notion that I formalized in 1975 in a paper entitled "Foreign Capital, Dependence, Destabilization, and Feasibility of Transition to Socialism," keeping in mind the experience of Chile where aid was indeed cut off because of President Allende's ideological shift in policies to the left, raising the ire and precipitating the retribution of Dr. Kissinger.[12] Lest you think that I point my finger only at the CIA, let me assure you that the precipitous withdrawal of the Soviet Union from China in 1971, leaving aid-assisted plants unfinished and untended, matches up the scores on this front between the two superpowers.

Our optimistic assurance about how to generate steady growth has thus been moderated by a sense of the complexity of the developmental process. In turn, it has provided a significant stimulus to those who seek to integrate analytically the political processes with more orthodox economic-theoretical formulations. It is hard, for instance, for development scholars to consider now the role of multinationals in development without considering the possibility that the optimal mix of tax-cum-subsidy policies on trade and capital flows, which we economists work out with such ingenuity and show to be the welfare-maximizing policy options for a host country, may simply be infeasible once we allow for the policy set itself to be constrained politically by a symbiotic relationship that may ensue between the host elites, the host governments and the multinationals. In short, development economists have become participants in the growing trend in economic theorizing towards analytically rigorous political economy.

But the major area of dampened optimism has been undoubtedly the link between growth and poverty. It would appear that, contrary to our expectations and certainly our hopes, the impact on poverty is a slow and indeed a rather tenuous process.

I must confess that I am not at all sympathetic to the claims of those who argue that the postwar development experience has substantially discredited those of us who pinned our hopes on growth to reduce (absolute) poverty. Such assertions have become commonplace in the documentation that has been emerging in the last decade from certain multilateral institutions; but they are false. This is not the occasion to examine these assertions fully. It should suffice to make simply the following observations.

First, the statistical evidence on which these assertions are based is often only marginally better than that for the post-Industrial-Revolution England which has kept distinguished historians such as Eric Hobsbawm and Max Hartwell in interminable conflict over whether the proletariat was immiserized by the Revolution or not! Three gloomy assertions need to be distinguished: (1) that the income distribution has worsened: Simon Kuznets has conjectured that the income distribution worsens initially with growth and then improves; (2) that, along with worsened income distribution, the absolute income at the bottom (say, 20%) has declined since the lower share has outweighed the larger total due to growth; and (3) that, furthermore, the per capita income within the bottom groups has fallen too.[13] These steadily worse outcomes from the viewpoint of poverty are conceptually easy to define in theory; they are equally difficult to measure in practice. I think that it is hard to conclude with confidence, on the available evidence for most developing countries, that any of these three outcomes represents a central tendency, or for that matter a *definitively established* outcome in any major developing country. One always hesitates to talk down the available data when they go against one's thesis: it can be terribly self-serving. But I must say that I find it very hard to side with quantitative economists against an imaginative developmental observer and analyst such as Gilbert Etienne who, in his recent provocative study of India's Green Revolution, scathingly attacks the veracity of studies which utilize national statistical sample surveys (as in India) to reach melodramatic doomsday statistical conclusions.[14] These, he argues convincingly with his firsthand evidence from extended stays in the very same villages that are sample-surveyed, are at variance with the properly observed facts. He has done what I call "doing in India what you do in China," that is, disregard the numbers which in any case are hardly available for China and simply see what your naked eye shows you. He has gone back over time to several villages that he had looked at intensively, often more than a decade earlier. And he observes, asks, examines, and records: much like Jan Myrdal in his celebrated Report from Liu Ling but with more anthropological, sociological and economic discipline and less poetry. The results are what we did expect: growth has indeed pushed several of the poor on in life.[15] Doubtless, some poor have been left behind; others have been even impoverished further. But then, as Arthur Lewis has wisely remarked, it is inherent in the developmental process that some see the opportunities and seize them, leaving others behind until they wish to, and can, follow. Politics and economics can both constrain the capacity of the laggards to follow. Thus, for instance, the Green Revolution in some instances may well have polarized the distribution of property in the countryside, enriching the farmers with access to credit, fertilizers and irrigation and immiserized those who did not: in so doing, the radicals had dared to hope that the Green Revolution would help usher in the Red Revolution. But, if Etienne is correct, this has not happened in anything like a significant degree in his cross-section of villages in India.

Second, as many skeptics of the doomsayers have remarked, India is hardly an impressive example of the failure of the pull-up thesis since growth itself has been rather lacklustre. On the other hand, data on real wages plus much indirect evidence would suggest that a few fast growing countries such as South Korea and Taiwan have had a noticeable impact on poverty. We simply cannot hang the defendant on current evidence; the jury is still out.

On the other hand, I would say that the policy of redistribution, as an *alternative* to growth, as a means of attacking poverty (the so-called "China option") has finally come to be seen in the skeptical light as some of us had viewed it in the 1950s. If China was not growth-oriented, we believed that it had triumphed in eliminating poverty, at least in the first round, in the first flush of the Maoist regime. Now, after the window has opened a tiny crack in the aftermath of the Cultural Revolution and the failures of the Great Leap Forward, we know, as one of my colleagues remarked in 1975 in a trenchant and telling early critique of the Sinological romance with the Chinese experiment, that if one has seen one commune one has *not* seen them all.[16] We also know that the barefoot doctors generally wear shoes; that their professional competence occasionally exceeds only marginally that of the average grandmother; and that doctors have dragged their feet almost as successfully as elsewhere when assigned to go to the countryside, indeed to the point where Liu-shao-chi's major crimes were declared by the Red Guards and official pronouncements to include the sabotage of the campaign to carry doctors to the rural areas. We have known now for several years, and tentative and tenuous demographic analysis seems to accent this, that China could not cope with a substantial famine in 1959, though it is probably foolish to jump to the conclusion, as some late converts to skepticism regarding China's successes have done, that this famine took as many as 15 to 25 million lives! We are further told that the Chinese concept of equality is intra-commune, *not* between communes: the rich communes do not generally share their affluence with the destitute ones. And we are now told by the new regime that more than 10% of the Chinese population may be below a rather austere poverty line: recall further that such governmental confessions often provide only lower-bound estimates. These tantalizing glimpses into China's assault on poverty will almost certainly not be allowed to develop into a fuller picture, since careful and unfettered scholarly scrutiny is no more possible here than in other socialist nations. I am afraid therefore that we shall have to reconcile ourselves to the uncomfortable situation where we do not know the extent to which China's *ex ante* egalitarian methods failed *ex post*, and whether the failures were due to a discordance between their announced and their true objectives or rather due to the limitations of the methods used to achieve the announced objectives. Indeed, for the foreseeable future, we will face the paradox that, in China's case, the more we learn, the less we will know.

On the other hand, the proposition that a more equitable distribution of assets at the start of the growth process will generally imply that the new incomes will, in turn, be distributed better is of course quite plausible. In the end, over a longer period, the forces that generate inequality will tend to unequalize the outcomes; but over a generation or two, the net outcomes would be more equal than if we were to start with unequal distribution of assets. The experience of South Korea and Taiwan, where Japanese occupation is credited with having brought about the initial asset-ownership equalization, underlines this near-truism well. Also, the experience in India, where several micro-level studies have shown the link between asset-ownership and new-income distribution to be a significant factor in a fair number of cases, only underlines the wisdom of supplementing the growth-oriented approach with policy measures that counter this bias.[17]

A policy of "redistribution with growth", where the redistribution of assets *precedes* the growth that is designed to impact on poverty, has therefore been advocated by several distinguished economists, among them Irma Adelman and Hollis Chenery.[18] If such redistribution can be undertaken politically, and is also not disruptive economically (as was the case with Soviet collectivization),[19] we can only rejoice. However, where such advantages do not obtain, the only logical and important policy option is to reorient and supplement the institutional mechanisms that channel the means of new income creation, so that the poorer classes have some access as well to them. I think that it is only fair to say that, at the start of the postwar period, we had not realized sufficiently the significance of these supplemental measures; we do now. But the dilemma still remains that the successful implementation of such institutional changes in the teeth of unequal asset distribution and hence unequal political power at the grassroots level is an extremely difficult task. And the degree of success of policies aimed at such changes is, to a large extent, a function of the extent to which "countervailing power" is available to the poor through the presence of social action groups and politically viable opposition parties.[20]

I would be remiss if I did not add that development economists have been equally surprised, but pleasantly this time, in discovering that they had exaggerated their early fears about the trade-off between "consumption" expenditures (such as on primary education and health) and investment expenditures aimed at growth and hence ultimate impact on poverty. It is difficult today to appreciate the widespread notion in the 1950s that primary education was simply a "natural right", whose implementation reflected the availability of resources. That it was possibly an important means for raising productivity and hence growth and therefore reducing poverty, and that it could therefore be justified on consequentialist ethics, was a later phenomenon. This holds equally for health expenditures that were viewed with inhibited enthusiasm also for fear that they would exacerbate population growth and only later were considered to have a possibly productivity-enhancing effect on populations that could otherwise be working at impaired efficiency or even to lead to a lowering of birth rates if, by reducing infant mortality and increasing survival rates, they enabled parents to produce fewer babies to wind up with their target family size in steady state. Much has indeed been written, though little is known with firmness, to wean us away therefore from the fear that such educational and health expenditures are *necessarily* at the expense of growth.[21] What is equally pleasurable is that many of these arguments apply with yet greater force when the expenditures are addressed to the poorer segments of the population. The case for undertaking *some* such expenditures, with focus on the poor, consistent with being engrossed in the growth strategy, is therefore now seen to be stronger than ever before.[22]

I must confess, however, that even when the growth process has been modified in ways suggested by postwar experience and experimentation so as to ensure that the incomes do get through to the poor, we must confront the dilemma that has long been known to the sociologists of poverty: that the poor may spend their incomes on frills rather than on food. As the Japanese proverb goes: "Each worm to his taste; some prefer nettles." Seamen's folklore recounts a story of the sailor who inherited a fortune, spent a third on women, a third on gin, and "frittered away" the rest!

For those of us who feel that certain basic needs ought to be satisfied, this tragic assertion of what economists have come to call rather extravagantly "consumer sovereignty" leaves us confronting a familiar moral-philosophical issue. Should we impose a paternalistic solution so that the poor are seduced into better fulfillment of what we regard as their basic needs? Ulysses chaining himself as the sirens approach and sacrificing his future freedom to choose in knowledge that otherwise he will choose to err is not quite the analogy we can employ here; for, in this instance, the chains are being imposed by someone else! The libertarians will see shades of "1984" in such paternalism, the thin end of the wedge, the road to serfdom. I do not. In fact, I see great virtue in paternalistic moves to *induce*, by supply and by taste-shifting policy measures, more nutrient food intake, greater use of clean water, among other things, by the poor. In thus compromising the important principle of freer choice, simply for the poor and not for others, evidently I adopt the moral-philosophical position that I do not care if the rich are malnourished from feeding on too many cakes but do if the poor are malnourished from buying too little bread, when their incomes can buy them both proper nourishment if only they were to choose to do so. In this, I am in the ethical company of Sofya (Sonia) Marmeladova in Dostoevsky's *Crime and Punishment* who, in turning to prostitution to support her destitute mother, sacrifices virtue for a greater good.

How should I conclude this retrospective essay? Doubtless, we have now become aware that development is a complex process, and that the removal of poverty, our central objective, is yet more arduous. Paradoxically, however, this realization may conceal in itself the seeds of a new optimism, provided we have an appropriate time frame. Let me illustrate, ending my essay as I began with a Japanese story that I owe to Dr. Goh Keng Swee, the architect of Singapore's economic success. An Australian expert, invited by the Japanese government, had this to say in his Report, as excerpted in the *Japan Times* of August 18, 1915:

> Japan commercially, I regret to say, does not bear the best reputation for executing business. Inferior goods, irregularity and indifferent shipments have caused no end of worry . . .
>
> My impression as to your cheap labour was soon disillusioned when I saw your people at work. No doubt they are lowly paid, but the return is equally so; to see your men at work made me feel that you are a very satisfied easy-going race who reckon time is no object. When I spoke to some managers they informed me that it was impossible to change the habits of national heritage . . .
>
> First class managers . . . are required to wake things up and get out of the go-as-you-please style that seems universal at present.

Few of us today would recognize in this account the modern Japan. In fact, the apocryphal story today goes the other way, with a tourist in Tokyo asking for a cheaper camera after looking at the high-quality display in a Japanese camera-shop in Ginza and being told: just go down to the American store at the end of the block; *they* sell shoddy stuff!

Perhaps our mistake was based in idealism: the poverty and backwardness in the underdeveloped countries left us psychologically with little choice except to be optimistic. But now that we have experienced sobering reality, it would be inexcusable to yield to despair. I feel, as always, that true wisdom here lies, if I may purloin Sir Geoffrey Crowther's famous witticism about the politics of *The Economist*, in the extreme center.

Notes

1. The quotation is from Edwin McClellan's *Two Japanese Novelists: Sōseki and Toson* (Tokyo: Charles E. Tuttle Company; and Chicago: University of Chicago Press, 1971), page 5. The original account is in Sōseki Zenshu Kankokai, (Ed.) Sōseki Zenshu, 20 Vols. (Tokyo, 1928), 20: 507–8.
2. P. N. Rosenstein-Rodan, "Problems of the industrialization of Eastern and South-Eastern Europe," *Economic Journal* (1943); and Ragnar Nurkse, *Problems of Capital Formation in Underdeveloped Countries* (Oxford: Basil Blackwell, 1953).
3. Albert Hirschman, *The Strategy of Economic Development* (New Haven: Yale University Press, 1958).
4. Walt W. Rostow, *The Stages of Economic Growth* (Cambridge University Press, 1960).
5. W. Arthur Lewis, "Economic development with unlimited supplies of labour", *Manchester School*, Vol. **22,** No. 2 (1954); and Jagdish Bhagwati, "W. Arthur Lewis: An appreciation," in Mark Gersowitz *et al.*, *The Theory and Experience of Economic Development: Essays in Honor of Sir W. Arthur Lewis* (London: Allen & Unwin, 1982).
6. Jagdish Bhagwati, *The Economics of Underdeveloped Countries,* World University Library Series (London: Weidenfeld and Nicolson, 1966).
7. Pitambar Pant and I were of the same mind on these questions, having both reflected over the available income-distributional data that I had assembled at the time for this purpose, having joined the Indian Statistical Institute in January 1962 as Professor of Economics in the Planning Unit assigned to work as a "think tank" for Pant in the Planning Commission. The main formal document that emerged from these deliberations under Pant's authorship was circulated in August 1962 by the Perspective Planning Division of the Planning Commission and part of it has been reprinted in T. N. Srinivasan and P. K. Bardhan (eds.), *Poverty and Income Distribution in India* (Calcutta: Statistical Publishing Society, 1974).

8. The quotes are from J. H. Parry, *The Spanish Theory of Empire in the Sixteenth Century* (New York: Octagon Books; 1974), p. 6.
9. Theodore W. Schultz, *Transforming Traditional Agriculture* (New Haven: Yale University Press, 1964).
10. David Landes, *The Unbound Prometheus: Technological Change and Industrial Development in Western Europe from 1750 to the Present* (Cambridge University Press, 1970).
11. David Morawetz, *Twenty-Five Years of Economic Development, 1950–1975* (Washington, DC: The World Bank, 1977).
12. Jagdish Bhagwati and Earl Grinols, "Foreign capital, dependence, destabilization and feasibility of transition to Socialism", *Journal of Development Economics,* Vol. **2** (1975); reprinted as Chapter 65 in my *International Factor Mobility: Essays in International Economic Theory* (Vol. **2**), Robert Feenstra (Ed.) (Cambridge: MIT Press, 1983).
13. There is, of course, the familiar problem, *within* the bottom group of (say) 20% of the population or households, that income distribution within that subset might have worsened alongside with the fall in the (mean) per capita income! That could be regarded as a yet further calamity, with justifiable consequential outrage.
14. Gilbert Etienne, *India's Changing Rural Scene: 1963–1979* (Delhi: Oxford University Press, 1982).
15. Jan Myrdal, *Report From A Chinese Village,* (New York: Signet Books Edition, 1966). Also see Jan Myrdal and Gun Kessle, *China: The Revolution Continued,* (New York: Vintage Books, Random House, 1972).
16. Cf. Padma Desai, "China and India: Development during the last 25 years", *The American Economic Review* (May 1975).
17. Cf. Suresh Tendulkar, "Rural institutional credit and rural development: A review article", *Indian Economic Review,* Vol. **XVIII**, No. 1.
18. Irma Adelman and Cynthia Morris, "Who Benefits from Economic Development?" (Washington, D.C.: World Bank) Mimeo; Irma Adelman and Cynthia Morris, *Economic Growth and Social Equity in Developing Countries* (Stanford University Press, 1983); and Hollis Chenery, Montek Ahluwalia, Clive Bell, John Duloy and Richard Jolly, *Redistribution with Growth* (London: Oxford University Press, 1974).

    In this generic class of strategies, I would also include an altogether different kind of proposal that I made for Indian planners to consider in 1973 in the Lal Bahadur Shastri Lectures (*India in the International Economy: A Policy Framework for a Progressive Society* (Hyderabad: Institute of Public Enterprise, 1973). I argued for a fractional nationalization of land in each village (or similar unit), which could be set apart to form a Chinese-style commune. Those destitute who wished to follow the slow and protracted route offered by the Indian strategy of predominantly relying on growth to impact on poverty would take their chances there; but those who wished to gain employment and some income right away would have immediate access to the commune *à la* China. The combination of both strategies, and access to either *by choice,* would mean that the destitute were not forced into the Indian option of freedom but slow poverty-alleviation *or* into the Chinese option of loss of freedom through forced removal to the communes but more rapid removal of abject poverty.
19. These are two hazards that do not seem to have afflicted China since the elimination of the kulaks seems to have occurred principally during the long civil war itself. Cf. Padma Desai, *op. cit.*
20. It should not be forgotten that, in the longer run, substantial growth itself is a factor generating the necessary countervailing power through the market place: by raising the demand for labor and increasing its opportunity cost.
21. For a splendid review of the literature on this issue, see Paul Streeten *et al., First Things First* (Oxford University Press, 1981), especially Chapters 4–6. It is perhaps ironic that the thesis that education is productive, and indeed a key to growth, principally came from "neoclassical" economists such as Theodore Schultz and Gary Becker! I should add, however, that the idea of education as "human capital", while very powerful indeed, is only a partial reflection of the full reality. Education as a screening device (Spence), education as a "socializing" technique (Bowles-Gintis) and education as an instrument of "job competition" in a job-ladder context (Bhagwati-Srinivasan-Fields-Thurow) are three other theories that legitimately compete for attention and have their empirical counterparts in developing countries. For a review, and a formal statement and analysis of the "job competition" theory, see Jagdish Bhagwati and T. N. Srinivasan, "Education in a 'job ladder' model and the fairness-in-hiring rule", *Journal of Public*

Economics, Vol. **7** (1977), pp. 1–22, reprinted in my *Essays in International Economic Theory* (Vol. **2**), Robert Feenstra (Ed.), (Cambridge: MIT Press, 1983), Chapter 49.

22. On the other hand, the difficulties of directing the expenditures on primary education and on health effectively to the poorer classes when the elites control the political system need to be recalled again. Questions such as the relative priority attached to primary and higher education in State spending and its relationship to the class nature of the State have been discussed at length by economists such as Samuel Bowles and myself; see the extended analysis in my "Education, class structure and income equality", *World Development* Vol. **1**, No. 5 (May 1973), pp. 21–36.

The Crisis: Some Reflections

GAMANI COREA

It is fitting that a volume of essays dedicated to Surendra Patel should cover a wide range of issues. His analytical contributions and his intellectual interests have spread over many subjects; although the transfer of technology occupied him for many years in UNCTAD, his approach to his work reflected these wider concerns. It is also fitting that this volume should begin with some views on the current crisis, not just because it has been an exceptional phenomenon in the postwar period, but also because it has thrown into relief underlying and longer term changes of significance to the workings of the world economy.

The recession of 1980 to 1983 had many of the familiar hallmarks of earlier recessions—the declines in output, employment and world trade; the collapse of commodity prices; the instability of exchange rates; the recourse to protectionism—all reflecting the global transmission of deflationary forces. Yet, there are at least two important respects in which the recession differs from its predecessors. One of these is now well recognized—the coexistence of inflation with declining employment and production. Indeed, the recession was itself the result of a policy response in the industrialized countries to this problem of "stagflation". The other difference is less widely perceived, but also significant. It relates to the potential of the developing countries in moderating and aggravating the global recession. In the late 1970s—in contrast to the experience of earlier cycles—the developing countries contributed to dampening the downturn in employment in the industrialized countries by maintaining a high level of import demand. This was made possible by the continued access to credit from the banking system by some of the newly industrializing countries and the high level of investment in the petroleum exporting countries. Subsequently, in the 1980s, the decline in import demand from the developing countries aggravated the downturn in activity and employment in the industrialized countries. Although this latter process was common to earlier recessions, the magnitudes were greater than before. For the first time, crisis conditions in developing countries—their debt service problems and cutbacks in investment—were affecting the developed countries and indeed the international banking and financial system.

These differences between the current recession and those of earlier times reflect changes that have been taking place over a longer period. The phenomenon of stagflation itself emerged in the context of a slowing down in the long-term trend of growth and productivity in the industrialized countries accompanying shifts in the sectoral composition of output and political and social changes that affected the distribution of real incomes, as well as international developments affecting prices in the field of energy. The changed context ruled out the "Keynesian" remedies to falling output and employment and induced governments to undertake and intensify deflationary policies despite the background of unemployment and declining growth.

The changed role of the developing countries was also the result of longer term changes. The transmission of forces from the centre to the periphery—so clearly depicted by Prebisch many years ago—continued to be dominant, but the developing countries taken as a whole had acquired a greater importance in the world economy. This is true in respect of their share of global output (around 20%), of the markets they provide for developed country exports (around 30%), and their influence in such key sectors as energy, finance and debt. The course of the development process and the economic fortunes of the developing countries had become, more than ever before, of consequence to the functioning and evolution of the global economy. What is significant now is not just the fact of interdependence, but also its changed character.

There is still, however, no consensus on the implications of this new interdependence for international policy. At UNCTAD VI in Belgrade, the developing countries tried to pursue its logic by proposing, as a response to the crisis, simultaneous and direct actions in the three key areas of finance, commodities and trade that would have served to reactivate development as well as reinforce recovery. But, although the theme of interdependence was widely acknowledged, there was no acceptance by the developed countries of the need for parallel or simultaneous actions to restore momentum to the development process in the Third World. The line of argument was essentially that the recovery of the industrialized countries was the first priority, that such recovery would benefit the developing countries, and that any direct action to reactivate development might even hamper the management of the recovery. In other words, 'trickle down" was the essential remedy for the developing countries and Recovery and Adjustment rather than Recovery and Development remained the order of the day.

The outcome of Belgrade and the continued stalemate in North-South negotiations in virtually every forum suggests that until a consensus emerges on the implications of the "new interdependence" there can be no real progress in the dialogue between developed and developing countries. At best, the actions taken on a multilateral basis will be of a minimal kind that would serve as gestures of goodwill to prevent a worsening of the atmosphere in the field of North-South relations. But is it realistic to seek such a consensus? Could there not be some validity to the trickle down concept? Is it practical, in the context of the constraints facing them, for the developed countries to endorse and engage in direct actions to reactivate development? In large measure, the answers to these questions depend on the view that is taken of the urgency and the desirable tempo of the development process in the countries of the Third World—not only in the immediate situation, but in the longer term as well.

There can, of course, be no doubt, given the present linkages between developed and developing countries, that a recovery of the industrialized countries is a necessary condition for the reactivation of development. The crisis that has engulfed the developing countries in recent years is the direct outcome of the recession in the developed countries and quite clearly a recovery in these latter countries will improve the prospects for growth in the Third World. Commodity prices, export markets, as well as aid and other financial flows to developing countries can certainly improve in the wake of an upturn in the economies of the industrialized countries. The crucial question, however, is to what extent? The social, economic, and indeed political imperatives of the Third World call for a substantial tempo of growth and transformation—even higher than that achieved over the 1960s and the 1970s. In an increasing number of these countries, the latent growth impulses are now sufficiently strong to support such a tempo of change. This was amply demonstrated during the brief period of the 1970s when several of these countries had access to financial resources of magnitudes that were unprecedented. The poorer and weaker countries may not as yet be poised for a similar performance, but they, too, require major investments and other reforms to break through their chronic and deep-seated problems.

It is doubtful, to say the least, that a recovery of the industrialized countries from the current recession will suffice to provide the impetus needed for an acceleration of development at the pace that is both necessary and possible. In the first place, doubts will continue to persist about the strength, durability and spread of the recovery in the industrialized countries. High interest rates, unstable exchange markets, continuing budget deficits and the ever-present danger of a revival of inflationary pressures continue to cloud the prospects for a strong and durable recovery. No less important, however, is the question of what rates of long-term growth the industrialized countries would attain once the recession has been overcome. Most projections discount the possibility—especially for Western Europe—of a return to the growth path of the 1950s and 1960s, If this were true, it would be unlikely that growth in the industrialized countries would serve as a sufficiently strong locomotive for accelerated growth in the Third World at the tempo that is needed on political and social grounds and that is possible in terms of their inherent potential.

A further consideration supports the case for a more direct approach to accelerating development. This is the present critical situation in developing countries and its political and social implications—both national and global. In pursuit of "adjustment", the developing countries have been obliged to undertake severe cuts in investment, imports and social programs that have aggravated unemployment and reduced living standards for the mass of the population. While some policy and management reforms may yield positive results, the cutbacks in investment and development spending cannot but generally weaken the prospects for future growth. In fact, the political, social and developmental consequences of the recent experience of contraction and adjustment may strike the developing countries with a time lag—after they begin to emerge from the most serious difficulties of the recession.

The case for the "immediate actions" proposed at Belgrade remains, therefore, valid. These actions were intended to cover a series of measures which, when taken together, would serve directly to reactivate development. It is difficult to see that they would be "burdensome" for the developed countries or would impair the management of the recovery rather than reinforce it. The most urgent need is to resume the flow of financial resources to the developing countries and to strengthen their liquidity position. This need not be achieved by increasing aid budgets. A sizable issue of special drawing rights, larger Fund quotas, and an increase in the borrowing potential of The World Bank are among the alternative modalities that could be utilized. A greater commitment in practice to commodity price stabilization through the establishment of commodity agreements and the Common Fund is a further avenue that could in no way impair the management of recovery in the industrialized countries. In the same way, the widening of developing country access to markets in these countries would help rather than hinder recovery by expanding the import demand of the developing countries and by helping them to service their debts. Many studies have demonstrated that trade with developing countries creates more jobs in the industrialized countries than it cancels.

The logic of the new interdependence has, however, to go beyond immediate actions to promote recovery and accelerate development. It is equally relevant to long-term policies and to the systems that support global economic growth and stability. As already noted, a recovery of the industrialized countries from the recession is unlikely to impart sufficiently strong impulses for a high and sustained tempo of development in the Third World. In the context of relatively slow growth in the industrialized countries over the medium- to long-term the development process in the developing countries will have to rely increasingly on two other supplementary supports—the diversification of their linkages through cooperation with each other and through trade with the socialist countries and greater support from the international institutional system.

But here again, there is hardly any recognition of this in the context of North-South negotiations. The importance of economic cooperation among developing countries is widely acknowledged, but no solution has yet been found regarding the manner in which the universal institutions of the UN system could support negotiations among these countries in pursuit of this goal. The idea of systems reform continues to meet strong resistance and is often caricatured as a bid to replace or destroy existing institutions. Yet, the institutional frameworks and systems in the areas of money, finance and trade that were established in the early postwar years have been seriously affected and undermined by both the longer term changes that have since taken place as well as by the recent crisis. If they are not adapted—or reformed—in time they will be further weakened and lose their relevance. The lessons of the current crisis point both to the need for urgent actions of a short-term kind and the longer term reform of the systems.

The international monetary system established at Bretton Woods was essentially concerned with the problem of preventing short-term disequilibria from disrupting the maintainance of full employment in the industrialized countries. The question of direct systems support for long-term growth was not ignored—the World Bank was set up to finance reconstruction and development—but such support was not based on any view of a desirable tempo of development and its implications for total resource flows. If a strong and sustained process of development in the Third World is now an imperative not only for the countries concerned but also for the global economy, there is a need for a new "development consensus" to underlie the design and operation of the international systems in the areas of money, finance and trade—parallel to the "full employment consensus" that guided their creation.

The supportive role played by the international monetary and financial system is particularly crucial. Short of some dramatic, but unlikely, change in the relative prices of their exports—as happened in the case of oil—the developing countries would need to incur substantial trade deficits to reach and sustain the tempo of growth that they are capable of and require. Some have argued for a new "Marshall Plan" to make this possible. But the recent history of bilateral aid and the economic environment in the developed countries does not encourage this prospect. Neither is it realistic to expect the private commercial banking system to resume the role it played in the late 1970s. Indeed this would not be desirable even if it were possible. Direct private investment, with a maximum of encouragement, may play a greater role than at present, but it is unlikely to provide either the magnitudes or the types of investments that are needed. There are also limits to direct borrowings from long-term capital markets.

What, then, could be the possible sources of finance? The answer must lie in large part at least in a reformed international monetary and financial system that lays greater stress on long-term financial flows than at the time of Bretton Woods. This is not the place to explore how this could be done. A vastly enlarged role for the long-term lending institutions such as the World Bank and the regional development banks; a better adaptation to the needs of developing countries of the IMF concept of short-term financing; the establishment of interest subsidization mechanisms; the creation of a "link" between international reserve creation and development finance; and the setting up of facilities to mobilize and channel structural surpluses are among the approaches that are relevant in this context. The debate on the reform of the international monetary and financial system has traditionally embraced a number of issues—the international reserve system, the regime of exchange rates, the magnitude and use of IMF resources, the nature of the adjustment process and so on. These issues are no doubt of vital importance. But of equal, if not greater, importance from the point of view of the dynamics of development is the question of long-term resource flows and the role of the international monetary and financial system in this regard. This, too, must be one of the key issues in the debate on reform.

There are, similarly, important issues that concern the evolution of the international trading system. Among these are the treatment of trade in services; the question of protectionism in agriculture; the best means of applying the principles of nondiscrimination and competitiveness to trade flows that are not always sensitive to tariff regulation such as state trading and the intra-trade of transnationals; the accommodation within a universal system of subsystems such as regional integration arrangements, free trade areas, and common markets; the question of a safeguards code; and, above all, the crucial link between protectionism and structural adjustment. Of special importance is the question of how the trading system of the future could reflect or incorporate a "development consensus". The Generalized System of Preferences provided for the preferential treatment of developing countries in an open system regulated only by tariffs. Its relevance, however, has been reduced both by the lowering of MFN tariff margins and by the increasing resort to non-tariff barriers to trade. For many developing countries with growing capabilities to export manufactured goods, the central issue has become one of access to markets let alone preferential access. At the same time, some of the basic premises of earlier understandings such as nondifferentiation between developing countries and nonreciprocity in trade negotiations are being called into question or even modified unilaterally. The time has surely come for a restatement or even a new understanding of the treatment of developing countries in world trade. This has to provide for the rapid and continued transformation of the economies of these countries in a dynamic setting of global growth and development. The link between the accommodation of developing countries in a global trading system and the process of structural adjustment in the industrialized countries has acquired a greater urgency than ever before.

A policy framework for the treatment of the commodity question should also form part of a tripod of international support systems along with money and finance and trade. This was well recognized in the Havana Charter but never really implemented. There are, however, several factors that make this important in the present and future context. The persistence of inflationary pressures in the industrialized countries has made movements in commodity prices an important factor in the management of macro-economic policy in these countries. At the same time, these movements are of consequence to the way in which developing countries sustain their import demand and their debt-servicing capacity and hence the way in which they support growth and stability in the world economy. In addition, long-term trends in commodity markets call for dynamic processes of diversification and transformation that could be assisted by a framework of international support. All this suggests that the "development consensus" that must underpin the evolution of the international systems of the future must also embrace the commodity issue and the policies and mechanisms to deal with it. The Integrated Program for Commodities and the Common Fund could be a starting point for this process.

The Winners and the Losers: The Rise of the South-East, the Decline of the North-West

JOHAN GALTUNG

The thesis is very simple: the New International Economic Order (NIEO) is a strategy for removing the North-West from its control position; it says nothing about where the new center is going to be. As a matter of fact, it sometimes reads as if the actors behind it think there is going to be no new center. But that is unrealistic. Whether by design or by the workings of the system, centers—more or less coordinated—will tend to materialize. Thus, whereas the point in the preceding section was that nothing is happening to challenge the capitalist structure of the world capitalist economic system, the thesis of the present section is that there will be a new center, and that center is in the South-East part of the world. Five reasons will be given for this proposition.[1]

First, the triangle described by China-Japan-Southeast-Asia has almost all the productive assets needed. Roughly one and a half billion people, among them a 600 million-strong Chinese labor force, Japanese technology and capital (with the highest saving ratio in the world: 19%),[2] raw materials from all over the area, particularly from Southeast Asia, including Indonesian oil (and also Chinese and Japanese offshore oil)—all of this together should satisfy a production function sufficient to meet a substantial portion of world market demands, particularly with Australia/New Zealand and Oceania added.[3]

Second is the underlying ethos. China and Japan and the part of Southeast Asia influenced by them exhibit that particular characteristic of Oriental civilizations: their additive or eclectic nature.[4] There is a Confucian/Buddhist combination common to both, although the tonality is different when Taoism is added in the Chinese case and Shintoism in the Japanese case. This combination makes not only for a considerable cultural communality, but also for a spirit of corporate endeavor in which Confucianism defines the vertical component (discipline, respect for authority, for non-manual work and so on), and Mahayana Buddhism the more horizontal, organic, solidarity aspect. Both of them are adding, in the usual eclectic manner, Western components:

Christianity for sure, liberalism, and, in the Chinese case, also Marxism. The Eastern component makes for a universalism in space, combined with an idea of progress in time, not found in the more Asian components of these civilizations. It is difficult to see how all of this combined could fail to make for an almost perfect ethos for a world-encompassing capitalism. Weber has explained the significance of Puritan Protestantism for the rise of capitalism, but at best his was only a theory giving a rationale for the capitalist entrepreneur himself, offering little or no comfort to the worker.[5] Marx explained what the worker's situation was and motivated the worker for revolt and distribution, but had very little to say that could inspire and justify entrepreneurial activity. The net result was, as we know, the two Norths, with entrepreneurial growth without much distribution in the West, and distribution without much entrepreneurial growth in the East.[6] The contention here is that the South-East world, not only Japan, is now transcending this apparent incompatibility because of its particular civilizational amalgam, and will run a very effective capitalism,[7] much more effective than that of the North-West.

Third, and partly as a reflection of the preceding theme, is the way Chinese and Japanese are organized. A Chinese name like Tang, or a Japanese name like Tanaka, stands for millions of people. In the Chinese case, there is a certain amount of solidarity among the Tangs, the Lims, the Changs, bordering on crystallizing them into an actor. To the extent that there are Chinese abroad—and this is a considerable extent[8]—a Chinese family name may *de facto* stand for a transnational corporation. Where Westerners tend to avoid very common names and prefer names that can better clothe their individualism,[9] Chinese and Japanese may prefer just the opposite, to sense the belongingness in a greater collectivity as expressed in a very common name, although there are counter-processes in the Japanese case. This expression of the famous collectivism of the Orient, among other factors, probably makes for less alienation in connection with the industrialization process than in the West. In a factory, the collectivity of workers and managers produces for the collectivity of customers, and both collectivities make sense in that particular non-Western setting. It is only in an individualizing culture that an individual worker producing for an individual customer in a direct individual-product-individual relationship makes so much sense,[10] and departures from that pattern create alienation, which may lead to some apathy, even withdrawal.

Fourth, the politics of the South-Eastern triangle is taking shape. There are weak points, or points that have to be played with great delicacy. China and Japan share the history of a terrible imperialist war only 50 years ago, and both of these major Powers are viewed with apprehension in the region. Chinese minorities everywhere may serve as points of contact but may also be seen as bridge-heads for expansionism. The people trained all over Southeast Asia by Japan during the *Dai-to-a* exercise (Great East Asia Co-prosperity—the Japanese wartime slogan) learned to respect Japanese efficiency and patterns of organization in general.[11] Although the feelings of wartime animosity have by and large withered away, a skepticism may still prevail, welling up as incidents of Japanese cars being burnt in Bangkok and Jakarta.[12] Nevertheless, it may safely be said that business circles in South-East Asia are heavily rooted in something Chinese, and something Japanese, in addition to their knowledge of local and United States business practices. It should be noted, however, that both the Catholic Philippines and Muslim Indonesia are Occidental where civilizational belongingness is concerned, and Oriental only geographically; and the same applies to the Malay majority of Malaysia—factors of some importance in the longer run. The countries and areas most easily integrated into this triangle would be the four mini-Japans: Singapore, Hong Kong, Taiwan Province of China and the Republic of Korea; in the longer run perhaps also the Democratic People's Republic of Korea and the countries of Indochina, and even Buddhist Thailand. Much delicate diplomacy would be needed, as witnessed by the recent (1982) textbook controversy: a major Japanese *gaffe*.[13]

Crucial, however, is the Tokyo-Beijing axis that the Japanese apparently started forging as soon as they realized that the United States and China had rediscovered each other back in 1971. The match of Chinese labor to Japanese capital and technology, within a shared ethos and geo-political potentiality, would seem perfect, but the details nonetheless have to be worked out. It had to take some time before Chinese workers could produce Japanese products, under license in the usual fashion, but this is now happening (the case of Sanyo products, for instance, "Made in the People's Republic of China").[14]

Fifth, to produce for the world market, and not merely for a national or regional market as would be the case for other newly industrializing countries (Brazil and Mexico, for instance, possibly also India), something very persuasive is needed: products that are competitive both in terms of quality and in terms of prices. It is hard to see why relatively inexpensive Chinese labor, coupled with Japanese and Chinese ingenuity, should not be able to produce a great many of these goods.[15] At this point, some particular reasons why Japanese products seem to be so good should be mentioned. There has, of course, always been an element of truth in the standard formulas used by Westerners for debunking Japan and to comfort themselves: the products are imitations of Western products only, and the technology is even "stolen"; the quality is "shoddy"; the prices are low because the proletariat is exploited; the prices are subsidized in one way or another through dumping practices; and—the 1970s version—Japanese methods involve a high cost to nature in terms of depletion and pollution. That element of truth is decreasing, however. The sixth point, that the Japanese market is not easily penetrated by foreign goods, remains true. Could the reason be partly that foreign goods are not really very good—as the Japanese often claim?

Today's Japanese products are often ingenious combinations, in an additive, eclectic manner, of components also known elsewhere—as when one of the excellent Japanese watches is equipped with a computer, or can be converted into a mini-TV just by pushing a little button and raising a small antenna. And they are certainly far from "shoddy", given that the Japanese seem to have reversed the tendency of Western capitalism towards "planned obsolescence" and make products to last, precisely because others do not. Moreover, the typical Japanese enterprise has a low productivity when everybody engaged is counted, because the products, in addition to being capital- and research-intensive, are also labor-intensive: they are made by thorough, cooperative, no-nonsense, serious craftsmanship with every detail attended to by many people. Westerners may point out that the total productivity is low; the Japanese may point out that the quality is high—but they seem to prefer the customers to do it for them. And the customers oblige: Japan having conquered the camera market to a large extent, the TV, radio, and music-machine market and watch markets are next on the list—just as in the sector of transportation the motorcycle market was the first one to be Japanized, now followed by personal cars, trucks and so on.[16] Integrated circuits, computers and all the related products are next in line, to be followed by airplanes and telematics.[14] Again, there is no reason to believe that these products will necessarily be inferior if they bear the stamp "Made in the People's Republic of China".

To the extent that what has been described above is taking shape in the 1980s, obviously the economic conflict formation is not a North/South one but a NW/SE formation. It is, moreover, already perceived as such: the complaints about the penetration into the Western part of the classical triangle of the old economic order—US-EC-Japan (the OECD countries, the Trilateral Commission countries)—are numerous. There is the usual admonition to "exercise restraint".[18] But why should the Japanese or anybody else in the South-East do that when the North-West never exercised restraint? Of course, the interesting thing about Japan is that it is a member of both triangles, and this is used to communicate, forcefully, the apprehensions of the North-West.[19] Sooner or later Japan will probably have to leave the first triangle in favor of the second one, though it may continue to keep it as a listening post. The initiative for this departure would probably, however, come from the North-West, not from Japan. When "exercising restraint" still fails to work for the reason, among others, that customers prefer Japanese products, the response of the North-West will have to be to impose tariff and non-tariff barriers. Psychological preparation for the latter may take the form of statistics about more serious accidents with small Japanese cars than with big American cars, although the probable reason is that they are small rather than that they are Japanese (and the Japanese answer may be to make bigger cars). Another preparation, obviously, is through entrapment.

Before considering what the South-East world would do in the event of a trade war one has to ask why the South-East is engaged in the entire exercise at all. The answer is probably very simple: the countries concerned know from experience how the North-West has been operating, that there are gains to be made if one can dominate the market—domination can lead to vast riches. In the final analysis, the objective is to achieve security, although they also know perfectly well that this road to security may become a road to insecurity.[20] Raw material sources and markets may be closed to them; hence one has to secure a minimum base of both, and the South-East triangle is already such a base. Even for the Japanese, it will take time to saturate the Chinese market with Japanese goods made in China. Should everything else fail there is always the *Dai-to-a*, the South-East world, the secure backyard. And should Japan decline and fall—not in the 1980s, but in the 1990s?—then its decline would offer little comfort to the North-West: ASEAN and China may take over.

About the rest of the world both Japan and China probably take a relatively detached view.[21] They do not think of themselves—as does the West—as being in the center and of the rest of the world as a periphery to be converted into a second-rate copy of the center. On the contrary, to the Chinese what is outside their own center are the Barbarians of the northern (Soviet), eastern (Japanese), southern (Vietnamese, Southeast Asia), and western (the Occidental) varieties. Of all of these, the northern Barbarians are considered the most dangerous, against whom protection should be sought through alliance with their enemies in the rest of Barbaria. The Barbarians to the East are regarded by the Chinese as a *nouveau riche,* slightly vulgarized variety of themselves. The southern Barbarians are thought of as a mixed lot, many of them extensions of China, whereas the Barbarians to the West are held to be more like a lost cause, infantile but clever in their onesidedness. They are like children playing with matchboxes, and one has to know how childlike they are, how dangerous are the matches, and how close the two are to each other. About Barbaria one has to be informed. To convert the Barbarians is a hopeless and even meaningless endeavor, however much they might need converting.

For the Japanese, the world image is not very different: they visualize themselves in the center, then outside that an East Asian zone of the countries mentioned, and beyond that zone the world as a resource. An evaluation of the world in purely utilitarian terms—positively and negatively from the self-centred point of view—is by no means unknown to Western theories of State: this is after all the essence of the concept of "State interest". The Orient might, however, go farther than the Occident in not abiding by universal norms of justice or other ethical rules if their observance should interfere with its own interests, not even seeing the world in universalist terms of Christian compassion or Roman-inspired law, or pretending to do so as Western Powers often do. This form of conduct has the advantage for the non-South-East countries of the world that there will be no attempts to convert them to talking Japanese or Chinese, to adopting cultural patterns of either Japan or China or both, or anything else of the kind. All they are asked to do is to pay for the goods offered in terms of natural resources, capital or technology. South-East capitalism, at the world level, is much more specific, more contractual than the diffuse, all-embracing Western imperialism was. The latter left peoples all over the world with languages not their own, with roots in a political, cultural and spiritual capital not their own. China and Japan will neither attempt nor want to do anything of that kind. They want a distance, and they will get it.[22] By contrast with Americans and French people, who feel flattered and confirmed when somebody far away speaks their own language perfectly, many Japanese and Chinese feel uneasy when somebody is getting under their skin that way. It is like being spied upon, like having nowhere to retreat.[23] It may be added that they have little reason to worry: theirs are very difficult languages, taking at least 5 years of concentrated study to learn. Those 5 years should preferably be at a very early stage in life—but in that case, one *is* a Chinese or Japanese. As the two understand each other up to 80% and can easily acquire the remaining 20% of at least their written languages, they have for all practical purposes a quite effective secret language among themselves with which to conduct their operations—an asset to be added to the others mentioned earlier. In addition, the sun rises eight hours earlier in Tokyo than in Europe, meaning that the Japanese can conclude transactions before a Zurich banker is even in his office!

However, what would happen if the North-West took harsher measures? For a start, such measures would take the form of a boycott, the erection of a tariff/non-tariff wall that would be unscalable, but may invite smuggling. The Japanese probably have no illusions that this would not be a possibility, and for that reason are doing everything to acquire the best possible allies within the framework of a liberal economy: the consumers. Any such wall, low or high, would mean that the consumers would have to subsidize employment for the workers, and profits for the managers and stockholders, by buying more expensive cars than they otherwise could buy—a dubious proposition from the point of view of a market economy. The Japanese may then call the bluff, but that would not decrease the agony of the dilemma.[24] In the meantime, the countries of the South-East would have to secure markets for their goods all over the Third World and possibly also in the North-East world, in the State socialist countries that are still under the spell of North-West capitalism, oriented towards American, French, German, Italian and even British products.[25] Needless to say, any trade embargo measure would be a blow to the South-East, as the North-West is by far the best paying customer, although the Third World *bourgeoisie* is now also coming up quite quickly. The question is whether the governments of the countries of the North-West would be prepared to pay the price of having their citizens subsidize jobs in factories that are unable to stand up to competition from the South-East, and how they would deal with such obvious effects as large-scale smuggling of goods from the South-East. It should also be noted that the Japanese Government may organize a consumers' boycott of North-West goods even if Japan as a country should lower tariff/non-tariff barriers. It is doubtful whether the governments of the North-West have a corresponding leverage on their own peoples.[26]

But what if the tension should worsen and even lead to warlike measures? Let us consider two possible scenarios. In the first scenario, the principal loser in consequence of the expansion of the South-East, the United States and possibly also some Western European allies, initiates warlike measures against the South-East. Today such an eventuality may seem not only hypothetical, but even a crazy figment of the imagination; yet there are, after all, very real interests at stake, and precedents for such a contingency have occurred in history in the 1850s and the 1940s. And it is also useful to realize that in such an event the Japanese would immediately establish closer ties with the Soviet Union (e.g., through projects in Siberia) and try to persuade the Chinese to do likewise.[27] It is important to realize that Japan and the Soviet Union have a bargaining counter that they can push back and forth between the two of them—the northern islands occupied by the Soviet Union after the Second World War. The Soviet Union might on some occasion hand them back, Japan might on some occasion recognize that they form part of Soviet territory; in either case, a considerable price would be paid. In any case, the mere prospect of this type of triple concert—Japan-China-Soviet Union—is hardly likely to make the United States move far in the direction of strong measures, again one instance of how powerless "the most powerful nation on earth"[28] actually has become.

Let us next consider the second possible scenario. The United States and the Soviet Union one day together define the world in terms of "Occident and Orient", or even "white and yellow", finding that there is more to unite them than to divide them in the face of the "yellow peril". This second scenario exhibits a more equal balance; demographically the Orient would be superior, economically Orient and Occident might be more even, militarily the Occident would be superior. A possible counter-argument might be, of course, that all the Soviet Union now has to do is to sit on the sidelines and wait for the South-East to outcompete the North-West sufficiently economically so as to weaken it also politically and militarily. It is not so obvious, however, that the Soviet Union wants to do this. It may fear that a militarily strong North-West in its economic decline might take the desperate action that might come out of a crumbling empire. It might want the North-West to remain strong, partly to scare its own population, partly to have a scapegoat, partly to have something to "catch up" with and "surpass", and partly because it admires the North-West as a source of technology and has not yet overcome the anti-Japanese prejudices referred to above. For such reasons the scenario is not at all implausible, and if it materialized it would be fraught with disastrous implications because of the civilizational and racial cleavages involved.[29]

The obvious Japanese/Chinese counter-strategy would be to see to it that the cold war between East and West is kept alive. As there is such a heavy ideological and political investment on either side, this should not be too difficult; in fact, what is needed would be only something like one Afghanistan every 5 to 10 years. The Russian theories of Chinese provocation in this particular conflict may therefore not be very far-fetched. And there is also another speculative perspective that could be interesting. Whereas the Chinese may produce an enormous amount of economic goods, they may prefer to use the Japanese as their economic agents abroad, their salesmen, turning Japan into a macro-Hong Kong for their purposes. The Japanese may prefer to let the Chinese do the political work for the South-East triangle. In that way, the Chinese would remain economically pure and more free to pursue their internal zigzag course and the Japanese more free to look politically neutral, solely engaged in economic pursuits. This does not exclude the use of the Chinese abroad as a general network, but much more important, in this perspective, is Japan as a macro-Hong Kong.

One may now ask: if the South-East triangle is that strong, why did not this process start earlier? For a few very simple reasons:

1. China was down, weakened by internal processes and external imperialism, a worn-out country; Japan came out of Tokugawa isolation as *the* modern and leading country in the Orient, tempted therefore into military adventurism, now facing a choice between a symbiotic or subservient relation to China, obviously preferring the former—as dominance over China is out of the question.
2. The Occidentalization process (in the sense of adding Occidental elements to the civilization) had not yet gone far enough to provide that forward and outward thrust so characteristic of the Occident.
3. The opportunity had not yet come: an Occident weakened by decolonization and the part of NIEO initiated by the Third World.

As a conclusion: the particular historical occasion did not yet present itself. Today it does; the 1980s will show to what extent it is made use of. Let me only add that this is not a "Japanese 21st Century" hypothesis. It is not about Japan only, Japan may not last that long. And it is a fact rather than a hypothesis, foreseeable for a long time,[30] about the South-East world. So let us call it the fourth world, and remember the Bible: The last shall be the first.

Notes

1. The reasons given below are good reasons why the South-East triangle is a strong one. But there is a deeper underlying assumption: that the world economic system organized in a capitalistic way will have to have some kind of center, for it cannot be really polycentric. This is explored to some extent by the author in his Millennium Lecture, London School of Economics, January 1981. A center is based on the law of uneven development: by and large, it has more of what is needed than is found elsewhere, above all control. At the same time, there is probably also a mental *Gestalt* at work: a notion among sufficiently many people around the world that this type of activity is of a kind that begs the question where the center is, and then consciously or unconsciously looks for evidence as to the whereabouts of that center. It's a little bit like a sports competition: given the patterns of individualism and competition, it would be very difficult for most people raised in that mental configuration not to feel dissatisfied, incomplete, unless a clear answer is given to the question, "Who won"? (See " 'A structural theory of imperialism'—Ten years later", *Millenium*, Vol. **9**, No. 3 (Winter 1980–81).)
2. Corresponding saving ratios in Western world might be between 3 and 7%; for the US it is about 6%. See *Newsweek* (2 August 1982), p. 48: interview with the Minister of Economic Planning of Japan.
3. Australia and New Zealand enter as rich dependencies, to be maintained adequately by the

center, the South-East triangle. Historically and culturally, racially and ethnically, they are different. I am indebted to Reginald Little, an Australian diplomat with much experience of Tokyo and Beijing, for informing me about how China and Japan may look from Australia and New Zealand, and vice versa; see his unpublished thesis (Institut Universitaire de Hautes Études Internationales, Geneva, 1978): "Economics, Civilization and World Order".

4. See Johan Galtung, "Five cosmologies: An impressionistic presentation", *GPID Papers* (Geneva: 1980), *Alternatives*, 1981. Surprise to Westerners: the Japanese and the Chinese combining orientations that do not combine in the West.
5. In fact, it is almost surprising how little Weber has to say about workers in his famous book on the spirit of capitalism—it is as if they simply did not exist. Weber's work was a rebuttal of Marxism, or intended as one—certainly not an adequate one when the problems Marx points to are left uncovered.
6. This point is explored, to some extent, in Johan Galtung. "On human-centred development", *GPID Papers* (Geneva: 1980).
7. And yet Max Weber, in his book *The Religion of China*, predicts that capitalism will not have much of a possibility in Japan!
8. I am indebted to Professor Ungku Aziz for pointing out so clearly the importance of this particular factor during a seminar in Kuala Lumpur, September 1979.
9. Of course, a Westerner with a very uncommon name might find protection, a source of solidarity, and a focus of identity in the group of people with that name. But this is hardly the case for the Smiths and the Joneses in the Anglo-Saxon world, for the Hansens and Jensens in the Nordic countries, and so on.
10. It should be noted that it is in that context that the theory of alienation is developed; one more example of how culture-bound a theory may be.
11. It is very difficult to ascertain how many people in Southeast Asia in fact were collaborators with the Japanese during the occupation, but the number must be considerable. Today one can still find Southeast Asians in their late middle age singing Japanese military songs and praising the efficiency of the Japanese war machine. And what one finds above all would be the sprawling growth of the Japanese economic machine, no doubt making use of the groundwork done during the occupation. For some impressions about this, see N. I. Low, *When Singapore Was Syonan-To* (Singapore: 1973).
12. Both Thai and Indonesian students seem to be good at organizing ritualistic burning of Japanese cars and flags. The question is: how much real resistance to Japanese consumer goods (better than others, cheaper than others) there really is in the population. At any rate, the Japanese fear such incidents.
13. See *The Daily Yomiuri* (15 August 1982), for an excellent summary.
14. One has a sense of *déjà vu* when one discovers that type of gadget in Mongolia, for instance (1980).
15. It should be pointed out that this association is not the same as trade between Japan and China, nor is it exactly the same as the Japanese building a factory in China, under a limited-time contract. It should rather be seen as a long-lasting partnership, as a type of transnational corporation with all kinds of transfers within the corporation and a high level of coordination when it comes to relationship to the outside.
16. As an example, see the article "Welche Autos kaufen die Deutschen?" *Die Welt* (30 January 1981), with the following table (in percent) of the makes of cars which were doing better than the year before, and those which were the losers:

The winners (in %)		*The losers (in %)*	
1. Mitsubishi	+100.7	1. Lada	−43.7
2. Toyota	+ 81.1	2. Leyland	−42.6
3. Datsun	+ 61.0	3. Alfa Romeo	−35.1
4. Honda	+ 50.0	4. Volvo	−31.7
5. Mazda	+ 44.2	5. Talbot	−20.4
6. Mercedes	+ 2.6	6. Ford	−19.9

17. For an excellent article on this particular subject, see "Integrated circuits in industry: How Japan bred a winner", by the *Financial Times* Far East editor in Tokyo, Charles Smith, *Financial Times* (29 January 1981). The article as usual points to the excellent bureaucracy-corporation-intelligentsia-cooperation in the Japanese system, and the Japanese genius for "improving on

the original". For the emphasis on "low-noise and energy-saving aircraft", see *Asahi Evening News* (9 February 1981).

18. A typical quote from the Western press: "Japan must alter its trade role, disillusioned US advisers agree—Japanese leaders will find that to try to continue upward mobility of their people through excessive penetration of others' markets is risky" (*International Herald Tribune*, 5 January 1981); but cf. slightly contradictory headlines about the same mission: "EEC sees 'glimmer of hope' over Japan trade", *Financial Times* (29 January 1981), and "Lawine aus Japan rollt weiter. Mission der EG gescheitert", *Die Welt* (30 January 1981).
19. It would be interesting to have a clearer view of Japan's past role in the Trilateral Commission; see a partial analysis in Holly Sklar (Ed.), *Trilateralism* (Montreal: Black Rose, 1980).
20. The question, of course, is how well they know that. A personal experience: Giving a talk about international political economics in general and the Japanese situation in particular to the editors of a leading Japanese newspaper early in the 1970s, I concluded that there would be considerable tension in Southeast Asia, and rightly so, because of the relationship that was emerging. The inference drawn by the editors, and they were not right-wing, was the following: This is probably correct; Japan should for that reason rather move to Africa!
21. See the newspaper article referred to in note 13. It should be pointed out very clearly that the Japanese attitude to East and Southeast Asia is very different from the attitude to the rest of the world, however—in a sense a more occidental attitude. It is not my impression that the Chinese have a corresponding trichotomy; the concept of barbarian seems to start right outside the doorsteps although there are barbarians of various shades and degrees.
22. What is being said here is that the distance between the internal sector and the external sector is greater than in the West. The external sector is even more remote, even less the object of empathy; and the internal sector is even more consolidated, at least potentially more an object of internal solidarity—although more in the Japanese case (Shintoism) than in the Chinese case.
23. The paradigmatic example is the Western room provided in so many Japanese homes: a room with Western furniture in which Weterners are received—thus far, but no further!
24. In this, there is something corresponding to another famous case in United States history: the contradiction between norms of equality and mobility and the practice of prejudice and discrimination against non-whites. Gunnar Myrdal referred to this as an American dilemma; this is likely to become an American dilemma II. One may also recall the history of the famous Navigation Act in British history.
25. It should be remembered that the State socialist countries still suffer from the Marxist blind spot: the enemy is private capitalism in its international manifestations, capitalist imperialism; it has its center in the North Atlantic area; this is the one to be beaten, not something emerging from a non-white, non-Occidental corner at the other end of the world. The model for the North-East is in the North-West, not in the South-East. It is the North-West the socialist countries want to emulate, even overtake. That this task may become considerably easier because of the way in which the North-West is being shaken by the emerging South-East will create tremendous problems ideologically, politically, psychologically. The racial/ethnic dimension of these problems should not be underestimated either.
26. It is possible that the United States Government might have that type of leverage on its own population; in times of crisis, the American people have a tremendous ability to respond collectively. The same might be true of the British. But would it be true of the French, of the Italians? Will they forego some individual advantage just because the government tries to persuade them to buy home-made cars?
27. Thus, it is my contention that what today looks like some kind of working triangle, the US-China-Japan connection, in fact is a very weak one. The US would do well not to underestimate the kind of wound left on the Japanese soul by the war, what happened before it, and its aftermath—and particularly the nuclear bomb attacks on Hiroshima and Nagasaki, with their racial overtones. And the Chinese have suffered even more humiliation from the West.
28. Of course, it is not wise of a country to use such expressions about itself, nor to refer to the US President as "the most powerful person on earth". The contradiction between such expressions and the reality during, for instance, the hostage crisis with Iran, becomes only too tangible.
29. See Johan Galtung, *The True Worlds: A Transitional Perspective* (New York: The Free Press, 1981), sect. 6.4, particularly p. 292.
30. I am thinking of the book by Hermann Kahn, *Emerging Japanese Superstate: Challenge and Response* (New York: Prentice-Hall, 1971).

Neither Prince nor Merchant: Citizen—An Introduction to the Third System

MARC NERFIN

Tell me: what would be worse than
not to be a citizen?
Dante, *The Divine Comedy*

Contrasting with governmental power—the Prince—and economic power—the Merchant, there is an immediate and autonomous power, sometimes patent, always latent: the power of the people. Some among the people develop an awareness of this, associate themselves with others and thus become citizens. Citizens and their associations, when they do not seek either governmental or economic power, constitute the third system. Contributing to make patent what is latent, the third system is one expression of the autonomous power of the people.

Associations are legion. Words are never innocent. The phrase "nongovernmental organization" implies that government is the center of society, and people its periphery. To insist on the autonomy of people also requires some semantical cleaning up. Except in quotations, we will use, instead of "NGO", the expressions "association", whose sense is wide enough, and "third system", in the precise acceptation proposed in this introduction. The 19th edition (1981) of the *Yearbook of International Organizations* describes 5955 international nongovernmental organizations *plus* 3443 internationally oriented national bodies. The 1981 *Directory of Non-Governmental Organizations in OECD Countries Active in Development Cooperation* includes the profiles of 1702 such bodies. There are many times more in the national and local spaces. In India, in 1978, a questionnaire on their participation in development could be sent to 1400 "NGOs"—of which more than 90% had no effective international links (Szuszkiewicz, 1979). In France, associations were in the hundreds in the 1930s, in the thousands in the 1950s, 10,000 in 1960, 30,000 in 1977 and anything between 300,000 and 500,000 in 1981, a year during which some 100 new associations were established every day (Passaris, 1981).

The game of numbers offers only limited pleasure and, in this case, even less relevance. In this magnitude, numbers only invite and at the same time defy typologizing. For the purpose of this introduction, a snapshot of that part of the galaxy of associations which could be considered as third system may suffice to discern some structure and some trends.

Associations are formed by citizens moved by their situation in society and/or some personal or moral reason. Social history suggests that the former mobile is more permanent, the latter more ardent, and the combination of both the best. A worker usually remains a worker, and his/her reasons to be active in a trade union are part and parcel of her/his social existence. The same holds true for members of ethnic minorities (or majorities). A woman has even deeper reasons to be a lifelong feminist activist. But not all workers or all women, become citizens, and the personal/moral motivation is always essential. Motivations are many, but observation of the third system as it currently unfolds suggests that there are a few deep-seated mobilizing themes, notably, local self-reliance, peace, women's liberation, human rights, environment, solidarity with the people of the Third World and consumers' self-defence.

The multiplicity of forms under which these associations appear corresponds to the diversity of mobiles and circumstances. Because they reflect the autonomy of the people, associations are often allergic to the forms defined by the establishment. The term is therefore used here, on purpose, in a rather loose sense.

Many associations are officially recognized and/or registered, with formal constitution, membership, organs, reporting, and so on. Others are just *ad hoc* gatherings of like-minded individuals who occasionally share ideas and experiences through a round-robin letter. In between, the spectrum includes all other possible configurations: some are underground; others do not care about their legal status and just exist; a few even resemble political parties, such as the Greens in Western Germany, and still are part of the third system as long as they remain a minority and keep out of the executive power. There may also be groups of marginal shareholders who try to voice social concerns in a transnational corporation. Quite a few, especially in Buddhist and Christian cultures, have a religious connotation. Others constitute vast movements assembling on some occasion several hundreds of thousands of people. In short, third system associations are as diverse as societies themselves.

Citizens and their associations usually act in a determined space—local, regional, national, multinational, global—but also, and increasingly so, in several spaces simultaneously. Amnesty International, to take only one example, acts in the global space through representations to United Nations bodies, in the national space through pressure on a government, and in the local space through the many groups that "adopt" a political prisoner and campaign for his/her liberation.

Whatever makes citizens join forces and wherever they take action, third system associations/activities can be considered under a few broad, non-mutually exclusive and non-comprehensive clusters.

Some are geared to the realization of an immediate *project* intended to solve a specific problem confronting the members of the association, or pursuing a more general objective: organizing a peasants' army (Bhoomi Sena) to enforce land reform in a district, improving a neighborhood, sharing appropriate technologies, ensuring equal access to jobs, decreasing working time, opposing the construction of a nuclear facility or the deployment of missiles, reconverting the manufacture of arms into that of socially useful products, preventing the export of dangerous drugs to the Third World or the dumping of toxic wastes, extending financial or technical support to a local initiative, facilitating the exchange of experiences, cross-cultural dialogues or the search for alternatives. The activity of the association often takes the form of a publication, the *Tribune* or the *ISIS Bulletin* to choose an example in the feminist movement.

Advocacy activities may be seen as constituting a second cluster. Associations may be formed to advocate a new world order or the New International Economic Order or a federalist world; a new approach to international security; better terms of trade for Third World countries; the recognition and effective respect of minority rights; breastfeeding; consumption of local products; education with production; another development in health; equality in opportunities among individuals and societies; protection and enhancement of the environment; eco-development; a reform of the United Nations . . .

A third cluster of associations deals with *accountability* ("those who hold power must be held accountable for the consequences of its exercise") and the necessary mechanisms. The Permanent Peoples' Tribunal and its predecessor, the Russell Tribunal, are examples as far as the Prince is concerned, the International Organization of Consumer Unions or Ralph Nader's Corporate Accountability Research Group and its monthly *Multinational Monitor* as far as the Merchant is concerned. The European Environment Bureau monitors environmental affairs in the European Economic Community, and the Environment Liaison Bureau in Nairobi the activities of UNEP, the United Nations Environment Program.

The three types of activities usually imply some underpinning policy-oriented research and have a broad educational role, and some associations devote themselves specifically to such functions. This hazy picture may perhaps, at this stage, be complemented by a more focused look at three specific spots in the third system constellation.

First, *peace*. The largest post-1945 demonstrations took place in Western Europe in the Fall of 1981. Over 50,000 people marched in the streets of Paris and Brussels, 100,000 in Rome, 250,000 in London and Bonn to oppose the deployment of Pershing 2 and cruise missiles in Western Europe. A million Americans overwhelmed New York Central Park and mid-town Manhattan on June 12, 1982 on the occasion of the United Nations General Assembly Second Special Session on Disarmament—a demonstration far larger than any during the 1960s anti-Vietnam war protests, and possibly the largest ever in New York City.

Even before the New York demonstration, George Kennan, a senior US establishment figure and diplomat, has this definitive comment:

> The recent growth and gathering strength of the anti-nuclear war movement here and in Europe is to my mind the most striking phenomenon of this beginning of the 80s. It is all the more impressive because it is so extensively spontaneous . . . At the heart of it lie some very fundamental and reasonable and powerful motivations: among them . . . a very real exasperation with their governments . . . They are the expression of a deep instinctive insistence, if you don't mind, on sheer survival . . . Our government will ignore this fact at its peril. This movement is too powerful, too elementary, too deeply embedded in the natural human instinct for self-preservation to be brushed aside (Kennan, 1982).

Second, *women's liberation.* Social orders created and dominated by men have failed. Another development, calling as it does for new structures, implies the rejection of any mimetism. Competing with men to run the same society, women would only reproduce outdated and inappropriate patterns. The chance of the women's liberation movement, which asserts itself everywhere, in all spaces and all battlefields, is to be imaginative, innovative, alternative. Like the Third World, the young, the alienated, the exploited, the marginals, women represent at the same time the difference pregnant with change and the totality that prefigures a better world. They may be the midwives of another development.

Hazel Henderson (1976) has noted that "these non-governmental organizations formed over the past fifty years by women, their proliferation in many countries and subsequent convergence on world problems and the restructuring of policies to address them, are a prototype for international action".

"Women", writes Marilyn Ferguson (1980), "represent the greatest single force for political renewal in a civilization thoroughly out of balance. Just as individuals are enriched by developing both the masculine and feminine sides of the self (independence and nurturance, intellect and intuition), so the society is benefiting from a change in the balance of power between the sexes." And she quotes Gandhi who once said, "if satyagraha is to be the mode of the future, then the future belongs to women."

Third, *accountability.* The principle of accountability and its enforcement emerges as a central theme in the efforts to reassert the autonomous power of the people *vis-à-vis* that of the Prince and that of the Merchant. Neither appropriate mechanisms nor, even less, its formal recognition would by themselves transform the world and its societies, but its exercise could be a critical step in this direction. An instrument of democracy (that is, strictly speaking, the power of the people), accountability may progressively circumscribe the power of those who hold it. It may instill a new sense of self-confidence among the people. It is a natural battlefield for citizens' associations. A good example is provided by the Consumers Association of Penang (Malaysia), which is linked with the IOCU mentioned above. CAP monitors prices, advertisements, marketing techniques, sales of dangerous goods, inadequacy of health care, public transportation or housing. It alerts authorities and people on any abuse. It carries out research into basic needs satisfaction, fights environmental deterioration (e.g., chemical pollution, deforestation, over-fishing), provides support to local initiatives, organizes educational programs, publishes a monthly paper in Malay and English and disseminates statements in four languages as well as books, pamphlets or postcards. More specifically, it handles complaints from people against abuses they suffer from either the governmental or the economic powers.

A CAP comment is relevant to this discussion:

> In ex-colonial societies where the people have for too long been used to the passive acceptance of life's injustices, the successful lodging of a complaint changes the perception and attitude of individuals who now see that redress can be obtained if one is willing to do something positive about it. The Complaint Service thus becomes an effective means and channel through which the public is able to exercise its rights to fight business malpractices and to press for fair and better services from companies and government departments as well as to demand protection of these rights from authorities (Khor Kok Peng, 1980).

The phrase "third system" in the sense accepted here and in the practice with which it is associated was coined in the Fall of 1977. It was first embodied in the title of the "third system project" carried out between 1978 and 1980, by the International Foundation for Development Alternatives as a contribution to the elaboration of the United Nations International Development Strategy for the 1980s (IFDA, 1980; Nerfin, 1978). As implied in the definition used here, the concept extends well beyond the modest pretext of its origin.

The association with the phrase "Third World" is not only deliberate: as a matter of fact, both phrases come from the same source; both are meant to evoke *le Tiers État,* the "third estate" of the French *ancien régime*. Before the 1789 Revolution, French society comprised three "estates": the nobility, the clergy and the third estate—that is, the vast majority. Alfred Sauvy was the first, in 1952, to use the phrase "Third World" to refer to the periphery, or the South, a phrase which has since then gained wide acceptance. That Sauvy (1982) implied that the OECD countries (the West) and the CMEA countries (the East) were the other "worlds" does not justify the use of the expressions "first" and "second" world. In this context, ordinal numbers are historically, conceptually and politically misleading. However, third system is conceptually closer to third estate than third world to either. The latter concept is geo-political; it concerns countries. The former two are socio-political; they concern people, and that is what the third system is about.

For in the beginning were the people. As history unfolded, various social groups and individuals emerged as self-appointed rulers and leaders which, to put it schematically, managed to extract from the people, the producers, a tribute or plus-value. To this effect, landowners, warriors, aristocrats, clergymen, merchants, moneylenders, capitalists, generals and bureaucrats established various forms of government and economic organization. As a result of this evolution, people are dominated, at this particular point in history, by governmental power, the Prince, and by economic power, the Merchant, sometimes united, sometimes antagonistic, always present and both unable by themselves to offer workable solutions to the two major problems confronting humankind today: peace and development.

Government still fascinates, understandably when it is sought as a tool for change at the service of people, less so when it is opposed. The Prince may well be the object of hate or sarcasm, yet it is somehow admired. Even when a particular prince is really bad, the very manner in which the criticism is construed implies the possibility of a good prince, a prince-philosopher (me? or the friend who may listen to me?). Princes of the literary kingdom themselves, though princes in their own right, often bow in a way to the political prince. Consider for instance three of the great contemporary writers of Latin America, a continent which has had and still has its lot of bad princes: Miguel Angel Asturias with *Mr. President,* Alejo Carpentier with *The Recourse of the Method* and even Gabriel Garcia Marquez with *The Autumn of the Patriarch*: the prince is torn into pieces, and yet somewhere floats the incense of mythification.

The alternative to the Prince, for a handful, is age old anarchy. The Prince does not really matter, it is the institution which is bad, so let's get rid of it. Neither God nor Master, let us join forces, and tomorrow the International will be Humankind.

The Merchant used to be close to us. The growing geographical and economic distance between producer and consumer made it at the same time more mythical in our perception and more real as a determinant of our daily lives. It has now grown into a monster/hero whose misdeeds are proportional to the services it renders. We resent its influence and its riches because we are under its influence and because its riches come from us as participants in the process of production and as consumers. For instance, we oppose transnational corporations—to a point: I don't like Nestlé, but I don't give up coffee. I don't like ITT, but I need to communicate with my co-worker in Chile. I don't like Hoffman-La-Roche, but what may happen to my child without this irreplaceable drug?

One alternative to the Merchant is workers' self-management, peasants' cooperatives and equal exchange among them. But not everyone likes to be a manager, and almost everywhere self-management has begotten new managers and restarted the process of alienation. Another alternative is the Plan as a people-serving agent of rational production and distribution of socially useful goods and services. Whilst it alleviates many of the Merchant's shortcomings, the Planner still has to outdo the former's capacity to deliver the goods.

The permanent fascination of the Prince, the resilient reliance on the Merchant and the elusiveness of utopias suggest not only that utopias need to be revisited, but also that neither Prince nor Merchant have outlived their usefulness.

On the present level of productivity, with its implications in terms of surplus extraction, ideological manipulation through the mass media, economic, social and political organization, and weaponry for both external and internal use, the fact is that some form of government, in the boundaries of the current nation-states system, is unavoidable, and to a point, necessary. In a world dominated by two superpowers, smaller or less powerful states still offer some protection against total subservience. Similarly unavoidable and to a point necessary is some form of economic organization, private or public, guided by the invisible hand or by the "Plan", mixing one way or the other transnational corporations, state enterprises, national and local capitalism. When the most basic human needs remain unsatisfied, those able to foster production as well as those able to ensure some social security through redistribution still have a function.

This is to say that we have to recognize that the Prince and the Merchant, as alienated from us and as alienating as they are, remain parts of the realm of necessity. What is at stake today is neither the withering away of the state nor generalized producers'/consumers' self-management. But understanding necessity never prevented liberty.

The next fact to reckon with is that the Prince and the Merchant control only parts of power and that their power is subsidiary to that of the people.

Whether they exist *de facto* or *de jure*, whether legitimate or not, whether serving the general interest or not, they cannot possibly represent society in its totality and plurality. Moreover, they exert their powers *upon* society. Their very existence depends on the people. They are, in a sense, our creatures.

People—the men and women we are—also have power, which we can exert both on them and on ourselves. The power of the people is thus *autonomous*.

Beyond the principles, as essential as they are, stands reality. People—societies—are not homogeneous.

To come back to an earlier analogy, the French third estate was not homogeneous. Opposing the nobility and the clergy were the merchants, the artisans, the peasants, the first industrial capitalists, the intellectuals. Their interests, beyond the abolition of absolutism, could hardly be the same. The ideology, formulated by Jean-Jacques Rousseau, Denis Diderot and others, was essentially democratic—but the bourgeoisie, assuming leadership, merely used it to take over from nobility and clergy, which disappeared as both governmental and economic power. By and large, the people simply changed masters. A similar scenario occurred, in a different context and for different reasons, after the 1917 socialist revolution in Russia. Bureaucracy, not bourgeoisie, took over and one had to wait until the summer of 1980, in Poland, to see the people—the workers—get together as an autonomous social force in a society born out of the quest for justice and socialism.

This suggests two observations: First, the people, save in exceptional circumstances, do not act politically as such, *en masse*. Not all people act as citizens. Minorities, self-propelled but society-borne, either act on the behalf of people or, better, perceiving a problem and outlining a solution, formulate a *project* in which people recognize themselves. They join forces and thus give meaning and dimension to the *movement* through which the latent power of the people becomes patent. Second, whatever the ideology, the social origin or the motivations, no minority, group or individual seeking or exercising power can be given a blank check.

Whatever provisional or lasting benefits the people may get from a change of power, or from power or counter-power exercised on their behalf, they will remain in a subservient situation if they do not retain their autonomous power. This is not to blame those who aspire to or get governmental power—many are genuinely honest and try in earnest to achieve what they had set out to do—but a reflection of the nature of the Prince and a reminder that in the polity as in the economy, the division of labor breeds "disabling professions" (Illich, 1978).

The third system is not coterminous with the people. It is only those among the people who are reaching a critical consciousness of the role they may play. It is not a party or an organization, but the movement of those, associations or citizens, who perceive that the essence of history is the endless struggle by which human beings grope to master their own destiny, a struggle which is in itself the process of humanization of man. The third system does not seek governmental or economic power. On the contrary, its function is to help people to assert their own autonomous power *vis-à-vis* the Prince and the Merchant. It endeavors to listen to those never or rarely heard and to offer, at least, a tribune to the unheard voices.

Realities, telecommunications and perceptions progressively confirm the oneness of this planet. The risk of nuclear holocaust and the combination of underdevelopment and maldevelopment make us one. Environment and health hazards make us interdependent. Two thirds of the planet—its oceans—are open to global management. Citizens and associations struggling in local spaces need to share experiences with others and to feel they belong to a larger whole. Those working in the global space, beyond their "sectoral" differences, also need to share experiences and to articulate their actions with those working in the other spaces.

The other systems of power do have their own linkages, mechanisms of consultation, concertation and joint action. Princes, whether they somehow represent their people or oppress them, have their fora, regional—like the Arab League, the Association of Southeast Asian Nations, the Council for Mutual Economic Assistance, the Economic Community of West African States, the European Economic Community, the Sistema Economico Latino Americano—or global—the United Nations system of agencies, programs and conferences. Those of the Merchants who operate beyond their national borders have formed transnational corporations. The third system as such has nothing of this sort. Anyhow its links are bound to be different.

There seem to be two possible approaches to third system global relations. The first gravitates around the United Nations system; the second, which is *networking*, while not necessarily excluding the first one, is more in line with the nature and goals of third system associations.

The United Nations approach itself is at least twofold. The first branch involves some improbable journey towards a remote utopia but nevertheless deserves to be mentioned. After all, the Charter opens with the famous "We the peoples of the United Nations", and not with "We the governments of the United States of the World". The EEC Commission, which has more power than the UN Security Council, coexists with a European Parliament which may have a limited role but whose members, elected as they are, somehow speak for the people of the member countries. Another possible analogy is offered by the International Labor Organization, whose General Conference is composed of four representatives of each member-state with individual voting power. Two are delegates of government, one of the employers, and one of the workers.

This may not go very far, especially since the "non-government" delegates are chosen by their respective government in agreement with each country's most representative industrial organization, but something along these lines would be a significant improvement in the United Nations proper, where some 700 "non-governmental organizations" are in some sort of consultative status with the Economic and Social Council. They are a mixed bag of organizations, ranging from the International Association of the Soap and Detergent Industry to the Christian Peace Conference. Most are essentially Western, and whether their relationships with the ECOSOC are "obsolete and unproductive", as some say, does not really matter (Andemicael, 1978). Furthermore, there is a persistent habit, in the "public information" sectors of the United Nations, to consider NGOs as mere conveyor belts of intergovernmental or bureaucratic wisdom distilled from above to the "public opinion" seen as a passive receptacle.

It is, of course, not prohibited to dream of another United Nations. Overcoming the consequences of the original sin of an organization of governments of nation-states, it may reflect better the real world through, for instance, some representation of the three powers—the Prince, the Merchant, and the people in their diversity. But its constitution, selection/election procedure and the rest is better left to experts to work out. Time is not pressing.

In the meantime, and leaving aside both utopia and the boggy bottom of ECOSOC NGOs, a *major departure from the traditional relationship between the United Nations and non-governmental organizations* must be mentioned. In October 1979, at a meeting on infant and young child feeding which took place *at the center of the WHO/UNICEF decision-making process,* various groups of participants were involved on an *equal* footing: representatives of governments, scientists, health workers, executives of infant food manufactures, representative of the United Nations system *and* constituent-based associations from both South and North. The composition of that meeting helped making it a *qualitative leap forward in the approach to infant feeding* (Lemaresquier, 1980). And there should be no surprise that the Executive Director of UNICEF, Jim Grant (1980), is of the opinion that "we have had a remarkable amount of structural change in the past 30 years . . . most of this change has been brought about by public pressure, with people ahead of governments . . . The outstanding example is the national liberation movements, which have all been against governments. The civil rights movement in the US was another case of people being ahead of the government and forcing change. And the environmental and women's movements . . .".

The 1979 event, unique happening or precedent, this remains to be seen, came after almost a decade of a new presence of associations in United Nations meetings. Perhaps the turning point was the 1972 Stockholm Conference on the environment, where as many interesting and far-reaching things happened in the adjacent Forum as in the inter-governmental assembly. Since that date, associations have been present and active in most major United Nations conferences, advocating their views, exchanging information, organizing debates, monitoring the position of governments, lobbying delegates, publishing journals or linking up with people at large. Examples are the World Food Conference (Rome, 1974), the Seventh Special Session of the United Nations General Assembly on Development and International Cooperation (New York, 1975), Habitat (Vancouver, 1976), UNCTAD IV (Nairobi, 1976), the Conference on new and renewable sources of energy (Nairobi, 1980) or the Second Special Session on Disarmament (New York, 1982).

Whatever the immediate impact of such activities, they serve another, and far-reaching purpose, not without similarities to the International Exhibitions of the last century. The First International—the International Workingmen's Association—resulted from a meeting of workers in London, in 1862, on the occasion of the International Exhibition. Perhaps the United Nations conferences of the 1980s will be seen in retrospect as having played a similar role, facilitating contacts between people otherwise scattered, opening new space to networking.

Networking is the other approach to third system linkages. There is nothing new in its practice: since the beginning of history, some people have always been in touch with others on the basis of common values and interests. What is new is that networking becomes progressively *global* and is more and more used not out of mere necessity, but as a conscious alternative to institutional arrangements.

It becomes global because of the new perception of the oneness of humankind, and because technology makes it possible: air travel and the Xerox machine, and the tapes, and now, in a new revolution, telecommunications. "During the coming decade," says the Chancellor of the University of Moratuwa, Sri Lanka, "more and more businessmen, well-heeled tourists and virtually all newspersons will be carrying attaché-case-sized units that will permit direct two-way communication with their homes or offices, via the most convenient satellite. These will provide voice, telex, and video facilities (still photos and, for those who read it, live TV coverage). As these units become cheaper, smaller and more universal, they will make travellers totally independent of national communications systems" (Clarke, 1983). And this need not be limited to Princes, Merchants and affluent people: it may and will be used by third system associations.

Networking already offers a concrete alternative to conventional institutions. The latter are usually designed and operated in a pyramidal manner so as to provide for hierarchical relations between a center and a periphery, a leader and those led (even when centers or leaders result from some consensus). They rest on a vertical division of labor between the bureaucrats and the membership. They nurture disabling professions. They are internally and externally competitive and foster bigness. They seek and dispense information rather than facilitating communication. They breed conformism. They are change-resistant and self-perpetuating. As a whole, they hinder rather than enhance freedom.

In sharp contrast, networks operate horizontally. Their centers are everywhere, their peripheries nowhere. Networking simply means that a number of autonomous, equal and usually small groups link up to share knowledge, practice solidarity or act jointly and simultaneously in different spaces. Based as they are on moral (as distinct from professional or institutional) motivations, networks are cooperative, not competitive. Communicating is of their essence. They ignore coordination as a specialized task. Leadership, if and when needed, is shifting. The *raison d'être* of networks is not in themselves, but in a job to be done. When there is one, they set themselves up. They adjust quickly to changing circumstances. They are resilient in adversity (for instance, that one entity is co-opted by the establishment does not affect the whole). If they are no longer useful, they disappear. They are transient. Moving outside mainstreams and beaten tracks on somewhat marginal paths, networkers learn from each other to look elsewhere and beyond the conventional and the immediate. Being multidimensional, networks stimulate imagination and innovation. They expand the sphere of autonomy and freedom.

Discussing the task, substance and strategy of the social movements in trying to stop the apocalypse, Rudolf Bahro (1983) writes:

> There are various seemingly irrational responses in vogue: the New Age Movement or the Aquarian Conspiracy. One thing about them is correct: what is required really is a world-embracing counter-movement, and there is no archimedian point within the existing institutions which could be used to bring about even the smallest change of course. Without forces which attack from outside, the atomic holocaust is not to be staved off... Only the most basic social movements can bring about that break in cultural continuity without which we shall be unable to save our very existence.

This also applies to the daily holocaust brought about by underdevelopment and maldevelopment and other results of the same misdirection of human affairs.

This identifies a critical problem. The counter-movement, the conspiracy, the third system, however named, still lacks full universality. Its source is the same everywhere—the autonomous power of people, and so is its most immediate goal—survival.

But the latent power of people materializes only here and there. This is what happens when hundreds of thousands of Europeans assemble to oppose nuclear weapons or when the consumers of Penang lodge complaints against the

Merchant's abuses. But these remain worlds apart. Networking may now be part of North American reality, and the basis of the work of the International Baby Food Network Action, but things still look as if these were only isolated islands emerging in the still unconnected archipelago of another development.

The movement and the networks discussed here are never created from above. They are only self-generated, but if self-generation is not somehow stimulated, the risk would be great that past and present inequalities will once more survive and negate the very future which is dreamt of. The way to consciousness passes through global information, so as to make every significant event in any local space an event in the global space. This points out to a crucial battlefield of the third system, communications.

There is, of course, a tremendous risk embedded in any communication structure, that of being manipulated by those who run or control it. No institutional mechanism will ever suppress the risk. Protection may only come from a technology that would be both appropriate and capable of appropriation by those using it.

What matters, in communications as in all other matters, is to make available to people the instruments through which they may exercise their autonomous power. In the process of realizing this potential, the role of citizens and/or associations is both critical and temporary. They may be the seeds of change, but if seeds do not flourish, there is no harvest. The medium may not be the message, but the process is certainly the policy, if the process means *enabling* people to become citizens, empowering them to act autonomously, to hold Prince, Merchant, and third system associations accountable for the consequences of the use of whatever power, opposed or complementary, they have. For the last thing to do would be to exonerate the third system from what it requires from the others. Only full accountability will help the third system to avoid bureaucratization, resist co-option, keep its role of countervailing power, preserve its capacity for permanent renewal, strive to run itself by imagination—in a word, remain the servant of the people.

References

Alger, Chadwick F., "Empowering people for global participation", in *International Transnational Associations*, No. 12 (December 1980), pp. 508–510.

Alger, Chadwick F., *The organizational context of development: Illuminating paths for wider participation* (Tokyo: United Nations University, 1979). Doc. HSDR GPID 17, mimeographed bibliography.

Andemicael, Berhanykun (rapporteur), *Non-governmental Organizations in Economic and Social Development* (New York: UNITAR, 1975), 44 pp.

Bahro, Rudolf, "Who can stop the Apocalypse", in *IFDA Dossier 34* (March–April 1983), pp. 51–64.

Clarke, Arthur C., *Beyond the Global Village,* Address on World Telecommunications Day, United Nations, New York, 17 May 1983. 14 pp., Mimeo.

Dag Hammarskjöld Foundation, *What Now—Another Development* (Uppsala: *Development Dialogue*, Special issue, 1975), 128 pp.

Elgin, Duane, *Voluntary Simplicity: An Ecological Lifestyle that Promotes Personal and Social Renewal* (New York: Bantam Books, 1982), 264 pp.

Ellis, William N., "New age world governance", in *IFDA Dossier 34* (March–April 1983), pp. 43–50.
Ferguson, Marilyn, *The Aquarian Conspiracy: Personal and Social Transformation in the 1980s* (New York: St. Martin's Press, 1980; and London: Granada Publishing Ltd, 1982), 496 pp.
Galtung, Johan, "Nonterritorial actors and the problems of peace", in Saul Mendlowitz (Ed.), *On the Creation of a Just World Order* (New York: The Free Press, 1975), pp. 151–188.
Gran, Guy, *Development by People: Citizen Construction of a Just World* (New York: Praeger, 1983) 482 pp., bibl., index.
Grant, James, "Achieving social and economic goals for the year 2000", in *Compass,* No. 8 (January–April 1981), p. 1.
Henderson, Hazel, "Citizen movements for greater global equality", in *International Social Sciences Journal,* Vol. XXVIII, No. 4 (1976), pp. 773–788.
Hine, Virginia, "The basic paradigm of a future socio-cultural system", in *World Issues* (April–May 1977).
Illich, Ivan, *The Right to Useful Unemployment* (London: Marion Boyars, 1978).
International Foundation for Development Alternatives, "Building blocks for alternative development strategies: A progress report from the third system project", *IFDA Dossier 17* (Nyon: May/June 1980), 150 pp.
Judge, A. J. N., "The associative society of the future", in *International Transnational Associations,* No. 6 (1979), pp. 259–265.
Kennan, George, "On nuclear war", in *The New York Review of Books* (21 January 1982).
Khor Kok Peng, "Value for the people: The potential role of a consumer movement in the Third World", in *IFDA Dossier 18* (August/September 1980), pp. 1–13.
Lemaresquier, Thierry, "Beyond infant feeding: The case for another relationship between NGOs and the UN system", in *Development Dialogue,* No. 1 (1980), pp. 120–125.
Lipnack, Jessica and Jeffrey Stamps, *Networking: The First Report and Directory* (New York: Doubleday, 1982).
Meister, Albert, *Vers une sociologie des associations* (Paris: Les Editions Ouvrières, 1972), 306 pp.
Michanek, Ernst, *Role of Swedish Non-Governmental Organizations in International Development Cooperation* (Stockholm: SIDA, 1977), Mimeo., 32 pp.
Nerfin, Marc, "A new United Nations development strategy for the 80s and beyond: The role of the third system", in A. J. Dolman and J. van Ettinger (Eds.), *Partners in Tomorrow: Strategies for a New International Order,* presented to Jan Tinbergen on the occasion of his 75th birthday (New York: E. P. Dutton, 1978), pp. 71–82.
Passaris, Solange, "Les enjeux de la vie associative en France", in *IFDA Dossier 24* (July/August 1981), pp. 15–24.
Sauvy, Alfred, *Mondes en marche* (Paris: Calmann-Lévy, 1982), 228 pp.
Sheth, D. L., "Grassroots stirrings and the future of politics", in *Alternatives,* Vol. **IX**, No. 1 (March 1983), pp. 1–24.
Szuszkiewicz, J. S., "Can they do it: Participation of NGOs of Third World countries in NEIO-oriented projects", in *IFDA Dossier 10* (August 1979), pp. 83–95.

2

SOME DIMENSIONS OF THE DEVELOPMENT DEBATE

Economic Distance and the Living Standard

AMARTYA K. SEN

Introduction

In an important paper published in 1964, called "Economic Distance between Nations: Its Origin, Measurement and Outlook,"[1] Surendra Patel made skillful use of comparative data on national products, but also expressed some deep skepticism about the general procedure of using such data:

> So far, the economic distance has been usually measured by the conventional yardstick of per capita income—this handy expression, in easily remembered numerals, of a country's economic well-being. Economists and statisticians have now produced such estimates for nearly every country. Politicians and planners have seized upon this mystic entity. They have used it as an infallible indicator for diagnosing ills, prescribing remedies, and for measuring progress. But the wider the use of these estimates, the more subject they have become to critical evaluation.[2]

Part of the object of this note is to continue that critical evaluation. Another objective is to make some illustrative international comparisons of aspects of living standards of a few developing countries.

The next two sections are devoted to discussing the conceptual framework for comparing living standards and measuring economic distance. In Section 4, some comparative analyses of Brazil, China, India, Mexico and Sri Lanka are presented. Some particular features of the China-India contrast are further discussed in Section 5, including the relevance of comparisons of growth rates. The paper ends with some concluding remarks.

Utility or Commodity Index?

It is fair to say that in theoretical welfare economics, the notion of the standard of living would be typically identified with utility or welfare. Some economists such as Pigou (1920) and Robertson (1952) may make a distinction between "economic welfare" and "total welfare" in the more inclusive sense,[3] but whether or not that distinction is made, the common welfare-economic view of living standard is in terms of some notion of utility.

Another common view of living standard—typically more used in the empirical literature but also in the theory of index numbers of "real income" or "real consumption"—identifies it with some scalar measure of the vector of goods and services commanded.[4] This can be seen, in some sense, as an index of "opulence"—reflecting what one has, rather than what one gets out of what one has.

It is not uncommon in economic theory to see "utility" as an index of "opulence." Possessing a "preferred" (or "higher indexed," or "larger") commodity bundle is then identified with having "more utility." Thus seen, the two approaches coincide. But this identification is illegitimate. The commodity bundle is only one source of utility. There are other sources such as environment and even the person's psychology.

The view that identifies opulence with utility tries to rule out these other influences by special assumptions. It is sometimes simply assumed that "everything else is the same." Or specifically, environmental factors get the chop with the assumption of "no externalities," and the psychological features are thought to be rendered innocuous by the assumption of "constant tastes." The latter does not work, for a reason that is worth spelling out.[5]

Constancy of tastes only guarantees that the indifference map remains unchanged. In terms of market behavior (i.e, what is bought and what is rejected), constant tastes coincide with no change in such behavior. But that does not, of course, guarantee that the happiness (or the intensity of desire) associated with each bundle has not changed. To illustrate, consider a person who has a manic-depressive fluctuation between being "high", when he is generally very elated, and being "low", when he is generally unhappy. Assume that he shuttles between these two mental states with an unchanging indifference map, i.e., with unchanging tastes and market behavior. He prefers x to y in both states, but while he is "blissful" having x and "jolly well" having y in his "high" state, he is "miserable" having x and "in terrible agony" having y in his "low" state. Now compare situation 1 in which he is low but has x, with situation 2 in which he is high but has y. In terms of opulence (i.e., comparison of commodity bundles as such), he is better off in situation 1, since he then has x, and he prefers x to y (in both states). But it would be absurd to claim that he has more happiness or utility in situation 1 since he is "miserable" in that situation in contrast with being "jolly well' in situation 2. The important thing to note is that this conflict arises despite *absolute constancy of tastes,* and the market behavior of the person would be identical in the two states.

The point is that a commodity index cannot, in any way, be seen as also an index of utilities to the possessor, unless assumptions are made that are a good deal stronger than those of no externalities and constant tastes. In particular, we have to assume that all other influences on utility, including mental characteristics, are completely unchanged between the two situations. This is not only a very demanding requirement, but also deeply problematic in terms of the dominant methodology of neoclassical economics, which led in the first place to identifying utility with choice over commodity bundles. Differences in psychology that are not reflected in differences of tastes are not revealed by observed data in terms of market behavior. So, the assumption of no change in psychology is not "testable" in the way that inconsistency of market behavior is. This is not a fatal problem for an approach that would not take such a limited view of observability and testing,[6] but that limited view is the one that has provided the theoretical underpinning of market-data-oriented consumer theory in the positivist tradition of neoclassical economics.

Once the crude assumption of constancy of everything else other than the commodity bundle is dropped, the coincidence of the utility index and the commodity index collapses. The case for dropping that assumption is strong enough when dealing with the same person at two different points of time, but it is altogether overwhelming when dealing with two different persons. Indeed, the works of such writers as Scitovsky (1976), Hirsch (1976), Leibenstein (1976), Elster (1979), Basu (1979), Hirschman (1982), Akerlof (1983) and others make it clear how odd it would be to assume no psychological variations in dealing with one person and another even within a particular society, not to mention between different cultures.

Capabilities and Partial Orderings

In view of the conflict between characterizing the living standard in terms of utility and that in terms of commodity index, we have to face the question, which one should we choose? What the analysis of the last section shows is that we cannot answer "both". But it does not indicate that we ought to choose one of the two. Indeed, the right answer would seem to be: *Neither!*

Living standard cannot be the same thing as utility or happiness. If an immensely rich, able-bodied, healthy, well-educated person were to tell us that he is not happy, we would have no *prima facie* reason to disbelieve him (he *could be* really unhappy), but we would scarcely think that his unhappiness indicates that his standard of living is low. He is unhappy *despite* his high standard of living. The fact that we see nothing unbelievable or absurd in such a possibility indicates quite clearly that the notion of living standard and that of utility or happiness are not identical. The former may be typically very important for the latter, but they are not the same.

But does that mean that living standard is best seen in terms of the commodity command of the person, that is, the bundles of goods and services that he or she can buy? That seems quite implausible, too. Consider a case in which the difference between two persons arises from some tangible physical factors. In particular, consider person 1, who is a cripple, having the same income and facing the same market as person 2, who is able-bodied. Their commodity commands are the same; but are their standards of living the same? The latter can walk freely without help; the former cannot. The latter goes swimming in the sea, or cycling in the country in the holidays; the former cannot. The failure of person 1 to be able to do what person 2 does is not due to a lower command over commodity bundles; it arises from person 1's inability to convert commodities into functioning (walking, swimming, cycling) in the way person 2 does. It would be not at all unreasonable to see person 1 as having a lower standard of living, since he cannot live in the way person 2 can, and the loss may be important for him.

The standard of living must surely be a reflection of what kind of living the person can achieve, and that is neither just a matter of owning commodities, nor the same as how happy the person turns out to be. The standard of living is best seen as an indicator of the *functioning* a person can do. If a person cannot have enough food to be well-nourished in the way that another person is, then his standard of living is, in that respect, lower. This is not because he is less happy (he may not be—he may have "inner" sources of strength and joy). Nor necessarily is it because he can buy less food than the other person (he may be able to buy as much, or more, but still be forced to be less nourished due to having a larger body size, or high metabolism or stomach parasites, or because he lives in a colder country).

Commodity commands are contingently—though importantly—related to standard of living, and happiness may also be closely associated with living standard, but neither commodity command nor happiness is the same thing as living standard. What we have to look for is what the person in question can or cannot do in terms of his material existence. Can he meet his nutritional needs, can he escape avoidable morbidity, can he live long, can he read and write, can he move about as he wants, and so on? It is a set of *capabilities* that we have to look for—the respective abilities to do various things that give him freedom in his material existence.

I have tried to present the capabilities approach more fully in the context of equality (Sen, 1980), judging personal advantage (Sen, 1982a), and assessing poverty (Sen, 1983a); and in a book—*Commodities and Capabilities* (Sen, 1983b), the capabilities approach to the evaluation of living standard is explored.[7] There are various technical problems that need careful handling, especially dealing with the identification and weighting of capabilities, in each specific context. These issues have been taken up in that volume (Sen, 1983b), and I shall not go further into those questions here.

I should, however, point out here that in talking about living standard, there is no absolute necessity to convert a *vector* of capabilities into one *scalar* index, reflected by one real number. Measurement of a concept that has inherent plurality (and corresponding ambiguities) is not helped by trying to see it as a one-dimensional measure, like that of weight or height. There is nothing illegitimate in saying that a person has a higher standard of living in one respect (e.g., nutrition), but lower in another respect (e.g., freedom to move about). Whether his living standard *altogether* is higher may or may not be a very interesting question, and the answer should certainly depend on what use might be made of such an aggregate assessment.

The literature on the quality of life indices (e.g., Grant, 1978; Morris, 1979), is obviously relevant for studying standard of living. If one were to be critical of the emphasis in some of these contributions, perhaps the importance attached to getting one overall indicator (e.g., "PQLI" in Morris, 1979) may be seen to be somewhat misplaced. That approach seems to lead to a rather narrow list of functionings, excluding many relevant ones in the interest of simple indexing, and it also leads to a rigid set of fixed relative weights among the few variables chosen.

The comparison of vectors lead to a *partial* order of dominance without bringing in any weighting (except that the weights are positive and finite). As the ranges within which the weights might lie are narrowed, the partial order gets more and more extended, reaching completeness typically much before each range of weights has been respectively reduced to one number.[8]

Vector comparison and partial orderings seem the natural way of dealing with comparisons of living standard. It is worth bearing in mind in this context that even the traditional theory of real-income comparisons with all its simplifying assumptions does not yield a complete order on the basis of actual observation of market behavior, so that the partial-ordering approach has been the natural one to use even in those highly simple cases.[9]

To summarize, in assessing living standard, we are not looking for utilities, nor for commodity indices, but for capabilities related to material existence. The capabilities to function take the form of a vector, and comparisons of such vectors yield a partial ordering. That partial ordering can be consistently extended by narrowing the range of possible weights. There is no need to push this narrowing arbitrarily, and certainly none to reduce each range to a point. A complete ordering is not needed; and the natural mathematical form of the comparison of standard of living is that of a partial ordering, possibly more extensive than that of vector dominance, but not necessarily anywhere near completeness.[10]

Some International Comparisons

Given the limitations of comparable data, it is not easy to make extensive comparisons of achievements of different countries in the field of extending capabilities. One reason why the data tend to be relatively scarce in this area compared with, say, data underlying GNP and GDP estimates, is the lack of strong demand for such data. There is no reason why it should not be possible to get more comparative data on, say, morbidity or undernutrition, in different countries. The weaknesses in the theory of standard of living have been partly responsible for the underdevelopment of the data base, which could have been used for a more relevant empirical treatment of comparative living standards in different countries.

In the long run, the most important task may be to expand the data base, but this should not prevent us from making use of whatever data may be easily available already. In Table 1 data are presented for five developing countries,

TABLE 1. *Comparative data on specific achievements of five countries*

Country	GNP per head ($) (1980)	Life expectancy (years) (1980)	Infant mortality (1980)	Child death rate (1980)	Adult literacy rate (%) (1977)	Secondary schooling ratio (%) (1979)	Higher education ratio (%) (1978)
India	240	52	123	17	36	27	8
Sri Lanka	270	66	44	3	85	79	1
China	290	64	56	5	66	53	1
Brazil	2,050	63	77	7	76	32	11
Mexico	2,090	65	56	4	81	45	12

Source: *World Development Report 1982* (New York: Oxford University Press, 1982).

viz., India, Sri Lanka, China, Brazil and Mexico; the data are all taken from *World Development Report 1982*. There is an undue concentration here on matters only of life and death and of education, ignoring many other important capabilities such as prevalence of undernourishment, extent of morbidity, adequacy of basic clothing, ability to be housed and sheltered, etc.; but given the present data base, these other comparisons are less easy to make at this moment.

Even with these limitations, some rather interesting patterns can be noticed in the data presented in Table 1. In terms of GNP per head, India, Sri Lanka and China fall broadly in the same group—the difference between them is not really much—whereas Brazil and Mexico are in a different part of the spectrum altogether. On the other hand, in terms of life expectancy, infant mortality and child death rate, India stands on its own, whereas Sri Lanka, China, Brazil and Mexico form a different group, with Sri Lanka having the edge over the others. In the context, therefore, of one of the most important capabilities—that of long life—India stays behind, whereas Sri Lanka and China join a group of countries that are many times richer in terms of GNP per head.

Turning to basic education, using the indicators of adult literacy rates, once again India stands away from the others, which belong to roughly the same group, with Sri Lanka again having the edge over the others. The picture is more muddled with the ratio of secondary school students to the corresponding age cohort,[11] but, in fact, Sri Lanka and China—though much poorer than Brazil and Mexico—have substantially *higher* figures, with India once again at the bottom of this list. In terms, therefore, of life-and-death matters, and elementary and secondary education, Sir Lanka and China part India's company (indeed that of all other countries in the same income group[12]) and join or overtake the richer Brazilian and Mexican economies.

The picture, however, is quite different when we look at the universities and higher education. It is now India's turn to part with the company of China and Sri Lanka and join—or come close to joining—Brazil and Mexico. The elitist character of the Indian society and polity, which I have tried to discuss in more general terms elesewhere (Sen, 1982b), is well illustrated by these contrasts. In terms of chances of receiving higher education, the Indian upper and middle classes are not far behind those in Brazil and Mexico,[13] and way ahead of corresponding groups in China and Sri Lanka (indeed with a higher education ratio about eight times higher than in these countries). On the other hand, the capabilities of the Indian masses are enormously below the masses in China and Sri Lanka in terms of the ability to live long, the ability to survive mortality during infancy and childhood, the ability to read and write, and the ability to receive substantial schooling.

India and China

In terms of the basic capabilities of survival and education, Sri Lanka and China do stand out. In fact, India's record is not really exceptionally bad for the corresponding income group. The point can be brought out by comparing India's position with many other countries, e.g., Pakistan or Nigeria, each richer than India in terms of GNP per head (Nigeria a lot richer), and each having rather worse records than India in terms of *each* of the indicators considered. The policies that have led to the remarkable achievements of China and Sri Lanka in respect of basic capabilities have been discussed elsewhere,[14] and I shall not go into that question here. Both have pursued public policies geared to delivering food, basic health services and school education extremely widely, and they have reaped as they have sown.

The contrast between China and India arouses a good deal of interest for obvious reasons. The superiority of China's overall performance in expanding the standard of living of the people is decisive.[15]. It is important in this context to be clear that the contrast is not really one of national income and GNP, since the two countries are fairly close to each other in these respects. The point is of considerable policy relevance, since the contrast between China's and India's growth rates of GNP or GDP has received such a lot of attention. It is often taken for granted that China's achievements in the field of raising standard of living is largely due to its fast economic growth.

In fact, there are good grounds for doubting whether China's growth rates of GNP or GDP have really been very much higher than those of India. There is a bit of a puzzle here. China's GNP growth rates are not available, but the growth rates of GDP per head during the decades of the 1960s and 1970s, are widely accepted to be very much higher in China than in India. The figures given in *World Development Report 1982* bear this out, and are presented here in Table 2. *World Development Report 1982* also indicates that in 1980, the GNP per head of China was about 21% higher than that of India, and the GDP per head was about 22% higher. If we extrapolate backwards at the identified growth rates of GDP per head for the 1970s and 1960s this would indicate that China's GDP and GNP were, in *per capita* terms, more than 20% *lower* than India's at the beginning of the 1960s! This is scarcely credible, since all the estimates suggest that China's income per head was comparable with—possibly even a little

TABLE 2. *Estimated growth rates and corresponding relative incomes: China and India*

	1960–70 GDP growth rate (%): Estimated by WDR 1982 (1)	1960–70 growth rate of GDP per head (%): Estimated by WDR 1982 (2)	1970–80 GDP growth rate (%): Estimated by WDR 1982 (3)	1970–80 growth rate of GDP per head (%): Estimated by WDR 1982 (4)	1960 GDP per head as a ratio of 1980 GDP per head corresponding to WDR growth estimates (5)	1980 GDP: Estimated by WDR 1982 ($) (6)	Alleged 1960 GDP per head: Estimated from WDR growth estimates ($) (7)	1980 GNP: Estimated by WDR 1982 ($) (8)	Alleged 1960 GNP per head: Estimated from WDR growth estimates ($) (9)
China	5.2	3.3	5.8	4.0	0.488	260	127	290	142
India	3.4	1.1	3.6	1.5	0.772	211	163	240	185
China/India ratio	1.5	3.0	1.6	2.7		1.23	0.78	1.21	0.77

Procedure and Sources: The calculations here are all based on data as presented by *World Development Report 1982*. Growth rates of GDP per head are calculated from WDR estimates of GDP growth rates (Table 2) and population growth rates (Table 18). Using these compound growth rates, the estimated ratios of 1960 GDP per head to 1980 GDP per head have been calculated (column 5). The 1980 GDP per head for each country (column 6) is calculated from GDP figures (Table 3) and population figures (Table 1). The "alleged 1960 GDP per head" is deduced from columns (5) and (6). Similarly, the "alleged 1960 GNP per head" is deduced from GNP per head data (Table 1) with the assumption of the same GNP and GDP growth (since Chinese GNP growth rates have not been given by WDR 1982). In fact, the reference years in the case of China are 1980 and 1961 respectively (see explanation in *World Development Report*, Table 2, p. 113), and for India 1980 and 1960. This does nothing to reduce the implausibility of India being so much richer than China two decades ago.

higher than—India's when systematic planning started in the two countries in the 1950s.[16]

It is very likely that China's growth rate has been higher than India's, but not radically higher. China has remained a poor country in terms of GNP and had not decisively parted company with India in this respect. Where it has parted company is precisely on the matter of standard of living, which must not be confused with GNP per head. The capabilities of the Chinese masses are now immensely superior in many vital respects to those of the Indian masses. They live a great deal longer; have much safer infancy and childhood; can deal more effectively with illness and disease; can mostly read and write; and so on. The "economic distance" between China and India is much more telling in terms of living standard and capabilities than in terms of GNP and commodities.[17] In the context of policy debates in India and elsewhere, where the high growth rate is assigned an unduly hallowed position, this basic point about criteria and program is important to bear in mind.

Concluding Remarks

This has been a hybrid paper—some theory, some empirical analysis. While there will be no attempt to summarize, some of the issues may be conveniently highlighted in this concluding section.

1. Living standard is often—mostly in welfare economic theory—seen in terms of utilities. It is also frequently seen as opulence—reflected by an index of commodity bundle that the person has. In modern economic theory, utility and opulence are often confounded, in the process of seeing utility as a real-valued representation of preferences over commodity vectors. It is easily shown that utility cannot be identified with opulence, and if living standard is identified with one, it cannot also be identified with the other.

2. Is living standard to be seen as a utility index or as an index of the commodity vector? The answer that is difficult to escape is: Neither!

3. There is a good case for seeing living standard in terms of people's capabilities—what they can or cannot do. The capability view differs from the utility view and from the commodity view.

4. Capabilities are inherently heterogeneous, and in comparing them, the basic mathematical relation is that of a partial ordering. The least controversial partial ordering is that of vector dominance, but that minimal partial ordering can be systematically extended by narrowing the ranges of relative weights to be put on different capabilities. But there is no need to insist on getting a complete ordering. Forcing an arbitrarily narrow range of weights reduces the relevance and credibility of the derived ordering. The literature on PQLI and other aggregate indicators is important, but suffers from the defect of a very narrow coverage of aspects of the so-called "quality" of life and a very rigid weighting system (unique numbers rather than ranges of values). The perspective of capabilities as the basis of living standard provides the theoretical

underpinning of a less arbitrary approach, which is at the same time, not indifferent to the so-called "quality" of life (including the *quantity* of life—measured in years—lived by each!).

5. The relative position of different countries vary greatly with the focus of attention in comparing living standards. While India, Sri Lanka and China are quite close to each other in GNP per head and very far below Brazil and Mexico, when it comes to longevity, infant mortality, child death rate, literacy and schooling, China and Sri Lanka join—or overtake—Brazil and Mexico, with India remaining far behind. India, however, leaves China and Sri Lanka behind on the matter of higher education, coming close to Brazil and Mexico, and this contrast illustrates rather sharply the political and social character of India and the influence of upper and middle classes on the direction of Indian planning, which can be illustrated with other information as well.

6. In the contrast between the successes of China and Sri Lanka as opposed to the failure of India to raise the living standard of the masses, it is China and Sri Lanka that are exceptional. Indeed, there are many countries with comparable or much higher GNP per head than India, which have achieved even less in terms of the features that were used for comparison. The nature of the public policies followed in China and Sri Lanka indicate the causes of the contrast.

7. While it is generally held that the Chinese growth rates of GDP or GNP have been much higher than India's, there are good reasons to doubt this. The success of China *vis-à-vis* India is not so much due to faster growth rate—the difference has not been dramatic—but due to aiming public policy at expanding the capabilities of people in vital matters. The issue relates to the direction of planning. It is worth noting in this context that South Korea, which has been so successful in growing at an extraordinarily fast rate and has now a GNP per head more than five times that of China or Sri Lanka, has not still been able to overtake either of those countries in terms of life expectancy.

8. There is need for greater theoretical clarity as well as wider empirical exploration in assessing and comparing living standard. The approach of capabilities provides a better perspective for examining the standard of living and for doing empirical analysis. The importance of the question relates not merely to assessing the "economic distance between nations" (to use Surendra Patel's expression), but also to drawing policy conclusions from these comparisons. Good policy cannot really be founded on bad description.

Notes

1. Patel (1964). Reprinted in Patel (1965), pp. 183–199.
2. Patel (1965), p. 192.
3. Robertson (1952), pp. 14–15.
4. See, for example, Afriat (1972), Samuelson and Swamy (1974), and Deaton and Muellbauer (1980).
5. See also Sen (1979a), pp. 12–13.
6. See Sen (1973, 1982a). See also Hahn and Hollis (1979).

7. The capabilities approach can be traced to the writings of Adam Smith (1776) and Karl Marx (1887), among others. On this see Sen (1983a, 1983b).
8. The mathematics of the procedure is virtually the same as that in deriving an aggregate welfare partial ordering from an *n*-tuple of utility functions with partial cardinality or partial interpersonal comparability, on which see Sen (1970a, 1970b), Blackorby (1975), Fine (1975), Basu (1979), and Bezembinder and van Acker (1979).
9. See Hicks (1940, 1958) and Samuelson (1947). The issue is separately examined in Sen (1979a).
10. There is an analogy here with cost-benefit analysis with ranges of weights rather than a set of unique numerical weights, on which see Dasgupta, Marglin and Sen (1972) and Sen (1975), pp. 112–114.
11. The right data to use to reflect the impact of secondary educational facilities would have been the ratio of secondary educated people in the population (comparable with literacy rates), but comparative data of this type could not be found.
12. *World Development Report 1982,* Table 1, pp. 21–23.
13. In fact, members of the Indian upper and middle classes may have a better chance in this respect than the upper and middle classes in Brazil and Mexico, since the higher educational students in India come from a relatively smaller proportion of the population than they do in Brazil and Mexico.
14. I have discussed the policy issues in Sen (1981, 1982c, 1983c).
15. China's success is, however, qualified by the fact that it has not been able to prevent the occurrence of famine; indeed, there has been a substantial one, during 1959–61, with a great many millions dying, as has been acknowledged only recently (Zhu Zhengzhi, 1980: Sun Yefung, 1981; People's Republic of China, 1981; Feng Yang County Communist Party Committee, 1980; and Bernstein, 1983). India has succeeded in avoiding famines largely because of the active role of relatively free newspapers and powerful opposition parties in forcing the government's hand in undertaking relief operations and causing adjustment of policies. However, while the pressure of newspapers and opposition parties have helped to guarantee the avoidance of acute starvation and famine, they have not provided a strong force against regular, endemic malnutrition, which is rampant in India and relatively rare in China. Dramatic deprivations engage the attention of the media and political opposition, and become electoral issues in a way that the quiet continuation of persistent, orderly hunger does not. On this contrast, see Sen (1982c, 1983c).
16. Kuznets (1966) estimates the same GNP per capita for China and India, and a "product per capita" 20% higher in China in 1958 (pp. 360–361).
17. Note that this calls for a more fundamental departure than correcting GNP figures for distributional differences (see Sen, 1976a; 1976b).

References

Afriat, S. N., "The theory of international comparisons of real incomes and prices", in D. J. Daley, (ed.), *International Comparisons of Prices and Outputs* (New York: NBER, 1972).

Akerlof, G. "Loyalty filters", *American Economic Review,* Vol. 73, (1983).

Basu, K., *Revealed Preferences of Governments* (Cambridge University Press, 1979).

Bernstein, T. P., "Hunger and the State: Grain Procurements during the Great Leap Forward, with a Soviet Perspective", Mimeo. (East Asia Center, Columbia University, 1983).

Bezembinder, Th., and P. van Acker, "A Note on Sen's Partial Comparability Model" (Nijmegen: Department of Psychology, Katholieke Universiteit, 1979).

Blackorby, C., "Degrees of cardinality and aggregate partial orderings", *Econometrica*, Vol. **43** (1975).

Dasgupta, P., S. Marglin, and A. Sen, *Guidelines for Project Evaluation* (New York: UNIDO, 1972).

Deaton, A., and J. Muellbauer, *Economics and Consumer Behaviour* (Cambridge University Press, 1980).

Elster, J., *Ulysses and the Sirens* (Cambridge University Press, 1979).

Feng Yang Communist Party Committee, "An investigation into the household production contract system in Liyuan Commune", *New York Review of Books,* Vol. **30** (16 June 1982); translated from *Nongye Jingji Congkan* (25 November 1980).

Fine, B., "A note on 'Interpersonal aggregation and partial comparability'," *Econometrica*, Vol. **43** (1975).
Grant, J., *Disparity Reduction Rates in Social Indicators* (Washington, D.C.: Overseas Development Council, 1978).
Hahn, F., and M. Hollis. (Eds.), *Philosophy and Economic Theory* (Oxford University Press, 1979).
Hicks, J. R., "Valuation of social income", *Review of Economic Studies*, Vol. **7** (1940).
Hicks, J. R., "Measurement of real income", *Oxford Economic Papers*, Vol. **10** (1958).
Hirsch, F., *Social Limits to Growth* (Cambridge, Mass.: Harvard University Press, 1976).
Hirschman, A. O., *Shifting Involvements* (Princeton University Press, 1982).
Kynch, J., and A. Sen, "Indian women: Well-being and survival", Mimeo., forthcoming in *Cambridge Journal of Economics* (1982).
Kuznets, S., *Modern Economic Growth* (New Haven, Conn.: Yale University Press, 1966).
Leibenstein, H., *Beyond Economic Man* (Cambridge, Mass.: Harvard University Press, 1976).
Marx, K., *Capital,* Volumes **I–III.** English translation of Volume **I** (London: Sonnenschein, 1887).
Morris, M. D., *Measuring the Conditions of the World's Poor: The Physical Quality of Life Index* (Oxford: Pergamon Press, 1979).
Patel, S. J. "Economic distance between nations: Its origin, measurement and outlook", *Economic Journal*, Vol. **74** (1964).
Patel, S. J., *Essays on Economic Transition (*London: Asia Publishing House, 1965).
People's Republic of China, *Foreign Broadcast Information Service*, No. 58 (26 March 1981).
Pigou, A. C., *Economics of Welfare* (London: Macmillan, 1920).
Robertson, D. H., *Utility and All That and Other Essays* (London: Allen & Unwin, 1952).
Samuelson, P. A., *Foundations of Economic Analysis* (Cambridge, Mass.: Harvard University Press, 1947).
Samuelson, P. A., and S. Swamy, "Invariant economic index numbers and canonical duality: Survey and synthesis", *American Economic Review*, Vol. **64** (1974).
Scitovsky, T., *The Joyless Economy* (Oxford University Press, 1976).
Sen, A. K., *Collective Choice and Social Welfare* (1970a, reprinted Amsterdam: North-Holland, 1979).
Sen, A. K., "Interpersonal aggregation and partial comparability", *Econometrica*, Vol. 38 (1970b); also "Correction", *Econometrica*, Vol. **40** (Reprinted in Sen, 1982a).
Sen, A. K., "Behaviour and the concept of preference", *Economica*, Vol. **40** (1973) (Reprinted in Sen, 1982a).
Sen, A. K., *Employment, Technology and Development* (Oxford: Clarendon Press, 1975).
Sen, A. K., "Real national income", *Review of Economic Studies*, Vol. **43** (1976a) (Reprinted in Sen, 1982a).
Sen, A. K., "Poverty: An ordinal approach to measurement", *Econometrica*, Vol. **44** (1976b). (Reprinted in Sen, 1982a).
Sen, A. K., "The welfare basis of real income comparisons", *Journal of Economic Literature*, Vol. **17** (1979a).
Sen, A. K., "Personal utilities and public judgments: or what's wrong with welfare economics?", *Economic Journal*, Vol. **89** (1979b) (Reprinted in Sen, 1982a.).
Sen, A. K., "Equality of what?" in S. McMurrin (Ed.), *Tanner Lectures on Human Values*, Vol. **I** (Cambridge University Press, 1980). (Reprinted in Sen, 1982a).
Sen, A. K., "Public action and the quality of life in developing countries", *Oxford Bulletin of Economics and Statistics*, Vol. **43** (1981).
Sen, A. K., *Choice, Welfare and Measurement* (Oxford: Blackwell, 1982a).
Sen, A. K., "How is India doing?", *New York Review of Books*, Vol. **29** (1982b), Christmas Number.
Sen, A. K., "Food battles: Conflicts in the access to food", Coromandel Lecture (12 December 1982a). (Reprinted in *Mainstream*, 8 January 1983).
Sen, A. K., "Poor, relatively speaking", *Oxford Economic Papers*, Vol. **35** (1983a).
Sen, A. K., *Commodities and Capabilities*, Hennipman Lecture (1982) (Amsterdam: North-Holland, 1983b).
Sen, A. K., "Development: Which way now?", *Economic Journal*, Vol. **93** (1983a).
Smith, A., *An Inquiry into the Nature and Causes of Wealth of Nations* (1776), Everyman's Library Edition (London: Dent, 1954).
Sun Yefang, Article in *Jingji Guanli*, No. 2 (15 February 1981); English translation in People's Republic of China (1981).
Zhu Zhengzhi, Article in *Jingji Kexue*, No. 3 (1980) summarized in Bernstein (1983).

Energy and International Security in the 1980s: Realities or Misperceptions?

MIGUEL S. WIONCZEK

The volume of research being carried out in the international organizations with the sponsorship of the countries which import different sources of energy, and in centers of learning located in the industrialized Western countries, as well as the number of books and studies published between 1980 and 1982 on the relationship between energy and both international and national security, continues to be impressive. Even more, many countries would seem to give just as much importance now to this particular aspect of international relations as they did in the 1970s following the first "oil crisis" of 1973–74.

However, in the last 10 years the world energy markets have undergone very important changes. What, then, is the origin of this persistent fascination with the question of the strategic effects of access to energy at a time when energy supply has become greater than demand? The object of this essay, which was written before the collapse of international energy markets in the spring of 1983, is to demystify the conventional views still fashionable among many people at all levels of sophistication and expertise. There is a widely held belief that behind political and military insecurity and behind many international conflicts lies the problem of a lack of security regarding access to energy sources located outside the national territory of the energy-importing countries. The purpose of this essay is to prove that to consider the question of energy access as the main cause of world, regional or national insecurity is to fall victim either to misperceptions or propaganda.

Conventional wisdom has it that the links between energy and international security are two-sided. International tensions and conflicts can disrupt the production of energy-exporting nations or their willingness to supply energy to others, sever transport links between producers and consumers, and damage the domestic energy facilities of combatant nations. Because the energy demands of military activities are potentially very great, such disruptions that would very seriously affect the security of the combatant parties must be avoided at all costs.

Furthermore, recent literature on the subject suggests that problems regarding the control of access to energy (and sometimes the steps taken to achieve this) can influence the likelihood and severity of international conflicts. These connections include the contributions of energy problems to the root causes of conflict, the potential of energy problems to serve as the immediate stimulus or trigger for conflict, and the links between the energy problem and policies and the means of conflict (that is, the size and character of military forces).[1]

This emphasis on the links between energy and security was reinforced by the second "oil shock" of 1979–80 which brought to the fore in the Western industrial societies and particularly in the United States, for the second time in less than 10 years, the specter of presumably long-term worldwide energy shortages and the dependence of large energy-consuming countries in the North Atlantic area upon oil that comes mainly from the Middle East. Given the persistent military, economic and social instability of that part of the so-called Third World, due in part to the 1979 Iranian Revolution and to Arab-Israeli conflict, intertwined with the global conflict between the two superpowers—the United States and the USSR—one can hardly be surprised that until very recently the problem of access to Middle East oil figured very high on the American alliance's list of possible sources of serious regional conflicts, which, if unchecked, might spill over into a global conflict between the two superpowers. The Western mass media and academic circles, again particularly in the United States, were not alone in dedicating in 1980–81 a great deal of space and attention to the issue of national or regional security as measured by the possibility of denial presumably by the USSR, of access to Middle Eastern oil. The recent list of US national strategic priorities, elaborated by Pentagon officials for Defense Secretary, Caspar Weinberger, in May 1982 and presumably approved by President Reagan, put as the second priority (after Western Europe) for general strategic planning (in case of world war with the Soviet Union) "to guarantee access to the oil in Southwest Asia", and to "defend the Gulf's oil".[2]

Before going into an objective analysis of the correctness or incorrectness of the perceptions underlying the view of the relationship between security and oil in international politics, under the conditions of the pro-war hysteria that reigns in the United States, and to a slightly lesser extent in some of its European allies, we should draw attention to the fact that between the date of the second "oil shock" and the end of 1982, the world energy picture underwent a very substantial change when one compares the recent situation with the conditions prevailing during the 1970s. Since early 1981, evidence has been accumulating at the world level that the so-called "energy crisis", as reflected by the two oil shocks, the first in 1973–74 and the second in 1979–80, is over. Furthermore, no international energy experts, whether from serious academic circles, governments of importing and exporting countries or the private or state-owned oil industry expected by mid-1982, that any new "oil crisis" or "energy crisis" would take place at least during what remains of the 20th century unless completely unpredictable political, social or military events take place.

This sort of consensus emerged, among others, at an international colloquium on the prospects for the world hydrocarbon markets in the 1980s, organized in April 1982 in Mexico City by the Energy Program of El Colegio de México, with the participation of a large group of high-level experts from such different countries as the United States, Canada, Great Britain, France, Italy, Norway, the USSR, Japan, Venezuela and Mexico, and also a number of international organizations including the International Energy Agency and the OPEC.[3]

A few major points of the findings of this international colloquium might be mentioned here as relevant to the debate on energy and international security in which the main oil-importing industrial nations are involved.

First, starting with the aftermath of the first "oil crisis" in 1973–74, the growth rate for energy, and particularly for hydrocarbons, per unit of production, continued to decline in *all* industrial energy-intensive economies in response to the increased energy prices, which resulted in energy savings and conservation measures. There is no reason to suppose that this trend, which is reflected by energy-saving technological advancements can be stopped and reversed.

Second, the world economic crisis, whose emergence can be traced back to the end of the 1960s, that is to say, before the first "oil crisis", and whose end is not yet in sight, was another determining factor in the erosion of world energy demand over the past 10 years.

Third, in response to the increase in hydrocarbon prices and the uncertain outlook for supplies from traditional sources, not only did new producers of oil and natural gas appear on the world scene in the 1970s, but more new producers are expected to appear in the 1980s. This constant increase in energy supply is due to several factors: (a) There are more conventional energy sources on the planet than has ever been thought. (b) Technological progress in the hydrocarbon sector and also with regard to alternative energy sources is rapid, sustained and worldwide. (c) Under the depressed international trade conditions, oil producing and exporting countries consider this particular energy commodity to be easier to sell than any other. (d) The industrial countries which produce oil for domestic consumption have recently been implementing energy policy measures aimed at decreasing dependence on foreign energy sources for reasons of "national security".

Fourth, the present precarious balance that exists between world energy (and hydrocarbon) supply and demand has been achieved mainly thanks to the OPEC policies aimed at reducing supply to a much lower level than that of OPEC member countries' oil and gas production capacity. The restrictive OPEC oil export policies have not been easy to implement, however, because of the steady expansion of supply by the non-OPEC oil exporting countries. The

OPEC's participation in the world hydrocarbon markets has been declining in real and relative terms since 1979, even leaving aside Middle East oil which disappeared from the international markets as the consequence of the war between Iran and Iraq.

In view of the uncertainties surrounding the recovery from both the worldwide economic crisis and the radical changes that have occurred in the conditions of the world energy market (and therefore world hydrocarbon markets), whose structure became even more complicated with the emergence of production and trade in natural gas and, marginally, the appearance of heavy oils and synfuels, very few international energy experts venture to make projections or forecasts about the world energy situation beyond 1986. Moreover, the layman deserves to be warned that all global forecasts, including the most recently available (published in 1980 and 1981), are still based on the assumptions made in the 1970s which proved to be totally wrong if the latest world energy developments are duly taken into account. Under the most optimistic recent (early 1983) assumptions about the medium-term behavior of the world economy, world energy supply will continue increasing in line with a very slow increase in world energy demand. If these optimistic assumptions prove correct, energy prices would remain more or less stable at present prices, but would decline in real terms unless unexpected and dramatic developments take place, such as those postulated by pessimistic political and military analysts in some Western industrial countries.

Leaving aside the arguments of the pessimistic strategic-military school that assumes that disasters, which are never clearly defined, will take place in the Middle East, and will be followed by the violent intervention by the industrial energy-consuming countries seeking to guarantee access to energy sources in that part of the world, one is obliged to conclude that at the regional, national and international levels, the question of energy security has acquired completely different dimensions from those it had in the 1970s.

Under the present conditions of worldwide abundance of available and potential energy sources, the problem does not lend itself to the type of solution that employs military resources designed to defend or prevent access to another country's energy resources. It can be solved only by the elimination of the sources of conflict which have little to do with problems which are supposedly related to the (nonexistent) scarcity of energy resources and their uneven distribution over the planet earth. In other words, the links between energy and security are not locational, economic or technological in nature. They originate in the struggle for power, in correct and incorrect perceptions of the energy problems, international politics and in propaganda.

In mid-1982, it was common to hear in the United States two contradictory versions of the conflict between that country's national security interest in the field of energy and the "inimical Soviet intentions". The first version, mentioned previously, referred to alleged long-term Soviet plans to "take over" the Middle East so as to prevent the US and its allies from having access to Arab energy resources. The second was related to the long-standing Soviet decision to proceed with the construction of a giant gas line from Western Siberia to Western Europe which, once finished in 1984, is planned to guarantee the European allies of the United States an ample supply of energy for the next 20 years.

The most important argument in favor of US and NATO strategic and other activities aimed—it is said—at denying the Soviet Union the "opportunity to cut the energy lifeline connecting the Middle East with Western Europe and the US" was the Soviet invasion of Afghanistan in December 1979. The second argument to support the US Persian Gulf strategic doctrine was that it was supposedly becoming difficult for the Soviet Union to obtain its own energy sources and it might need the Middle East oil for itself. As far as the continued US offensive against the Soviet Union's agreements for selling its oil and natural gas to Western Europe is concerned, another argument was used: such expanded Soviet supplies would seriously diminish Western European security by offering the USSR the chance to cut off these supplies at any time.

One does not have to be a particularly brilliant, internationally recognized energy expert nor an unconditional admirer of the Soviet Union or a professional enemy of the United States, to reach the conclusion that both sets of arguments *cannot* be true at the same time, simply because taken as a whole they go against the principles of logic and common sense. Moreover, putting Soviet intentions in these terms amounts to accusing the USSR of working against its own interests, or, in other words, of acting against national Soviet security. Furthermore, more sophisticated minds, whether or not they are involved in the analysis of the world energy situations, but which are dedicated to the serious study of international affairs, will rapidly arrive at the conclusion that the two official US theories that the Soviet Union is bent upon taking over Middle East oil and that at the same time the Soviet Union is attempting to win the allegiance of the Western European purchasers of its natural gas in order to blackmail them at some future date cannot be taken very seriously either when considered separately.

Enough is known by now about the background to the problem of Afghanistan to be able to hold that it had nothing to do with Persian Gulf oil, the alleged Soviet "Drang nach Sud" to the warm waters of the Indian Ocean, etc.[4] More than enough is known from Western sources also to sustain that the extreme military, political, economic and social instability of the Middle East is not exactly of the Soviet's making. After all, no evidence exists that the Israeli Prime Minister is working for the Soviet Union.

Finally, Soviet energy planning and its pipeline project for supplying Western Europe with very considerable amounts of natural gas for a least the next 20 years shows that the Soviet Union has large exportable energy resources and, moreover, definitely suggests that it needs to export them for very real economic and financial reasons. According to such neutral sources as the UN Economic Commission for Europe, Soviet exports of energy to the non-socialist countries accounted in 1981 for almost one half of that country's convertible foreign currency earnings.

Moreover, there are reasons to take seriously the most recent official Soviet statement that United States' unilateral sanctions, which in the summer of 1982 took the form of Washington's banning the sales to the Soviet Union of high technology pipeline equipment not only by American industry and its subsidiaries in Western Europe, but also by European firms operating under American-owned licenses, will hardly delay the completion of the project by 1984 in which most of Western Europe is very much interested. This interest is due not only to the financial and economic advantages which such an agreement could bring the purchasers of Soviet natural gas but also to Western Europe's great concern over the growing overall instability in the Middle East and the possible subsequent reduction in oil exports from this area to Western Europe. In other words, if we assume correctly that the availability of uninterrupted energy supplies plays an important role in national and regional security considerations, Western Europe wants to diversify its sources of supply in the adjacent areas and to add the Soviet Union to the list of its permanent suppliers along with the North Sea, North Africa and the Middle East. In the conditions of a growing worldwide supply of energy sources which is expected in the remaining decades of the 20th century, Western Europe can easily fulfill its energy security objectives. The obstacles in the adjacent areas do not occur as a result of the "diabolic designs" of the Soviets, but are created by the overseas political leaders of the anti-Soviet holy nuclear crusade and their strategists who call for military occupation of the Middle East in the case of a "possible" Soviet invasion of the oil-countries in that region.

The explanation behind the continued fascination of most Western strategists and of some, albeit fewer, international oil experts with the possibility of new violent interruptions of Western access to Middle East oil lies in ideological and propagandistic motives in the case of the former group and in the misperceptions of the latter.[5] The international colloquium, held in Mexico City in the spring of 1982 and mentioned earlier, provided interesting evidence about the origin of these misperceptions. Top-level energy experts from a dozen Western oil-importing countries and oil-producing and exporting countries did not discuss in detail the question of a possible interruption in international hydrocarbon trade in the 1980s and 1990s, and only accepted that such interruptions may occur as a result of local conflicts on the periphery of the industrialized world (including the Middle East, of course) or of limited wars in which the two superpowers participate directly or indirectly.

It is interesting, however, to point out that in their discussions of the prospects for global energy supply and demand most Western experts used the concepts of the "Free World" as opposed to "other sources of hydrocarbons". The ideological and political connotations of this division of the world by experts who cannot otherwise be considered vigorous champions of the Cold War are far-reaching. Moreover, they reveal the importance of political or economic misperceptions in international relations. Some of these misperceptions originate in the failure to recognize the importance of pre-existing beliefs of which most societies, individuals, experts and governments are guilty but not necessarily conscious.[6]

As one of America's most prominent political scientists, Robert Jarvis from the University of California at Los Angeles, warned in the mid-1970s:

> Because people do not understand the degree to which their inferences are derived from their expectations, they tend to see their interpretation of evidence as "compelling" rather than "plausible".[7]

At the time of "crisis" or of "expected crisis", misperceptions exceed rational perceptions. As George Kennan, a long-time student of the relations between the two superpowers, argues Western societies respond to pressure in a very dangerous way:

> There is . . . nothing in nature more egocentrical than the embattled democracy. It soon becomes the victim of its own war propaganda. It then tends to attach to its own cause an absolute value which distorts its own vision of everything else . . . It will readily be seen that people who have got themselves into this frame of mind have little understanding for the issues in any context other than the one in which they are involved. The idea of people wasting time and substance on any *other* issues seems to them preposterous.[8]

Once in order to pursue this examination of the problem of misperceptions in the field of energy and international security, one accepts Kennan's description of the United States as an "embattled democracy" (or rectifies it by defining that the country as a power that is deeply torn by internal conflict and whose role in world affairs is declining steadily), it becomes relatively easy to understand why the United States is at the center of such great concern over energy supply security. On the other hand, in Western Europe, where dependence on foreign energy sources is much greater than in the United States, and which happens to be a relatively close neighbor of the USSR, such concern has never reached the same pitch as in the United States. On the contrary, Western Europe is willing to reduce dependence for energy on the Middle East, towards diversifying its supplies by turning to the Soviet Union and other countries, perhaps because it has more realistic view of the roots of the Middle East situation.

In a recent attempt to sum up the past two decades of domestic social conflict and the disillusionment that swept Ronald Reagan into the White House, Theodore H. White, a quite objective although conservative American observer, holds that the key factor responsible for the hysteria unleashed over the problem of energy security in the United States at the end of the 1970s and also for the outcome of the 1980 presidential elections was the shortage of gasoline suffered by millions of Americans following the Revolution in Iran.[9] Together with the United States impotence in the case of the Teheran hostages, this shortage led American citizens to feel not only that the country was badly governed, but also that its way of life and well-being were very seriously threatened by unfriendly and distant enemies, of which, the Soviet Union must, of course, have been not only one but the principal one. It did not matter that Iran had been "lost" not because of any type of Soviet intervention in the

Middle East, but as the result of the holy alliance between the United States Army and the corrupt and power-mad Shah of Iran. What really mattered to the Americans in 1980 was that the invisible "enemy", using its infinite conspiratorial ability, was not 12,000 miles away from the Atlantic coast of the United States but present in the United States itself, reducing the amount of fuel available at American gasoline stations from Vermont right across to California. Putting an end to catastrophes of this kind, whether in the far away Middle East or at a local gasoline station, became a major "national security" issue. The sense of alarm fed in the United States by the "nationalization" process of oil resources in most distant foreign countries in the late 1960s and 1970s also played an important part in creating an atmosphere of uncertainty over the question of American access to foreign oil.

In our search for the historical reasons behind this recent obsession with energy as a vital factor in the security of the North Atlantic superpower, we might recall the history of United States policy on foreign oil during and immediately following the Second World War, between 1941 and 1947. According to the excellent study which has been made on this subject:

> In the decade before Pearl Harbor the administration of Franklin D. Roosevelt showed no interest in developing national policy for foreign oil matters. New Dealers in charge of oil concerned themselves with the home front, where economic depression and industrial collapse threatened national survival as much as any war ever had. Only as depression gave way to real war did federal officials turn their attention to foreign policy, first as part of the organization for war and much later as part of the preparation for peace.[10]

By 1947–48, however,

> As national interest became more and more interwoven with the interests of private enterprise, the wartime effort to define and to carry out a coherent foreign policy fell to pieces. Even so, by the early postwar era that effort has produced a record of impressive accomplishment in the Middle East . . . Most important of all the basis for a sizeable Middle Eastern oil industry with strong American participation has been established. . . . The structure was basically sound. It had achieved the primary aims of security and order.
>
> So it seemed in 1947. Three decades later the irony of those decisions can be fully appreciated. The security achieved in the early postwar period brought only ephemeral safety. (Since US government officials adopted) . . . the assumption that commercial and national interests were identical . . . the possibilities for constant and responsible planning by both governments and oil companies vanished too soon, and in the end the men who sought security promoted its collapse.[11]

United States energy security collapsed in the Middle East in the early 1970s in spite, one may add, of the fact that Soviet policy in the region was far from successful either, judging by the events which took place in Egypt following Nasser's death. In the face of available historical evidence, it does not make much sense in the early 1980s to interpret the tricky Middle East situation in terms of the conflict between the two superpowers. If the United States could stop seeing energy problems, and particularly those relating to the international supply and demand of hydrocarbons, as a central part of the East-West conflict and if the United States could start to recognize the structural changes which have occurred in the energy markets and the policies of the major oil-producing and exporting countries, perhaps the links between energy and international security could be easily brought into line and become manageable.

But even if this were the case, another international problem which is rarely considered seriously would still be left unsolved. We refer to the question of supplying energy for the development of the poor non-oil producing countries on the periphery of the industrialized world, that is Latin America, Africa and Asia. For reasons which are far from clear, Western analysts of the international energy situation continue even at this stage to be almost unanimous in claiming that energy consumption in the remaining decades of the present century will increase rapidly in the developing countries, especially in those with large potential hydrocarbon reserves where domestic consumption of oil is heavily subsidized.

Serious reflection is needed at this point because the developing countries are not in a position to finance the development of alternative energy sources to substitute oil. The functioning of their economies will continue to depend largely on this conventional source of energy which is made up mainly by imported oil. However, at present, no Western energy experts are able to explain how these countries will pay their debts and continue purchasing more oil without further exacerbating their balance-of-payments problems, especially when their own economic recovery is so closely linked to the uncertain outcome of the crisis which the Western industrialized countries are currently undergoing.[12]

Long-term credits and preferential treatment by OPEC members, who are themselves facing a steady drop in oil revenues, are not enough to sustain a rapid increase in energy demand by the developing countries. As the borrowing capacity of the developing countries in the West's private capital markets becomes exhausted, it will be very difficult for the potential energy demand present in the developing countries to be converted into real demand. Strange though it may seem, such factors as the deterioration in the terms of trade of the developing countries, their growing current account deficits, the size of the foreign debt, among others, are not, however, taken into account when the future of the world energy sector is debated in the West under the assumption that there will be a steady growth in the demand for energy by the non-oil producing developing countries.

In conclusion, since the economic stagnation and impoverishment of the world peripheries, which are a most important part of the present international economic crisis, will only increase international political and social chaos and very possibly tie this into the East-West conflict, the problems concerning the links between energy and the international security of the advanced industrialized countries of the North become dwarfed by the size which the same problems acquire in the developing countries. These last aspects must be addressed most seriously by those concerned with them since, in the final analysis, they are no less important for the preservation of peace than the East-West strategic and political conflicts and the international security considerations of these two main blocs.

Notes

1. 31st Pugwash Conference on Science and World Affairs, Banff, Alberta, Canada, 28 August–2 September, 1981, Final Draft Report of Working Group on Energy.
2. George C. Wilson, "Pentagon document stresses priority of defending Gulf's oil", *International Herald Tribune* (Paris: 3 June 1982).
3. For more details, see Miguel S. Wionczek (Ed.), *World Hydrocarbon Markets: Current Status, Projected Prospects and Future Trends* (Oxford: Pergamon Press, 1982).
4. A series of thoughtful articles by renowned Western experts was written in 1981–82 on the Afghanistan problem. All of them differ from the official US propaganda line. The most recent contribution is Jagot S. Mehta's "Afghanistan: A neutral solution", *Foreign Policy* (Washington, D.C.), No. 47 (Summer 1982).
5. A book published in 1981 by such an august academic body as Harvard's Energy and Security Research Project (David A. Deese and Joseph S. Nye, (Eds.), *Energy and Security* (Cambridge Mass.: Ballinger Publishing Company)) starts with the presentation of its contents in the following terms: "The dependence of the United States and its allies on oil from unstable regions has erected a severe threat to world peace. The uncertainty of Soviet intentions in the Persian Gulf heightens the risk. Never have so many nations been so ill-prepared to meet a direct threat to their economic well-being and political power".
6. For a fascinating analysis of this point, see Robert Jervis, *Perception and Misperception in International Politics* (Princeton University Press, 1976).
7. *Ibid.,* p. 181.
8. George Kennan, *Russia and the West under Lenin and Stalin* (New York: Mentor, 1962), pp. 11–12.
9. Theodore H. White, "Summing up—A chronicle of two decades of social experiment . . . and the disillusionment that swept Ronald Reagan into the White House", *The New York Times Magazine* (2 May 1982).
10. Michael B. Stoff, *Oil, War and American Security—The Search for a National Policy on Foreign Oil, 1941–1947* (New Haven and London: Yale University Press, 1980), p. 11.
11. *Ibid.,* pp. 214–215.
12. Miguel S. Wionczek, "Reflections on the Prospects for the World Hydrocarbon Markets during the Eighties", Chapter VI of the book mentioned in Note 3.

The North–South Negotiating Process

MUCHKUND DUBEY

The Negotiating Cycle

North-South negotiations, like any other multilateral negotiation, constitute a multiphase and multifaceted process. The process can be said to start with the quest for ideas which become the basis of negotiations. The ideas get converted into issues which are joined by individuals and groups of countries. The formulation of individual country, subregional, regional and wider group (G-77) positions on the issues is an important phase of the negotiating process. Important trade-offs among the countries of the South and North take place during this phase. Then comes the crucial stage of trade-offs and compromises between the North and South, resulting in the adoption of resolutions, declarations and decisions, either by consensus or by majority voting. Finally, there remain the tasks of getting the ideas translated into concrete actions, maintaining their continuity and ensuring progress in their evolution to ensure the progression of international economic relations to higher and higher levels of excellence.

The nearly complete cycle of the negotiating process described in the above paragraph is a somewhat rationalized and idealized version of what happens in actual practice. We have resorted to such a rationalization mainly for the purpose of facilitating comprehension and orderly discussion. The negotiating process seldom unfolds in neatly divided phases as in the cycle described above. Quite often, one phase of the negotiation merges with the next. National and group interests sometimes constitute the mainspring of the evolution of ideas. The countries of the North inexorably exert pressures and use all the leverage at their command to influence the formulation of the country, subregional, regional and wider group positions of the South. Likewise, the countries of the South seek to influence the process of the formulation of the group position of the North. Even otherwise, in order to arrive at a realistic and negotiable group position, each group has to take into account the interests of the other groups. While introducing the first draft of the International Development Strategy for the Second United Nations Development Decade in the Second Committee at the Twenty-Fifth Commemorative session of the General Assembly and addressing the other groups of countries represented in the Committee, the then Chairman of the Group of 77 (the author of this paper) had said: "When in the early hours of this morning I was finalizing the text of this draft, you were more overwhelmingly present before me than ever during the last 2-year period of negotiations on this document". Intra- and inter-group negotiations are involved at each phase of the cycle. These negotiations do not always involve confrontation or the physical presence of the negotiating parties. They are so inextricably tied in the web of international relations that each party is constantly under the influence of the others while engaged in the process of formulating its individual country or group position.

Every negotiation process does not have as well-defined and comprehensive a cycle as the one outlined above. Some negotiating processes are much shorter and quite mechanical. But, then, many of these shorter processes can be linked to the larger cycle of North-South negotiations centered on certain key ideas that might be going on at the same time. Similarly, each country or delegate need not be involved with each phase of the cycle. The extent of the involvement with the cycle, of a particular country, depends upon its perceived national interest in the idea constituting the focus of the negotiation, its perception of its stake in the wider regional and global objectives sought to be achieved through the negotiation and the general alertness, intelligence and effectiveness of its delegate.

The final phase of the cycle—getting the ideas translated into concrete action, maintaining their continuity and ensuring progress in their evolution—is the most difficult and complex part of the negotiating process. Most of the resolutions/declarations incorporating the result of an important negotiating process, provide for a built-in review mechanism to assess progress in implementation. But, more frequently, to ensure implementation, it becomes necessary to initiate a whole series of new negotiating processes. Each of these processes may constitute a separate cycle of negotiation. Ensuring the continuity of ideas and progress in their evolution calls for an extraordinary degree of alertness, sensitivity and tenacity on the part of the delegates. A delegate must be intellectually equipped to handle this difficult and challenging task. And those who are so equipped find their tasks rendered difficult and quite often hopeless, because of the inability of most governments to perceive their national interests in the preservation of and further progress in the basic ideas and fundamental objectives of international development cooperation.

Sources of Ideas

Ideas that are subsequently converted into issues for negotiation emanate from a variety of sources, such as academia, individual governments, secretariats of UN organizations, study groups, committees and commissions set up by inter-governmental decisions or at the initiative of the UN secretariats. Ideas emanating from academia are generally of a basic conceptual nature, whereas those advanced by governments, UN secretariats, UN study groups/commissions, have greater operational content. Basic ideas from which many negotiating issues have been derived are indeed very few. Some of these ideas are: trade as an engine of growth; departure from the most favored nation principle; extension of the planning approach and models for formulating international development strategies; and international accountability. The Haberler Committee (1958) was one of the earliest group of experts to come out with ideas and proposals that became the basis of the subsequent negotiations within GATT. Ragnar Nurske was mainly responsible for propounding the concept of trade as an engine of growth. Jan Tinbergen and Raul Prebisch have been the pioneers in advancing ideas for the extension of the national development planning method and model to guide international economic cooperation. The concept of international development strategy was derived mainly from a book on international development planning written by Tinbergen in the mid-1960s, whereas Prebisch was the first person to use the term "international development strategy". Important ideas of international accountability were conceived, elaborated and advanced by individual governments with the assistance of UN secretariats. For example, India played an important role in the evolution of and negotiations on such concepts of international accountability as targets for aid flows and action programs for the removal of barriers to the exports of developing countries.

The performance of UN study groups, committees and commissions in advancing ideas has been mixed. Some have strikingly original or significant ideas like the "link" and "commodity reserve currency". Some have merely refurbished and presented in a systematic, elaborate and coherent framework, ideas that had been floating about for some time. A striking example in this category is the country programming approach recommended by the UNDP Commission headed by Sir Robert Jackson. A large number of study groups and committees/commissions degenerate into presenting a mere summary of the state of the art and providing justifications for the pet prejudices and vested interests of particular secretariats or groups of countries. There are examples of committees and commissions having been set up without first consulting or seeking the approval of governments. Quite often, the commissioners heading the commissions and chairpersons of the study groups and committees are selected so as to influence the outcome of the deliberations of the committees/commissions in particular directions. For the same purpose, there is also a great deal of behind-the-scene maneuvering to influence the composition of committees and commissions. Thus the notion of the objectivity of experts functioning in their personal capacity as members of UN study groups and committees, is illusory more often than not. Prejudice is introduced at various stages and levels of their functioning. It is not uncommon for independent experts to come fully briefed by their government and to function at their behest. Many expert groups have been aborted and have produced mere statements of differing positions of governments rather than ideas based on objective considerations, because of the ever-present propensity of governments to influence the opinions of their experts.

Converting Ideas into Issues

Once the ideas come to light, the negotiator's task is to convert them into concrete issues for inter-governmental negotiations. An important part of the negotiating process involves presenting the ideas and stimulating interest in them so that they become viable issues. Ideas generally assume the form of "norms". The task of negotiators is to ensure wide acceptance of the norm so that it becomes a moral obligation or contractual commitment. Thus, the negotiation process bridges the gap between ideas as desirable norms and ideas as obligations or commitments. Individuals who pioneer an idea are frequently surprised—at times pleasantly, but more often with a sense of anguish and frustration—when they see the shape their idea assumes when it is converted into an issue; often their surprise is even greater when they seen the final form their idea takes when it gains acceptance and is transmuted into an obligation or commitment assumed by governments following the negotiations.

Selection of Ideas

Ideas to be converted into negotiating issues must be selected with great care. All the ideas that clamor to be turned into issues for multilateral negotiations do not deserve to be elevated to this status. Many do not serve any wider regional or general interests, even though they may serve the peripheral national interest of one or a few countries. If proper discretion and judgment is not applied and peripheral ideas or issues are taken up for negotiation, then there is the danger of the entire negotiating process being discredited. In addition, attention will be diverted from more crucial North-South issues.

Quite often, a whole edifice of ideas is erected to justify the creation of additional posts or the upgrading of existing posts in a secretariat. As in the national context, so in the context of a multinational negotiation, there is often a person behind the idea, wanting to serve an individual or a national selfish interest. It has frequently been evident as negotiations reached the concluding stages, that the basic ideas and the country positions on them are readily sacrificed at the altar of acquiring high positions for particular individuals. Frequently, other delegates know what the entire game is about and yet they readily play for political reasons.

In the case of many of the recent global conferences centered on specific issues like environment, combatting desertification, water resources development, among others, the main ideas behind the conferences were given short shrift; and the entire negotiating process was geared to the creation of new institutions and new posts to staff them. It is true that institutions—either new or the existing ones duly strengthened—are necessary to monitor progress in the implementation of the decisions of a conference, carry forward the idea behind the conference, and mobilize support for it on a continuing basis. But the entire negotiating process gets distorted when it is obvious from the beginning that the ultimate purpose of the negotiation is going to be to accommodate the interests of certain individuals or the empire-building ambitions of some branch of the secretariat.

Selling the Idea

Sometimes, a long educational process among developing countries must precede and go side-by-side with the negotiating process. Many delegations are not equipped to grasp easily the significance of the idea to be converted into a negotiating issue. Nor are they able to understand the ramifications of the approach that should be adopted in the negotiations. For example, when the idea of an international development strategy emerged in 1968, most delegations of developing countries were not aware of its great significance and far-reaching implications for international development cooperation. They thought the strategy would be just a general resolution of the kind that launched the first United Nations Development Decade. It was left to a few delegates to conceive the entire idea of an international development strategy in the form of a projection at the international level of the basic national development planning model. These delegates then worked out the details of the model, explained to their colleagues from other developing countries the great significance of this model for making progress in North-South negotiations and as a comprehensive framework for international accountability, and defined the main negotiating issues, both among developing countries themselves and between developed and developing countries. The educational process started well before the negotiations commenced among developing countries and continued simultaneously with these negotiations and those between developed and developing countries.

Similarly, before commencing the negotiations on the international development strategy for the 1980s, it became necessary to launch a long educational process to resolve the conflict a number of developing countries perceived between a new international development strategy and the New International Economic Order. This conflict persisted in one form or another until the end of the negotiating process. And so did the educational process.

Preventing Deflection from the Main Idea

Another important challenge the negotiators of the South have to face is to be able to frustrate efforts designed to get the negotiations deflected from the original idea into the by-lanes of what may be called spurious ideas.

At the time the negotiation began on the International Development Strategy for the Second United Nations Decade, the Secretariat suggested that the Strategy should be a mere set of norms which governments could subscribe to by adopting a general resolution. This would have reduced the Strategy to a set of pious declarations without embodying any commitment of governments. At a later stage of the negotiations on the Strategy, another attempt was made—this time by the prestigious Committee on Development Planning—to divert the exercise by suggesting that the Strategy should consist of a general declaration followed by individual declarations of governments. The general declaration part of the Strategy would have been negotiated and subscribed to by all governments. Individual declarations would have reflected only the intentions of governments and not their commitment to adopt policy measures to realize the goals and objectives of the Strategy. Different countries would have declared their intention to adopt different sets of policy measures. There was no assurance that such an approach would have resulted in any progress from the prevailing position. Moreover, this would have given the international community no standard yardstick to measure progress. The negotiators of the South had to wage a Herculean battle to stick to the model of the Strategy originally conceived by them and make it a totally negotiated document embodying the commitment of governments.

Continuity of Ideas

It is important to maintain continuity of the ideas constituting the basis of North-South negotiations. This, in fact, is the only way to make progress in the evolution of ideas and to ensure the progression of international economic relations to higher and still higher levels of excellence. For this purpose, the negotiators must know the present state of the art. Otherwise, they will simply waste time rediscovering the wheel.

Unfortunately, during the last 10 years or so, continuity in many of the ideas has been disrupted. In several areas, negotiations have resulted in the erosion of the gains of the past and the adoption of retrogressive measures. *Ad hoc*-ism has replaced progressive development of ideas. For example, increases in financial assistance available for activities in specific areas have been erroneously regarded as progress without taking into account that there has simultaneously been a decline in the proportion of GNP of the developed countries made available as official development assistance. A series of *ad hoc* but qualitatively similar and marginal measures have been adopted in different areas to create a semblance of progress; there has, at the same time, been a retreat from the general positions reached long ago, particularly on fundamental and structural issues. The present process of the North-South dialogue has become general and repetitive without being cumulative.

Deterioration in the international economic and political environment sometimes makes it necessary to withdraw from the positions reached in the past. This has been particularly true since the onset of the decade of the 1970s when most of the developed market economies have been in the grip of frequent recessions and continuing inflation accompanied by socially unacceptable levels of unemployment. During this period, the world economy also entered a phase of selective scarcity of key natural resources brought about mainly by excessive consumption of these resources in developed countries. Concern with the environment has led to a conscious effort to change prevailing life styles, even at the cost of material prosperity. All this has led to drastic adjustments in the priority and importance of ideas which constituted the main negotiating issues during the 1960s.

In this unfavorable climate of international development cooperation, many of the gains achieved earlier in the progressive development of ideas have been eroded. The presently accepted position on the target for flow of official development assistance represents a retreat from the position reached in the International Development Strategy for the Second United Nations Development Decade. Many of the advances made in the 1960s for removal of barriers to the exports of developing countries have been drastically set back. The system of coordinated and unified programming of all technical cooperation activities of the UN System instituted in the early 1970s, is now virtually in shambles. The intrusion into multilateral programs of bilateral prejudices and influences has been carried to a point where it poses a threat not only to the very identity of the multilateral programs, but also to the ideals and objectives of the United Nations. In the GATT Ministerial Meeting held in November 1982, the developing countries devoted most of their effort and energy to ensuring that positions reached and principles accepted earlier were not eroded under the offensive launched by the developed countries. Success in this effort was regarded as the main achievement of the developing countries at this meeting.

Tactical withdrawals from previous positions become inevitable in certain circumstances. First, it may be desirable to stage a tactical retreat from positions reached earlier on issues which were regarded as important at that time but which are no longer so, to make progress on issues which have now become the burning concern of the international community. For example, in recent negotiations, major NIEO issues calling for structural changes in the world economic system have rightly been given priority over such issues as commodity pricing policy or *ad hoc* tariff reductions, that figured prominently in earlier negotiations.

Second, when the climate for negotiations is unfavorable, it is tactically important to preserve an idea even in a diluted form instead of allowing it to be totally discarded. For example, in the present unfavorable situation with regard to the flow of official development assistance, it is important to preserve the concept of commitment to a specified target for the ODA flow, even though it may not be possible to adhere to the deadlines previously agreed upon for achieving this target. Whereas tactical retreats from previously agreed positions may be desirable, important ideas on which consensus has been reached in the past should not be allowed to be diluted or abandoned for short-term *ad hoc*, opportunistic gains.

Formulations of Country Positions

Once an idea becomes a negotiating issue, it is necessary for individual governments to formulate their positions on various aspects of the issues. The education process that goes on in multinational groups is an important aid to the formulation of such positions. A country formulates its position basically in the light of its perception of national, regional and wider group interests involved in the negotiating issue. Subregional, regional or wider group interests may not be immediately perceived as national interests; but they have the potential for becoming so in the medium- and long-run. Few countries are, therefore, immediately able to see their long-term national interests in the regional or general interests to be articulated and safeguarded in the negotiating process. The vast majority of them come unprepared to the negotiating table, quite willing to be guided by and cooperate with those who happen to know better.

Even with regard to national interests, very few countries come to a negotiating conference with a clear perception of their interests and well-defined positions designed to safeguard and promote such interests. Many developing countries do not have at their disposal the necessary expertise to analyze and grasp the significance of an idea that has matured into a negotiating issue, to assess how it affects their national, regional and wider group interests and to formulate a clear-cut position reflecting these interests. They have, however, a broad notion of how their national interests are affected during the course of the negotiations.

Countries that go to a negotiating conference with a well-formulated position are the ones which assume the leadership role in such conferences. Generally a direct co-relationship exists between the degree of national or regional interests involved and the extent of preparedness and the clarity and elaborateness of the country position formulated for the negotiations. It was, therefore, natural that India was one of the leaders in the negotiations on liberalization of trade in manufactures, preferences and aid flows; Malaysia and Australia assumed the leadership role in the negotiations on commodities; and Upper Volta, the Sudan, and recently Bangladesh, were leaders on special measures in favor of the least developed among developing countries.

On issues such as the principles governing international economic relations and international development strategy it is not easy to perceive the national interest involved. Interest in such negotiating issues can be generated and sustained only through a process designed to inculcate the perception of enlightened and long-term national interest in the regional and wider group interests.

Sometimes, countries with no immediate national interest involved in an issue, are able to see the general interest and their own enlightened longer-term national interests more clearly than the countries directly concerned with the issue. In such a situation, the former countries can play an important role in the negotiations. Since their immediate national interests are not involved, they need not be timid in championing the general cause. They are also in a stronger bargaining position than the countries whose immediate interests are involved. Therefore, in appropriate circumstances, they can safeguard the national interests of the latter countries better than their own delegates can. The countries not directly involved with the issue can also ensure, to the extent possible, that no conditions or guidelines are imposed on the countries with a weak bargaining position, that can subsequently be used as a precedent for being applied generally, thus adversely affecting the national interests of the former countries and those of similarly placed countries.

Formulation of Regional Positions

The process involved and the considerations to be taken into account in formulating regional positions are not very different from those that go into the formulation of the position of the Group of 77. These processes and considerations will be discussed in the subsequent section. The present section is, therefore, confined to the discussion of some general aspects of the problem of formulating regional positions.

A regional position is not the total of the positions of the individual countries of the region. The regional position must reflect distinctive regional experience, aspirations and needs. However, all regional positions are not necessarily unique to the region concerned. Many elements of regional positions are common to those incorporated in the regional positions of other regions. There is also a commonality between a regional position and a wider group position. Distinctive regional positions prevail with regard to a very few issues; and these positions alone constitute distinctive regional contributions to the formulation of ideas of general interest to all developing countries. For example, in the UNCTAD forum, distinctive contributions of Asia have been in the field of shipping, state trading and planning; Latin American contributions have been in the sphere of regional and subregional integration and commodities; and the African countries have emphasized human resources development and development of institutional and physical infrastructure.

Regional positions, like country and wider group positions change with changes in the world economy and the economies of the region. By the early 1970s, regional and subregional integration no longer remained the battle cry for the Latin American countries as it was during UNCTAD I. This was because schemes of regional and subregional integration encountered numerous difficulties in Latin America and could not be made to work efficiently. Many of those schemes were either given up or have been stagnating. Recently in Asia, attention has shifted towards problems of poverty, food and energy.

Formulation of the Position of the Group of 77

The classical method followed for reconciling differences in the positions of individual and groups of developing countries in order to evolve the common position of the developing countries as a whole, has been to formulate and agree upon a package of measures some of which are of interest to all developing countries, while others are of interest to only some of these countries. An essential precondition for the acceptance of such a package is that it must be comprehensive enough to take care of the interest of each and every developing country. At the same time, the package should be realistic enough to serve as a basis of negotiation with developed countries.

Each member of the Group of 77 must accept and welcome the inclusion in the package of measures which are not of any direct interest to it but which have a vital bearing on the outcome of the development effort of the other members of the Group. Members of the Group should not only put up with measures in the package that may not be of any interest to them, but in the interest of Group solidarity, should actively work to get them included in the package. For example, a country like India must show an understanding of the deep interest of many African countries in technical assistance and human resource development. The fact that India is no longer at the receiving end of these measures of international development cooperation should, instead of engendering indifference, enable India to appreciate how important these measures are for accelerating development. Similarly, by virtue of its pursuit right from the beginning of its development effort of a prudent and pragmatic policy towards multinational companies, India did not have much of a problem with them. But India must understand the obsession of the Latin American countries, given their bitter experience of dealing with these companies, concerning the question of the nationalization of foreign companies. Likewise, India with its diversified export structure was in a better position to dispense with existing preferences and to agree to merge them in a generalized system of preferences from which it had more to gain than other developing countries. But in order to reach an agreement on a package, India had to show an understanding of the insistence of many of the Caribbean and African countries on preserving their existing preferences or demanding compensation for allowing them to be merged with a generalized system of preferences. With their monoculture economies and one or a few commodities accounting for the bulk of their national incomes and export earnings, these countries had practically everything to lose had they given up the existing preferences without adequate safeguards.

On the other hand, the least developed countries should realize that some of the problems faced by the relatively more developed among the developing countries, are not as irrelevant to their development needs as it may appear superficially. For the least developed countries are themselves going to face these problems at later stages of their development. Eventually all these countries will embark on the task of industrializing and diversifying their economies and generally making themselves self-reliant. If measures they will urgently need at later stages of their development are adopted right now in response to the demands of the countries at a relatively higher stage of development, if at least some of the rules of the game are already changed in favor of the developed countries as a whole, and if there is already some progress in restructuring the system of international economic relations, it will make the future task of development so much easier for the least developed countries. When they enter the subsequent phases of development, they will no longer be required to cross some of those hurdles some of the more advanced among the developing countries are at present struggling to overcome. Thus, a significant challenge of the problem of evolving a common platform of the developing countries as a whole is one of understanding and sympathy with others' problems and seeing in others' immediate interests, one's own long-term enlightened interests.

However, the real problem in harmonizing the interests of the developing countries in a common package of measures, arises from the fact that the package has to incorporate not only different measures of interest to different groups of developing countries, but also measures regarding which there is a clear conflict of interest among different groups of developing countries. This conflict must be resolved before the package becomes generally acceptable. We have discussed why in the negotiations on the generalized system of preferences, some of the developing countries were unable to countenance any erosion of their existing preferences without receiving compensatory benefits. Therefore, in the beginning they were not at all enthusiastic about the generalized system of preferences into which their special existing preferences were supposed to merge. Herein lay a conflict of interests that had to be resolved before the package could become acceptable.

A conflict also arises when measures for increasing flows of assistance to particular sectors or for particular activities are considered. This is because the overall resources are limited, and resources made available for activities in a particular area, or to particular groups of countries, are at the cost of those available for activities in other areas or for other groups of countries. Therefore, the groups of countries interested in the latter category of activities find their interests in conflict with those interested in the former category. Some of the least developed countries in their anxiety to claim a larger share of the existing level of ODA resources may be tempted not to join other developing countries in making a demand for an increase in the overall availability of such resources. Similarly, countries at earlier stages of development may sometimes be lured by selective commodity-by-commodity preferences and by proposals for introducing a system of graduation in the existing system of preferences because of their perception that in the short run, they stand to gain from such approaches even though it may be at the cost of the general interest of the developing countries.

The best chance of resolving such conflicts of interest lies in the ability and willingness of the least developed countries or countries placed in a similar situation to perceive their own long-term enlightened interest in the immediate interest of others and to eschew those of their short-term interests that conflict with the general interest of the developing countries and hence with their own enlightened self-interest. Thus even if it may not be in their immediate interest, the least developed among the developing countries and other countries suffering from special disabilities, must enthusiastically support other developing countries in their struggle for changing the rules of the game, in building the institutions they will themselves desperately need in the not-too-distant future, and in establishing permanently, systems, frameworks and facilities that will be indispensable at subsequent stages of their economic development.

The more advanced among the developing countries, on the other hand, must do for other developing countries what they expect developed countries to do for developing countries as a whole. For example, the more advanced developing countries should grant unilateral tariff preferences to the least developed in the expectation of implicit reciprocity. They should agree to the inclusion in the common package of measures designed to harness fully the productive capacity of the least developed. Prosperity is as much indivisible in the developing world as it is in the world as a whole. In the long term, the more advanced among the developing countries cannot but gain by agreeing to measures designed to promote the development of the least developed.

To put the package together, in the ultimate analysis, the principal parties involved in the conflict will have to make some sacrifice for the sake of the general cause of the developing countries. The long-term interest of the principal parties is bound to suffer if the package collapses. Therefore, in a conflict situation, the parties concerned should stop short of taking the differences of interests to a breaking point and make compromises even at the cost of short-term disadvantages, in order to evolve a common position of the developing countries as a whole.

Postponement of decisions on issues involving conflict of interest can work both ways. Sometimes it is best to leave the matter alone for the time being. Subsequent events have demonstrated that issues that have appeared more than life size because of their topicality and the passion of the moment, dwindle into insignificance with the passage of time. For example, at UNCTAD I, the Latin American countries made a big issue of the concept of regional and subregional integration. By the time of UNCTAD II, this issue lost much of its momentum as it became evident that schemes of regional and subregional integration in Latin America were proving to be at best a very limited success, if not a failure. On the other hand, the tactics adopted at UNCTAD II to postpone the issue of identifying least developed countries apparently misfired. The issue became so pressing that it was vigorously pursued in other forums where a set of not entirely scientific criteria was agreed upon and applied for identifying these countries. Thus, UNCTAD lost a very important initiative in its sphere of activities that it was able to regain only a few years later at the

United Nations Conference on Least Developed Countries in Paris in September 1981.

Sometimes, a strategic bargain between major groups of developing countries is necessary to evolve an agreed package as a basis of negotiation with developed countries. This happened at the Havana Summit Conference of the nonaligned countries in 1979 when the oil-importing developing countries agreed to lend their full support to the proposal of some of the major OPEC countries for a new global round of negotiations, in lieu of the latter countries agreeing to enter into a serious South-South dialogue as a preparation for and side-by-side with the global round of negotiations. The outcome was the adoption of a resolution on a global round of negotiations committing all the nonaligned countries to work for the launching of these negotiations and generally for their successful conclusion, and a set of guidelines for consolidating solidarity among the nonaligned countries, an important element of which was an implicit commitment by OPEC countries to supply, in periods of shortage, oil to oil-importing nonaligned countries, on a priority basis and through government-to-government contracts.

North–South Bargains

Recently, North-South bargains have been frequently analyzed in terms of game theory. It has been suggested that the developing countries should attach priority to those negotiating issues which make for positive-sum games, where there is a net positive gain to be shared by both the parties. *Ipso facto*, issues on which the negotiations involve a zero-sum game, where the gain for one country involves an equivalent loss for the other, should not be considered in North-South negotiation. Examples have been given of issues that can make the negotiations a positive-sum game and of those that make them a zero-sum game. For example, it has been suggested that negotiations on issues such as price stabilization measures, trade liberalization and multilateral balance-of-payments financing will produce a positive-sum game. On the other hand, bargaining on aid issues, price enhancement measures, debt cancellation, and so forth will involve a zero-sum game.

It is, of course, convenient to look at negotiating issues from the perspective of whether they make for positive-sum or zero-sum game negotiations. However, this is an over-simplified approach to selecting negotiating issues or establishing *inter se* priority among them. There is nothing intrinsic about an issue that makes the negotiation on it a positive-sum or zero-sum game; it is the perception of the parties to the negotiations, of interests involved in the issue that determine whether the negotiations on it are going to be a positive-sum or negative-sum game. And perception of interests depends on a variety of factors, such as the objectives to be pursued through the negotiations, the time horizon in which the objectives are seen, the economic and political circumstances in which the negotiating countries are placed, and the other party's capacity to inflict loss.

There could be three-tier objectives in the light of which interest involved in North-South negotiations can be perceived. First, there are long-term objectives of a broader general nature—peace, stability, harmony among nations, international social justice and human solidarity—to be achieved by such negotiations. These objectives, among others, provide a rationale for making the development of developing countries the common concern of the entire international community and support for measures for eradication of poverty and for equitable global redistribution of incomes and resources. Factors might emerge in the world economy or international economic relations that make some of these longer-term objectives a matter of immediate international concern. For example, development of developing countries is now taken more seriously as the concern of all countries than was the case in the immediate postwar years. Similarly certain recent developments have heightened the perception of interdependence among the economies of the North and South. It would, therefore, be too simple or cynical to assume that these broader long-term objectives do not provide any but a moral basis for bargaining because parties to the negotiations do not perceive their short- and medium-term interests in them.

Then there are what may be regarded as medium-term objectives like enhancing growth and prosperity through nonreciprocal measures for liberalization of trade and more rational utilization of the global resources by adopting international measures for harnessing the productive resources of the developing countries. It is now being increasingly perceived by the North that the short-term cost involved in pursuing these objectives can be more than compensated by long-term benefits to be derived by them. These are many other examples where perception of gain changes with a change in time perspective.

The third tier objectives are those which involve immediate loss or gain, such as preventing loss in employment, disruption of industries and balance-of-payments problems resulting from an increase in imports from developing countries and avoiding budgetary problems created by increased financial flows to these countries.

The whole question of bargaining on the basis of perception of loss or gain is further complicated by the fact that in negotiations on the same issue, some developed countries perceive a positive-sum game and others a zero-sum game. The perceptions depend on how the individual or groups of developed countries discount their general long-term, medium to long- and short-term objectives. The rate of discount will depend on the country's geographical position, the size and structure of its economy, the degree of economic autonomy achieved by it, its economic and political vulnerability to significant changes in the world economy or international relations and its geopolitical strategies and objectives. It is well-known that because of the above factors, many of the negotiating games the United States regards as zero-sum are regarded as positive-sum by the Nordic countries and the Netherlands. For the same reason, even among the so-called hardliners among the countries of the North, there are important nuances in the perception of interest and important shifts in negotiating positions of one or another among these countries taking place from time to time, thereby changing the entire complexion of the game.

It should also be remembered that the gains in a negotiating game sometimes are not confined to the area directly covered by the negotiations. The pressure exerted by negotiations in one area, need not affect that area alone; its spinoffs in other areas must also be counted in assessing whether the game is going to be positive-sum or zero-sum. For example, it is now generally recognized that the struggle the developing countries waged for the establishment of a Capital Development Fund within the UN led to the creation of not only the Special Fund within the UN, but also the International Development Association within the World Bank. Similarly, the pressure that the developing countries mounted within UNCTAD to get a Supplementary Financing Facility created, was, in no small measure, responsible for the improvement of the IMF Compensatory Financing Facility and the establishment of the Extended Fund Facility.

A negotiating game becomes positive-sum not only when the more powerful partners perceive that they will be better off in terms of the net benefit the negotiations will yield, but also when they can be made to perceive that in the absence of a compromise, they will be worse off. Herein comes the role of the bargaining power of the less powerful partners and what they can do to strengthen it.

Strengthening the Bargaining Power of the South

Essentially, the bargaining strength of the South derives from the feeling of solidarity among its members, their unity and cohesion. Unity and cohesion are based on a subjective perception of identity or convergence of interest. The developing countries adopt common positions in negotiations not only because they see that their immediate interests are involved in the negotiating issues, but essentially because they have a shared understanding of the dynamics of the present international order and they agree among themselves on both the necessity and the method of removing disparity in the wealth and incomes between them and the developed countries. There is generally a tendency to underestimate the capacity of the developing countries to perceive their interest in this fashion in spite of their having demonstrated it time and again through negotiations in different forums. But this is as far as it goes. For enhancing the bargaining strength of the developing countries, the subjective sense of unity must be backed by material basis for solidarity.

Evolving a satisfactory package of measures constituting the common positions of the Group of 77 itself provides a material basis for the solidarity of the Group. We have already discussed the best way of putting together such a package.

South-South cooperation is the most important means for providing a material basis for the solidarity of the Group of 77. In this context, an early implementation of such proposals as the Common Fund of the Developing Countries themselves, Council of Producers' Alliances, Non-aligned Solidarity Fund, and Project Development Fund of the Non-aligned and Other Developing Countries, acquires special importance.

South-South negotiations sometimes have demonstration effects that can influence the North-South dialogue in a positive direction. The impression that the developing countries create through these negotiations about their ability to discharge their obligations to each other can be a factor influencing the manner in which the developed countries will discharge their obligations to the developing countries. Similarly, by formulating and implementing some of the schemes of cooperation among themselves, the developing countries can demonstrate the feasibility of adopting and implementing similar schemes at the global level between developed and developing countries, and thereby exert moral pressure on the developed countries to follow the examples of the developing countries.

To strengthen their bargaining position, developing countries must also take advantage of differences among the developed countries. They should be particularly aggressive in doing so when gaining the widest possible acceptance of their ideas can lead to obtaining broader, long-term goals through North-South negotiations. It is on these issues where developing countries should resort to majority voting and where the size of the majority is crucial.

Decision by Consensus or by Majority Voting

Where the proposals of the South (aid flows, contributions to special funds, setting up of institutions requiring large funding) cannot be implemented without the North's active support and where acceptance by the North involves a direct sacrifice on their part, the South must not try to force the issue. In these cases, the developing countries must agree to enough concessions to make it worthwhile for the developed countries to offer their willing consent. It is impractical to legislate sacrifices to be made by the North by a majority vote without their consent. Of course, the South should use all its bargaining clout to keep the North under pressure until the end; but the decision ultimately must be by consensus.

On issues involving broader, long-term objectives (for example, where a principle or an idea on which there can be only step-by-step progress, is to be preserved and advanced), a degree of confrontation between the North and the South may become unavoidable and even desirable. It is necessary to go on agitating about these ideas and keep them alive by majority votes. For example, it is unrealistic to expect that at the present stage of international economic relations, the developed countries would accept such ideas as an international taxation system, drastic restructuring of the world financial and monetary systems, developing countries' control over institutions like the International Monetary Fund (IMF) and the World Bank, or drastic changes in the rules governing international trade. At the same time, these are key issues relating to the establishment of a New International Economic Order; and, therefore, the developing countries cannot afford to abandon them. Of course, maximum effort must be made to achieve the widest measure of agreement on these ideas, because this is how they will advance. But in the ultimate analysis, they must be voted on to be kept alive.

Format of North-South Negotiations

The functioning of the group mechanism is a practical necessity for negotiations in a large multilateral forum. In any negotiation among a large number of countries, coalitions of countries with similar interests inevitably tend to form. The question is whether groups should be constituted and operate on an *ad hoc* basis for different negotiations or on the basis of well-established and firmly institutionalized entities representing broad common interests. Practical necessity will point to the inevitability of the latter type of arrangement. If groups crystallize and start functioning only after the negotiations start, negotiation will become a highly chaotic, cumbersome and time-consuming process. Functioning through institutionalized groups helps to make the negotiations a continuing exercise of a more orderly and systematic nature.

Well-established groups have existed within the United Nations ever since its founding. Under the Charter, elections to UN bodies are provided for on the basis of equitable representation from different regional groups. Western and socialist groups began functioning soon after the UN was established. With respect to the developing countries, there has been from the beginning the Afro-Asian group. This coalition was concerned mainly with political issues, but occasionally, it also adopted specific positions on economic issues. The Afro-Asian group was soon supplanted by the Non-aligned group. The Group of 75 was established in 1963 as preparations were underway for UNCTAD I; it was transformed into the Group of 77 by the time the Conference concluded. Since then it has remained the Group of 77 because of the great symbolic value this title acquired, even though more than 125 developing countries are now represented in the Group. It was also during UNCTAD I that the group mechanism was formally recognized in a United Nations Resolution and thus firmly institutionalized.

When the Group of 77 emerged on the scene in 1964, it was hailed as one of the most important phenomena of the postwar years. It is only during the recent years when the climate of international development cooperation has deteriorated sharply that the functioning of the Group of 77 has been questioned and doubts have been cast on the very rationale for and validity of its existence. It is being alleged that the Group of 77 is a voting bloc and that the Group votes indiscriminately, irrationally and monolithically, thereby making a mockery of the resolutions adopted by such votes.

The main reason for the dissatisfaction with the functioning of the Group of 77 is not because it functions the way it is alleged to do, but because it functions, as it must, inexorably in the interest of the developing countries. During the early years of the UN, when its membership was limited, the major economic and political powers frequently manipulated the bloc votes of regional groups in their favor. At that time, these powers never raised objections to bloc voting.

While developed countries tend to say that the functioning of the group mechanism, particularly that of the Group of 77, has introduced rigidities in North-South negotiations, they themselves are very well organized into groups of their own. In the UN forums, they function as Group B countries as defined in one of the decisions of UNCTAD I (Annex. A.V. I of the Final Act). Moreover they regularly and systematically coordinate their policies and harmonize their positions through the institutions of the OECD and EEC. They have at their disposal very efficient and prestigious secretariats, the world's top experts, and sizeable resources.

The Group of 77, on the other hand, has not been able to establish a secretariat of its own, has hardly any organized group of experts working for it and chronically suffers from paucity of resources to carry out research work to support its negotiating activities or even to hold meetings to formulate its positions. Thanks to their organizational resources and regular habits of coordination, the developed countries generally come better prepared to negotiating conferences than the Group of 77. They are able to project a consistent position when the same issue comes up for discussion in different forums, and they do not miss a single opportunity to advance their objectives. The developing countries, in contrast, work at cross purposes in many negotiating forums, and they are known to have allowed important objectives to go by default out of sheer inertia, unpreparedness or for want of persistence.

However, on broader long-term objectives of a more fundamental nature, the developing countries have repeatedly demonstrated their ability and will to remain united and alert. This unity is based on a kind of class consciousness among them. They all perceive themselves in what Prebisch called the "periphery of the world economy" and at the receiving end of the present international economic order. They feel very strongly that they are the poor, undeveloped and underprivileged segment of the world, that they do not have full control over their resources, that they are left outside the main decision-making process regarding the management of the world economy, that they cannot influence very much the world financial and monetary systems that are such a key determinant of their position in the world economy, and that they are at the bottom of the ladder of technological progress. Apart from this subjective perception of their unity, they have now learned that they must provide a solid objective basis by undertaking negotiations of their own for promoting and strengthening their collective self-reliance.

Recently, there has been a lot of talk about the appearance of chinks in the unity of the Group of 77. It is said that because of conflicts of interest among various groups of developing countries, sharp contradictions have appeared in the Group of 77, putting heavy pressure on its unity. It is predicted that soon a stage will come—if it has not already come—when the Group will disintegrate under the burden of its contradictions.

The developing countries have repeatedly shown particularly in UNCTAD, where the Group of 77 has been most cohesive and active, that they are capable of transcending the narrow short-term interests of individual members to preserve its unity. It is indeed remarkable how, despite conflicts of interests, the Group has managed to stand together.

It is, therefore, futile to predict the disappearance of the Group. It will remain a reality so long as the outcomes of negotiations in this unjust and unfair world economic system continue to depend not on the intrinsic merits of the negotiating issues, but on the bargaining power of the negotiating partners. The Group will continue to exist for quite some time to come not only because the functioning of the group mechanism is a practical necessity for any negotiation, but also because it just does not make any sense for the developing countries to give up the only bargaining power they have: that is their unity, their standing together as a group. Of course, the Group will disappear when most of its members come out of the periphery and join the mainstream of the world economy. But that will be like Nirvana, beyond which, in any case, there is no existence.

Another important indictment of the Group of 77 has been that the unwieldy size of the Group makes it time-consuming and at times even impracticable to carry on negotiations on behalf of the Group as a whole. It is, therefore, suggested that to ensure the effectiveness and speed of negotiations, the participants in them should be only principal interest groups or countries directly involved in the issue under negotiation.

In actual practice, even when the negotiations are conducted on behalf of the Group of 77, the real participants are those whose interests are directly involved and those who represent principal interest groups. Hard bargaining and serious negotiations always take place in smaller groups. This is in spite of the recent tendency to make all important negotiating committees, committees of the whole and all *ad hoc*/working/drafting groups, open-ended. This development reflects the recent trend of increasing demonetization of all UN organizations and bodies and the keen desire of even the smallest country to seize every opportunity to assert its sovereignty, one of the most visible forms of which is representation in UN bodies.

However, once the committee of the whole or the open-ended working group is set up and starts meeting, a sizeable proportion of the countries represented in them do not turn up. This is mainly because they do not have at their disposal the personnel or expertise needed to participate. The proportion of drop outs in the open-ended groups is even heavier, because there, real expertise and almost full-time attendance is required. Thus, the negotiations are inevitably confined to the delegates representing the principal interests involved in the issue.

But the democratic form at the negotiations must be scrupulously maintained. Every country must be given the feeling that it has full and equal opportunity to participate. The process involved in arriving at the stage where the negotiations are confined to principal interests must be democratic. Again, the outcome of the negotiations in smaller groups must be democratically discussed and endorsed. These are, of course, time-consuming procedures. But they are far less time-consuming than would be the case if there were no pre-established groups continuously negotiating among themselves and with the other groups. Besides, the time devoted to the maintenance of the democratic form is worthwhile because it saves time in the actual negotiation that takes place in smaller groups.

A View from the South: Growth of World Trade in Manufactures

VIJAY L. KELKAR*

In the history of the world economy, the post-Second World War period especially up to the early 1970s stands out due to its remarkable performance in terms of growth in output and world trade. It is estimated that during the period 1950–73 an average growth rate was 5% per annum for world output, while world trade grew at an even more rapid pace, almost at 8% per annum. It should be noted that both these indices are much higher compared to an earlier experience of the world economy even compared to the "golden age" of capitalism, i.e., from 1890 to 1913 when world production increased at the rate of 4% per annum, while world trade grew at 5% per annum. Table 1 gives the growth rate in the world trade and output in the manufacturing sector between the years 1853 to 1973. With this historical perspective, one can make at least three observations for the post-Second World War period: (1) the industrial output grew at a much more rapid pace, i.e., 6.5% per annum; (2) the growth in world trade in manufactured goods was even higher and grew at a rate of 9.5% per annum; and (3) the relationship between growth in output and trade has shown a remarkable increase, or, in other words, growth elasticity of world trade has been even higher.

Since 1973, there have been sharp changes in the world economy—declining growth rates in output and trade, relative shares of different commodity groups in world trade, a shift in relative prices especially that of energy, and increasing rates of inflation coupled with a shift in the pattern of balance-of-payment surpluses among the various countries. Equally, there have been major changes in the international monetary system. Indeed, there seems to be little doubt that since 1973 the world economy has entered a distinctly different phase. It can be argued that these developments since 1973 further underline the uniqueness of the post-1945 experience of the world economy in terms of its dynamism in industrial output and trade.

*The views expressed are the author's own and need not reflect the view of the organization with which he is associated. The author wishes to gratefully acknowledge Mr. H. Brewster, Mr. F. Clairmonte, Mr. A. Cornford, Mr. P. Ginman, Mr. P Rodrick, Mr. G. Sampson and Mr. K. Kojima for their helpful comments on an earlier draft. Needless to say, any remaining errors are his own.

TABLE 1. *Growth in world output and trade in manufactures (in % per annum)*

	1 1853–72	2 1872–99	3 1900–13	4 1929–37	5 1958–73	6 1973–81
Manufacturing output	3.7	3.2	3.9	2.5	6–6.5	3–3.2
Manufactured exports	4.27	3.09	5	−2	9.5–10	4.2–4.5

Sources: For Columns, 1, 2 and 3: W. A. Lewis, "World production and trade, 1870–1960", *Manchester School* (1952).
For Column 3: A. Maizels, *Industrial Growth and World Trade* (Cambridge University Press, 1863).
For Column 4: Economic Commission for Europe, *The European Economy from the 1950s to the 1970s* (Geneva: 1972).
For Column 5: GATT, *Networks of World Trade* (Geneva: 1978).
For Column 6: This is a provisional estimate based on GATT and other sources.

As far as understanding of this rather remarkable growth process is concerned, there is now a considerable literature. The works of Dennison (1967) and Madison (1979) are important contributions in this category and there is a growing consensus among scholars in this area.[1] Various studies are available that have looked into the world trade expansion. Unfortunately, it seems to us that no real consensus exists in these studies in terms of offering a comprehensive explanation of the phenomena, and some studies in fact are "off the track".

In this paper, an attempt has been made to understand the determinants that sustained the growth in world trade. The conclusions of the study are summarized as follows:

(1) Multilateral tariff reductions carried out under the GATT system have not been the major factor in stimulating growth in world trade in manufactures, and there were other important processes working in the international economy especially in the international capitalist economy comprising advanced industrialized countries as well as developing countries.

(2) Growth of international capital markets and capital movements or factor mobility have been a major source for growth in world trade in manufacturers.

(3) Declining transport costs or transactions costs which reduced the costs of transport of goods over "time" and "space" have been as significant, if not more, as tariff reductions in stimulating world trade.

(4) Increased productivity gains in export industries through accumulation and technology diffusion and the operations of transnationals imparted a trade-promoting bias to the world economy.

For the reasons stated earlier, this study focuses attention on the period up to 1973 as the developments after that year are qualitatively of a different character. Equally, attention is also only on world trade in manufactured products. In other words, trade in agricultural products, minerals, fuels, and so on as well as trade in services is excluded from our discussion. This is mainly because of three reasons; (1) the growth in manufactured exports during this period has been the most rapid and accounted for more than 75 to 80% of the total growth in world trade; (2) the supply and demand conditions in agricultural commodities, as well as in natural resources have distinctly different characteristics; and (3) until recently trade in non-factor services has not been a major component of world trade. In other words, we will only look at the most dynamic as well as largest segment of world trade, that is, trade in manufactured products. Equally, we have chosen the sub-period of 1958–73 for greater scrutiny because by 1958 the Western European countries had completed their postwar reconstruction; and as argued by Maizels (1963) in his classic study of world industrialization and trade, the world economy had assumed "normalcy" by 1958.[2] This was also the year by which the Rome Treaty establishing ECM was implemented.

In analyzing postwar growth in exports, it is interesting to note that there are quite a few distinctive features. Unlike the earlier years between 1953–73, the share of manufactured exports in total world exports increased in a monotonic way from 45 to 61% by 1973 (see Table 2). As noted by Professor W. A. Lewis (1952), historically until 1938, this share was almost constant and was in the neighborhood of 38%.[3] Equally, there was a qualitative difference in the movement in the terms of trade between manufactured and primary products. In contrast to the earlier period, the terms showed a movement in favor of manufactured exports during the postwar period.

TABLE 2. *Trend in the share of manufactures exports in total world trade*

	1899	1911	1938	1953	1963	1973	1980
Share of manufacturing exports	0.38	0.39	0.37	0.45	0.52	0.61	0.55

Source: Same as in Table 1.

If one disaggregates manufactures into consumer goods, capital goods and intermediate goods, another important feature of this growth has been a comparatively more rapid expansion in the trade of "intermediate" goods or "middle" products.[4] For instance, the growth of exports of intermediate manufactures or middle products during the period from 1929 to 1959 (the period for which data are available) was only 1.7% per annum and this rate was much lower than the rate of growth in total exports. However, the recent calculations suggest that during the period 1958–73 average per annum growth rate in world exports of middle products or intermediate products perhaps exceeded 10%, a rate that was higher than the rate of growth in total exports. We should, of course, note that the growth rates for these two different periods may not be strictly comparable because of some differences in the definition of intermediate goods. To make these comparable, we have tried to recalculate from the original data of Maizels (1963) the growth rate of exports of intermediate products for the period 1929 to 1959. It is likely that a discrepancy may still exist. However, it should be noted that for the period 1958 to 1973 our calculations may have underestimated the growth of exports of intermediate products as large categories of products—for example, iron and steel products such as tubes and pipes, spare parts, and subassemblies—have been classified as capital goods; in reality, these should have been defined as intermediate or middle products. This feature of rapid expansion in exports of intermediate products is important in more than one way and is suggestive of the importance of the variables other than tariffs in determining trade flows.

We should, of course, note that there is an important caveat in relating the growth rates of industrial production to industrial exports, as well as in making historical comparisons over different time periods. This is because while the data for industrial output are measured in terms of "value added", usually data for exports or trade are made available only in terms of "gross" value. In other words, in the presence of internationally traded intermediate products, there could be a considerable amount of double counting in the computation of the total world trade. For instance, if a country exports yarn and imports cotton for the yarn production, world trade figures simply aggregate both of these exports. One can quote such examples from other sectors, such as leather, metal-based products and chemicals. Stated differently, as there are imported inputs in export production, trade data aggregated over countries to indicate the total world exports would lead to "double counting". This in itself would not have mattered in growth rate comparisons, provided underlying ratios of "value added" in the export activities were constant. But there is evidence to suggest that this was not so.

We have already noted that the share of intermediate products in world trade in the aggregate has been growing. This would suggest that there are good reasons that the world trade would be "inflated" for the postwar period compared to the earlier period. In addition to this macro-evidence, there is considerable evidence, both for the developed as well as for the developing countries, which shows that the value-added in the export sector as a percentage of total value of exports may have declined over the period 1958 to 1973. For instance, the calculations made on the basis of input-output data for some of the industrialized countries shows that this ratio has declined.[5] There is further evidence of this phenomenon as revealed by the data for the exports to the United States under the tariff item 806.30 and 807.00. According to the USA customs rules for products coming under this tariff line, import duty is only on that portion which represents "value-added" in the exporting countries, provided the inputs used are of USA origin. Table 3 gives the data for such imports. "Value-added" in these exports of manufactures is between 50 to 70% and it is lower in developing countries. It is also clear that the value-added

TABLE 3. *US imports under tariff items 802.00 and 806.30. 1966–76 ($ millions)*

	Total value		Dutiable value		Total of US products	
	Total	Developing countries	Total	Developing countries	Total	Developing countries
1966	953.0	60.7	805.5	31.4	147.5	29.0
1967	1035.1	99.0	837.2	42.6	197.9	55.8
1968	1554.4	221.7	1263.7	97.7	290.6	124.0
1969	1838.8	394.8	1396.7	177.3	442.1	217.6
1970	2208.2	541.5	1671.8	245.9	536.3	295.5
1971	2765.8	652.5	2105.9	314.1	659.9	338.4
1972	3408.8	1066.5	2540.4	547.3	868.3	519.2
1973	4247.1	1557.3	3238.3	845.4	1008.8	711.8
1974	5371.8	2350.1	4059.0	1303.0	1312.8	1047.1
1975	5161.2	2261.7	3895.5	1238.7	1265.7	1023.1
1976	5719.0	2807.0	4173.2	1548.6	1545.7	1258.4
1977	7188.1	3306.8	5112.0	1721.4	1976.1	1585.3

Source: United States International Trade Commission

in the exporting countries has been declining during the 1960s and early 1970s. Note that these calculations, as well as those made on the basis of input-output tables, do not take into account the contribution of imported capital goods. Thus, even these statistics, in fact, overestimate domestic value-added content in the industrial exports. That international trade in capital goods has increased even at a higher rate than the overall exports suggests that this overestimation may be a serious one, especially in the case of the developing countries.

The implications of all this can be quite significant. When one makes an appropriate adjustment to the export data for later years to take into account the increasing content of imported inputs in manufactured exports, the growth in manufactured exports declines. It is clear that, for the purpose of historical comparison or for the comparison between rates of industrial output and international trade, such an adjustment is necessary. For instance, for the period 1958 to 1973 the growth in manufactured exports would only be 8–8.5% per annum and not 9.5–10% per annum if one takes into account an increase in import content of exports of around 10% or so. If such an adjustment is made, it is interesting to see that the output elasticity of world trade for the period 1958–73 is not much higher than during the period 1899 to 1913.

In passing, we note two other points. (1) In the total exports of manufactured products the share of the developing countries increased only marginally from 6.5 to 7.8% between 1959 and 1973, while during this period the share of intra-industrial countries went up sharply from 37 to 60%. As far as the socialist countries are concerned, their share declined marginally. In these comparisons we should, of course, remember that these ratios may change if one takes into account a more appropriate measure, such as "value-added" in the export sector. However, it is clear that such a measure would also indicate the changes of a similar order. (2) It is interesting to note that within the industrialized countries there were sharp changes in terms of the individual country's shares in the total exports of manufacturers. For instance, the share of Japan more than doubled and increased to 13% by 1973, while the USA share declined substantially from 21.5% to around 15–16%. Needless to say, these shifts reflect profound changes in the relative power in international economic relations and have had implications for the developments in the international economic system.

Professor W. A. Lewis (1981) in his lucid analysis of the growth of world trade with a historical perspective of more than a century, has argued that the most important determinant has been the rapid expansion of manufacturing output in the world economy itself.[6] He seems to give little weight to developments in trade policy, such as tariff reduction for the stimulation of world trade.

Other analysts have attached considerably greater importance to the role of commercial policies in explaining expansion in industrial output as well as in the world trade. For instance, Blackhurst *et al.* (1977) argued in a GATT study that trade liberalization exemplified by the multilateral tariff reduction under the GATT system, changing demographic trends, technological innovations, including those related to transport and communications have been the main reasons for this expansion.[7] In this study, no qualification has been made to indicate the contribution of each of these variables towards promotion of world trade. However, they seem to argue that commercial policies, such as binding and reduction of tariff barriers, contribute to a higher growth in industrial output through a more efficient allocation of resources and thus contribute both directly and indirectly to the trade expansion. They also seem to argue that the international institutional structure as exemplified by GATT increased stability and predictability of the international economic system, and this facilitated new investments in the industrial countries. In their analysis, the neoclassical paradigm based on "factor-proportions" as a determinant of international trade plays a central role in explaining patterns of trade and of allocation of resources. In this study, however, there is no adequate discussion of the role of factor movements or factor markets in explaining trade expansion.

This is not really surprising as in the neoclassical synthesis *á la* Heckscher-Ohlin-Samuelson (HOS) there usually is a much greater emphasis on the static aspects. The severity of the limitations of the neoclassical approach has been well analyzed by the recent neo-Ricardian contributions.[8] In addition to these limitations, with the HOS framework, there is an inherent tendency to look at factor movements and product movements as substitutes for each other. Undoubtedly, in his own classic work, Ohlin (1968) is far more subtle and does recognize the complexity of these issues; but the recent tradition seems to be shaped more by the works of Mundell (1975), which given the very restrictive neoclassical assumptions although logically correct, are not fully consistent with the realities of international trade.[9] One of the key results of Mundell's framework is that factor mobility and trade in commodities are substitutes for each other.[10]

In addition to the important issues of factor mobility, there is yet another problem that is missed regarding the role of exports. This relates to the role of exports in stimulation of "effective demand", an issue of central importance to the advanced capitalist economies. In other words, as the neoclassical framework is usually a full employment framework, such macro-economic aspects are not adequately taken into account in the analysis of international trade.

A more recent study by Batchelor *et al.* (1980) of N.I.S.R. has also examined in depth the determinants of growth in the world trade of manufactures.[11] In this rather exhaustive study, there is a detailed discussion of the role of exports in economic growth in the advanced industrial countries. This major study also includes a substantial amount of sophisticated econometric work. A far as the issue of the contribution of exports to industrial growth is concerned, this study shows that the experience of advanced industrialized countries does not seem to support the hypothesis of "export-led" growth. However, the authors do bring out an intimate relationship between exports and industrial growth, where direction of causation is much more complex; that is, for some countries or for some period, the causal chain may be from exports to industrial growth, while at other times, it may be the other way around. This is not altogether surprising. In a modern economy, there will be interaction between demand and supply schedules as both are shifting through differential productivity gains, technical progress and accumulation.

It is this dynamic interaction which perhaps is the heart of the matter. That is why measurement of "static" gains of trade usually shows that these are rather insignificant. For instance, a well-known study by Dennison (1967) has shown that in explaining postwar growth in the OECD countries, the contribution of trade policies through better resource allocation of the trade policies has had rather marginal effects.[12] Even in the case of Japan, a country that is perceived to be propelled by "export-led" growth, careful studies have shown that such an hypothesis is not sustainable, at least for the period that we are discussing.[13]

The study by Batchelor *et al.* has also looked at the various determinants of growth in world trade in manufactures. It seems to conclude that the commercial policy measures, especially the tariff reductions, have been the major factor for trade expansion in manufactured products. Here we note that perhaps Balassa (1977) is the one who assigns maximum weight to trade liberalization of manufactured products. He has argued that, in the 19th century, decline in transport costs was a major factor in stimulating world trade. However, in the post-Second World War period the reductions in tariff and non-tariff barriers under various rounds of trade negotiations under the GATT system have been of greater importance in stimulating world trade.[14] In supporting his arguments, he has relied on a number of econometric studies that have estimated price, income and tariff elasticities.

There are many econometric studies that have estimated price elasticities or tariff elasticities for individual traded commodities in different countries. In a dynamic setting where both demand and supply schedules are moving, there are serious problems in these estimates. Not only are there statistical and methodological problems as argued by Orcutt (1950) in his classic paper, but there are also related issues of separating contributions of other factors, such as declining costs through technological progress, from the contribution of declining tariffs to the stimulation of industrial imports.[15] In any case, elasticity measurements within a "partial equilibrium" framework are not appropriate to understand overall movements in world trade. Further, there have been considerable changes in the factor markets as well as in the factor movements that have not been fully taken into account in these estimation procedures; and this is, in our view, a serious lacuna in such types of analyses.

As shown in Table 4 it is clear that during the 1958 to 1973 period the sharpest reduction in the average rate of industrial tariffs has been in the EEC countries where the reduction has been as deep as 10% *ad valorem;* but for the industrial countries as a whole, on an average, this reduction has been much lower. For instance, in the United States, average tariff reduction was only 4 to

TABLE 4. *Average tariff rate on imports of manufactured goods, 1960–1977 (in percent)*

Year	USA	Japan	Netherlands	West Germany	France	Italy	UK	Austria
1960	12.2	15.9	1.4	6.0	12.6	12.8	14.8	18.0
1965	12.2	12.2	15.9	4.6	6.0	6.1	12.6	15.1
1970	9.9	12.2	2.5	3.2	2.3	2.7	8.8	10.5
1977	8.7	9.7	1.1	1.3	1.0	1.2	3.1	1.1

Source: P. Franger *et al.*, *Tariffs in the World Trade Model* (Oslo: Central Bureau of Statistics, 1980), Table 3.1.

5% *ad valorem* and in Japan only 6% *ad valorem.* The main reason for the much sharper decline in the average tariff rates in the West European countries is, of course, the formation of the European Economic Community in 1958, when tariffs were progressively abolished on intra-EEC exchanges. Hence, *ceteris paribus,* the trade creation effects due to tariff reduction would be higher on intra-EEC trade than on the overall world trade. As seen in Table 5, the growth in intra-EEC trade in manufactures was more dynamic than the overall growth in world trade in manufactures. Fortunately, various estimates are available that compute the trade expansion due to the creation and enlargement of the EEC. Perhaps the most satisfactory estimate is that of Kreinin (1981).[16] His method is quite ingenious and is able to isolate better the effects of the commercial policy measures from the other developments on the stimulation of trade.

As pointed out by Kreinin in his study, it is crucial to "keep constant" influences on trade flows other than tariff reduction or integration (such as income growth, shift in supply, etc.) in order to isolate and "flush out" the

TABLE 5 (a). *Growth of trade in manufactures (US$ billions) between 1959/60 and 1977/78*

	(a) 1959–60 (annual) average	(b) 1977–78 (annual) average	Ratio col. (b): col. (a)
Imports of the US and Canada	10.4	123	11.8
Imports of Western Europe (internal plus external)	23.5	322	13.8
External imports of Western Europe	5.5	67	12.1
Internal imports of Western Europe (intra-European trade)	18	255	14.2
Total World Exports	63.4	718	11.3

Source: GATT, *International Trade* (annual issues).

TABLE 5 (b). *Estimated annual effects of European integration on trade flows in manufactured products (US$ billion)*

		Annual trade creation	Annual trade diversion
A.	Two states of integration combined (independent estimate)	43	8
B.	First stage integration		
	EEC	10	2
	EFTA	3	11
	EEC plus EFTA	13	3
C.	Second stage integration	28	5
D.	Two stages added up	41	8

Source: M. Kreinin, *op. cit.*, Table 5.

effects of tariff reduction or integration. In *ex post* analysis, it is necessary to construct hypothetical estimates of what trade flows would have been in the absence of tariff reduction or integration and compare them with actual trade flows. Kreinin does this by employing the growth of worldwide trade or of trade of other similar countries as a *normalizer* to estimate hypothetical trade flows in EEC countries in the absence of a common market. Comparing this with actual trade flows yields an estimate of trade creation due to economic integration, that is, common market. As the normalizer, he has used the combined trade of the US, Japan and Canada. Such a procedure of normalization enables him to isolate static trade creation effects of the tariff reduction achieved under the creation of the European Common Market. This study shows that the additional trade created in manufactures because of the creation and enlargement of the EEC is around 15% of the total manufactures trade. When one looks at the fact that the Common Market implied deeper tariff reductions than the ones under the GATT, and when one further realizes that the *Common Market* also implied greater reductions of non-tariff barriers, it is clear that the trade expansion due to the tariff reduction under various GATT tariff rounds, would be lower for the trade of other countries, provided other important parameters, such as growth in incomes, were comparable. As during 1960 and 1970, the industrialized countries other than the countries of the EEC were growing at aggregate almost similar rates, it seems to us that the contribution to world trade of various tariff measures undertaken under the GATT would be much lower than the contribution of the institution of the Common Market to intra-EEC trade.

In fact, in aggregate, the pace of industrialization may have been marginally lower in the countries other than the EEC. We have already noted that the depth of tariff reduction for industrial imports on average was less than half in the industrial countries other than the countries of EEC. For instance, *ad valorem* tariff reduction in the EEC exceeded 10–12%, while it was only around 4–6% in the USA and Japan. Taking into account these factors, it is suggested that the "trade creation" effects of commercial policies would be only 5 to 8% of the total trade in manufactures. We may note here that even this estimate is high in the sense that if one takes into account only "value-added", it may be only around 4 to 6% of the "value-added" in the export sector. Obviously, there were several other important factors that stimulated world trade in manufactures. We suggest that these other factors are: (1) factor movements, especially capital flows; (2) declining costs of exports due to a relatively more rapid increase in productivity in the export sectors and through declining transport or transaction costs. These factors improved relative competitiveness of exports. For instance, according to an OECD study, labor productivity during the postwar period was increasing in the exporting manufacturing sector at a higher rate, 0.35% higher per annum, than the manufacturing sector as a whole.[17] This would mean that over the period 1958–73 this factor alone would contribute more than the tariff reduction did in lowering import prices, a critical variable in stimulating trade.

As Table 6 shows, during the period 1955–73, the rate of growth in transfer of real resources across national boundaries, that is international capital flows, was even higher than the growth in world trade, whether measured for total

TABLE 6. *International trade and capital flows for selected years*

Year	World trade (T) (in billion dollars)	Capital flows (F) (in billion dollars)
1955	84.7	3.9
1958	96.4	8.1
1959	102.4	6.6
1965	167	14.0
1970	283	25.4
1973	524	51.1

Source: IMF, *International Financial Statistics* (Washington, D.C.: various annual issues).
Fi = current account surplus for *i*-th country
F = *Fi* for all where *Fi* = 0

trade or for trade in manufactures. How does one measure factor movements? Admittedly, this is not an easy question in the sense of empirical investigation.[18] In this paper, we have taken the surplus on current account as an indicator of resources transfer from a country to the rest of the world in that year. Perhaps the balance on goods and services account may have been more appropriate as it is not influenced by transactions under "unrequited transfers". Private unrequited transfers largely signify workers remittances due to labor flows across national boundaries. This is also a factor flow, and for a few individual countries, such as Turkey, it is an important one. Unfortunately, such diaggregated data are not available for all the countries. Hence, we have taken the current account sample as an approximation of the transfer of capital resources. As the major industrial countries such as the US and West Germany had a considerable inflow of "guest workers", this approximation only underestimates the capital flow from these countries to the rest of the world.

For any given year, one aggregates each country's current account surplus wherever it is positive. Then this aggregate amount indicates the total flow of resources in that particular year. To put it differently, it is the gross account that was borrowed by the capital-importing countries or was lent by the resource-transferring countries. These transactions indicate the process that assisted different national economies in carrying out macroeconomic decisions affecting allocation of their resources over time, that is, domestic savings and investment decisions. As seen in the Table 6, the share of the capital resource flows in the world's total exports has been growing from around 6% in 1968–69 to almost 10% by 1973. As can be seen, the implication of these transfers is that if each of these countries was operating under the constraint of "balanced trade" implying absence of any transfer of resources across the national boundaries, world trade would have been correspondingly lower, as resource transfers across the national boundaries can take place only through trade of goods and services. In other words, net lending or net borrowing decisions of various countries generate additional possibilities for international trade.

During this period which countries were exporting capital? As can be seen from the IMF balance-of-payments data, the bulk of these resources was coming from the mature industrialized countries, such as the US, the Federal Republic of Germany, the Netherlands, and so on, and a few oil-producing countries of the Gulf region. Which were the borrowing countries? During this period, the major borrowers were the newly industrialized countries of Asia, Latin America and Southern Europe and a few dynamic countries of Europe, such as France and Italy. When one takes into account the fact that, during this period, it is precisely these capital exporting countries that had a surplus in exports of manufactures in terms of the world trade matrix, the importance of the transactions on capital account in expanding world trade in manufactures can be readily seen. We recall that the share of manufacture exports in world exports was between 55 and 60% in the 1960s. Thus, the contribution of factor movement to increase trade in manufactures would be almost 20%. The estimated contribution of tariff reduction to increase the total trade in manufactures was only 5 to 8%. Therefore, the contribution of capital mobility or factor movements in promoting trade in manufactures was much greater than that of commercial policy measures, such as reduction in tariff barriers.

This discussion brings out the dynamic role of trade in promotion of industrialization and growth. It shows how through international trade different countries are able to pursue more effectively intertemporal allocation of their resources aided by the possibilities of international borrowing and lending. In a "dynamic setting", this is the crucial role of trade. Countries that are willing and capable of sustaining capital accumulation because of the higher productivity of capital in their economies are enabled to do so by the countries whose savings exceed the investment opportunities under the conditions of full employment. To state it differently, it is suggested that the difference in "time preferences" and consequent factor mobility among nations may be a sufficient condition for promotion of international trade. In our view, this also indicates that the causation is from the capital movements to export surplus and thus stimulation of international trade.

It is clear that capital flows may take different forms—public loans, suppliers credit, portfolio capital, equity capital and joint ventures—and transnationals and the governments of lending capitalist countries have been most active in this. In capital-exporting countries, export surpluses also play an important role in terms of maintaining "effective demand" in the macroeconomic sense. This role has been emphasized by many scholars, notably by Thirlwall (1981) in his elaboration of the Harrod foreign trade multiplier.[19] Hence, under certain conditions, these capital flows can only have a positive impact on accumulation, technology diffusion and growth and thus on the dynamism of world economy and trade.

One of the most important elements in accelerating factor mobility has been the operations of transnationals. Their spread has been of tremendous significance to the development of capitalism itself. Scholars have argued that this was the period when worldwide division of labor and a world-market oriented industrialization phase became dominant in the history of capitalism.[20] Transnationals based in the dominant industrialized countries, such as the US, Federal Republic of Germany and the United Kingdom, increased their international investments at a rapid pace during the period under consideration, as seen in Table 7. These investments led to the process of internalization of production of a more oligopolistic type. This also had profound implications for the pattern and levels of foreign trade as intra-firm trade accounts for a major portion of the world trade in manufactures. For instance, in the United States in 1970, transnationals' trade with their majority-owned foreign subsidiaries accounted for 21% of the total exports of manufactures increasing from 16% in 1966. In 1970, the US-based transnationals accounted for 66% of the total US manufactured exports.[21] According to one estimate, a quarter to a third of the developed countries' total trade in manufactures takes place on an intra-firm basis.[22]

TABLE 7. *Trends in transnational investments*

		Host countries					
		North America	EEC	United Kingdom	Other EFTA	Other	Total
Investments by:							
United States	1963	5.29	2.38	2.59	0.21	3.23	13.70
	1967	7.56	4.73	3.70	0.45	6.09	22.53
	1971	9.94	8.00	3.24	0.82	9.44	33.44
United Kingdom	1963	0.92	0.40	n.a.	0.09	2.11	3.52
	1967	1.23	0.63	n.a.	0.11	2.73	4.60
	1971	1.79	1.32	n.a.	0.19	4.09	7.39
Germany	1963	0.19	0.23	0.02	0.13	0.46	1.03
	1967	0.35	0.70	0.04	0.21	0.70	2.00
	1971	0.75	1.43	0.12	0.48	1.61	4.39
Canada	1963	0.85		0.28			1.52
	1967	1.57		0.37			1.93
	1971	1.39		0.43			2.59
Sweden	1963	0.13	0.24	0.04	0.05	0.09	0.54
	1967	0.16	0.42	0.07	0.09	0.17	0.91
	1971	0.22	0.55	0.09	0.10	0.22	1.20

Source: R. A. Batchelor, *et al.*, *op. cit.*, Table 4.14.

The important role of intra-firm trade could have meant that a considerable part of world trade was determined by locational-type decisions regarding production and investments by these transnationals with the motivation of maximization of their after-tax global profits. Obviously, these after-tax profits will depend not only on the commercial policies, but also on the fiscal policies, especially corporate taxation policies. Incidentally, it is also possible that the operations of transnationals may also account for some of the acceleration in the growth rates of trade in intermediate or middle products. Not only does this open further possibilities of increasing profits through "transfer pricing", it also enables transnationals to exploit economies of scale, cost reduction via longer production runs, etc. The operations of transnationals have, of course, different impacts on the developed countries compared to the developing countries. Transnational operations have been important in many developing countries, especially those of Latin America. Studies show that the operations of transnationals have an adverse impact on the development of developing countries due to, among other things, the asymmetry of the relationship and the consequent dependence, especially technological dependence of the developing countries. This issue is, of course, so crucial that it requires a separate discussion.[23]

The international financial system consisting of intergovernmental institutions, such as the International Monetary Fund and the World Bank, and private institutions such as transnational banks, have played a crucial role in actualizing resource flows. The growth and strength of international capital markets, especially the Eurodollar market, has facilitated the transfer of resources. The growth of these integrated international capital markets perhaps also reduced the costs of financial intermediation. During the 1960s and early 1970s, one significant feature of the world economy was declining interest rates in the international capital markets. In fact, these rates became negative in some years for a few major currencies. The decline took place for short-term as well as long-term credit. This reduction in real interest rates had profound implications for international trade. Given the time-lag in any international exchange, declining interest rates reduced the cost of international transactions as transporting "over-time" became cheaper. As noted by Grassmann (1977), during the 1960s on average, the duration of the export credit increased from a few weeks to a few months.[24] Undoubtedly, the extent of this increase in the duration was even greater in the case of the developing countries, including the NICs. Thus, for many newly industrialized countries where the domestic interest rates were even higher, foreign trade financing opportunities opened by the international capital markets at declining interest rates meant considerable improvement in their competitiveness. Terms and conditions of trade credit have a considerable impact on the prices of the traded goods. As credit terms soften in terms of the interest rates, repayment period, etc., it can lead to decline in export prices in a competitive market. This is of considerable importance to the developing countries where the real costs of capital in domestic markets have been very high, exceeding 10% per annum. Under these conditions, access to international capital markets where interest rates are only around 2 to 3% meant considerable cost reduction for the exporters of manufactures. Some calculations suggest that this factor alone may have increased the competitiveness of exports by a greater margin than the tariff reduction achieved under the various GATT rounds in the importing markets.[25]

We also note that during the 1960s there was a substantial increase in transportation of manufactured products by modes other than shipping. Exports by air as well as by surface transport (roads, railways, etc.) grew considerably. Recent UNCTAD calculations show that between 1964 and 1970 the share of seaborne trade in total trade declined to 60% compared to 80% or so in 1964.[26] That during this period the seaborne trade of fuel increased quite substantially suggests that the share of seaborne manufactures in the total exports of manufactures may have declined at an even sharper rate. The same UNCTAD study indicates a rapid decline in transportation costs, especially the air freight costs. Among other things, declining energy prices especially of hydrocarbons was an important contributory factor in this reduction. Once again, for many manufactured products, this reduction in transport costs may have been as important as a reduction in tariffs in terms of improving competitiveness of exports.

This discussion clearly brings out the operation of various other factors that imparted dynamism to world trade in manufactures and that have been of a far greater importance than the commercial policies " *à la* GATT". In fact, one could argue that tariff reductions were perhaps more an "effect" than the cause of trade expansion. Or, to put it differently, tariff reductions were more of an accommodating character, accommodating real economic changes occurring in the international economy. What is being argued is that in this phase of growth, international capitalism sought to remove barriers against internationalization of production. Tariffs were one such barrier, and they had to be lowered to accommodate the charges that were occurring in the advanced industrialized countries.

To sum up, growth in international trade is intrinsically linked with international capital movements, and the relationship between these two has been a positive one. It is interesting to note that our conclusion is perhaps reinforced by the historical experience of the capitalist development, as well as by the contemporary experience of the trade among socialist countries. As Professor Lewis noted, between 1900 and 1913 when the growth in world trade was at its highest rate, it was accompanied by a higher level of capital exports compared to earlier experience of the 19th century. According to his estimate, these capital exports accounted for one-third of the trade expansion. Such capital transfers were reduced to much lower levels during the later period, a period of stagnation in world trade. We will also note that the intra-CMEA trade, i.e., trade among socialist countries, in industrial products in aggregate grew at a lower rate than world trade, even when the industrial expansion in the CMEA countries was comparatively more buoyant. Admittedly, there were various reasons for this phenomenon; but in our view, one of the important reasons is the payments or financial arrangements for intra-CMEA trade. It should be noted that the intra-CMEA trade is governed by the principle of a "balanced bilateral trade". Hence, in this system, net resource transfers have not been significant.[27] This feature, in our view, may have imposed a constraint on trade expansion relative to the arrangements where possibilities of international borrowing and lending exist.

What are the implications of this discussion for the future? We suggest there are at least two important questions: (1) What is the perspective for future growth in world trade in manufactures? (2) What should the strategy be for expanding trade among developing countries? Undoubtedly, in the future as in the past, growth in world exports will depend on overall growth as well as on growth in industrial output. However, in the future, even if the growth rate in the world industrial production picks up, the rate of growth in exports of manufactures will be smaller compared to its dynamism during the decade of the 1960s. There are various reasons for this situation and these are reflected in the changes that have occurred in the world economy, especially in the fields of international finance, such as patterns of global savings and investments, capital markets as reflected in the sharp increase in the real interest rates, and increasing transport costs due to the considerable price rise in energy. These developments would mean that transport costs of international exchange would go up as transporting "over time" as well as "over space" becomes relatively more expensive. This would influence levels as well as the structure of trade in manufactures. In addition to these factors some evidence suggests that the growth in the operations of transnationals may also be waning. All these developments can lead to lower growth in the trade in manufactures. Equally, there would be a change in the composition of industrial exports, a move away from the "middle" products to final products.

Finally, our discussion also suggests that a mere tariff reduction as envisaged in the Program of Global Systems of Tariff Preferences (GSTP) among developing countries will not be adequate for promoting trade among them. Vigorous efforts will have to be undertaken in the field of financial cooperation among the developing countries to facilitate factor flows. In other words, complementary efforts in promoting investment cooperation seem to be essential for trade expansion among developing countries.

Notes

1. See E. Dennison, *Why Growth Rates Differ* (Washington, D.C.: The Brookings Institution, 1967). A. Madison, "Long run dynamics of productivity growth", *Banco Nazionale del Lavoro* (Rome: March 1979).
2. A. Maizels, *Industrial Growth and World Trade* (Cambridge: Cambridge University Press, 1963).
3. W. Lewis, "World production and trade 1870–1960", *Manchester School (1952)*.
4. The importance of this phenomenon has been incorporated only recently in the standard international trade theory. For a recent analytical work see K. Sanyal and R. Jones, "The theory of trade in middle products" (Stockholm: Institute of International Studies, University of Stockholm, n.d.).
5. An analysis based upon input-output tables for Western European countries, shows that an increase in import content of export production has grown quite substantially, e.g., for France from 10 to 16%, and in Italy, from 14 to 23%. See Table 2 in R. Blackhurst *et al.*, *Trade Liberalization, Protectionism and Interdependence* (Geneva: GATT, 1977).
6. W. A. Lewis, "The rate of growth in world trade, 1980–1973" in S. Grassman and E. Lundberg (Eds.), *The World Economic Order, Past and Prospects* (London: Macmillan Press, 1981).
7. R. Blackhurst, *et al.*, *op. cit.*
8. The basis of this critique is essentially a capital-theoretic one. Among other things, this critique has shown the importance of dynamics. See I. Steedman, *Trade Amongst Growing Economies* (Cambridge University Press, 1979).
9. B. Ohlin, *Interregional and International Trade* (Cambridge, MA: Harvard University Press, 1968).
10. R. Mundell, "International trade and factor mobility", *American Economic Review* (March 1957).
11. R. Batchelor, *et al.*, *Industrialization and the Basis for Trade* (Cambridge University Press, 1980).
12. E. Dennison, *op. cit.*
13. A. Boltho, *Japan: An Economic Survey, 1953–1973* (Oxford University Press, 1975).
14. B. Balassa, "Effects of commercial policy on international trade location of production and factor movements", in B. Ohlin *et al.* (Eds.), *International Allocation of Economic Activity* (London: Macmillan, 1977).
15. G. Orcutt, "Measurements of price elasticities in international trade", *Review of Economics and Statistics* (May 1950).
16. M. Kreinin, "Static effects of EC enlargement on trade flows in manufactured products", *Kyklos*, Vol. **34** (1981).
17. OECD, "The international competitiveness of selected OECD countries", *Occasional Studies* (Paris: OECD, 1973).
18. For a discussion of the various problems involved in this issue, see E. Veil, "Surpluses and deficits in the balance of payments: Definitions and significance of alternative concepts", *OECD Occasional Studies* (Paris: OECD, 1975).

19. A. Thirlwall, "The Harrod trade multiplier and the importance of export-led growth", Mimeo. (1981).
20. See F. Frobel, *et al.*, *The New International Division of Labour* (Cambridge University Press, 1981).
21. R. Batchelor, *et al.*, *op. cit.*, p. 88.
22. S. Lall, "Transfer pricing by multinational manufacturing firms", *Oxford University Bulletin of Economics and Statistics* (1973).
23. For an understanding of the issues involved, see R. Barnet and R. Mueller, *Global Reach: The Power of Multinational Corporations* (New York: 1974).
24. S. Grassmann, "Financial aspects of internationalization process" (Stockholm: Institute of International Economic Studies, 1977).
25. For a general discussion of these issues and illustrative calculations of reduction in export prices, see C. Helman,"The effect of trade credit on price and price level comparison", *The Review of Economics and Statistics* (November 1981).
26. UNCTAD, *Review of Maritime Transport* 1972–1973 (TD/B/C.4/117) (Geneva: 1975). According to this report (see Table 45), freight charges for ton kilometers declined by more than 20% even in nominal terms, suggesting that in real terms it may have been reduced by 50% between 1962 and 1971. During this period, air freight increased more than fourfold and was 11.5 billion ton kilometers.
27. Regarding details for CMEA trade data, see Vienna Institute of Comparative Economic Studies, *COMECON Foreign Trade Data*, 1980 (London: Macmillan, 1981).

India: Development of the Idea of Development

ALICE THORNER

The wealth of a nation or a colony has long been a subject of interest to scholars as well as rulers. But the concept of development (usually economic—sometimes also social or political) as a national goal has a relatively short history in India as indeed elsewhere.

British officials in Victorian India expressed their concern for the "prosperity" of the tracts which they governed. In census reports and gazetteers, they recorded with satisfaction the clearing of new land for cultivation, the progress of canal irrigation, the opening of markets consequent upon the extension of railway lines and the establishment of rice, sugar and cotton mills. They tended to deplore the decline of traditional artisanal production. Above all, they noted whether the harvests had been abundant or deficient, ascribing the results to meteorological conditions, the qualities of husbandry of the local peasantry, and the presence or absence of debilitating fevers.

Few of them doubted the beneficent effect of the British presence that assured the rule of law and introduced public health measures as well as undertaking irrigation works and constructing railways. One of the few was a former administrator, William Digby, several generations of whose family served in various capacities in India. In his book, *"Prosperous" British India: A Revelation from Official Records* (London: 1901), Digby argued that India was poorer in 1900 than in 1850, and *per capita* levels of living had declined throughout the second half of the 19th century.

Criticisms along these lines had already been voiced and continued to be put forward by Indian nationalists, notably Dadabhai Naoroji and Romesh Chandra Dutt. In its crudest version, the patriotic position attributed all of India's ills to foreign exploitation. Once the dead weight of imperialism was thrown off, and the "drain" stopped up, the garden of Eden which the early British adventurers had discovered in such places as deltaic Bengal and the forested Ganges valley would once again spring into bloom.

Typical replies are to be found in *Economic Development of the British Overseas Empire* (London: 1924) by Dr. L. C. A. Knowles of the London School of

Economics and *Economic Development of India* (1929) by Dr. Vera Anstey of the same institution. As we see, the word "development" was already in use by the 1920s, but in the sense rather of a statement as to what had occurred than of a program for future action.

From the time Gandhi rose to prominence on the Indian scene, the nationalist movement carried within it two main currents of economic thought. Both envisaged fundamental change; both were addressed to the problem of the poverty of the masses. Even before the first great civil disobedience movement, Gandhi had begun to elaborate his "constructive program" of village uplift through self-help. While refraining from any direct challenge to the local power structure, the Gandhians hoped to rebuild a more just society by the reformation of individuals. Gandhian volunteers fanned out to the villages to teach personal hygiene, sanitation, spinning, weaving and other cottage industries, and to promote adult as well as child education. Similar efforts of exhortation and aid for rural improvement had been pioneered by Gokhale's Servants of India Society in the Deccan. One might also mention the activities of F. W. Brayne, a British Officer in the Punjab; of Hamilton who established a colony in Bengal called Gosabha; of the poet Tagore who set up cottage industries in a village called Sriniketan as a sort of auxiliary to his university at Shantiniketan; of Spencer, who promoted fish culture and other innovations at Marthandam in Kerala. The crucial new element introduced by Gandhi was an ideology—a principled belief that the villages could reconstruct a former almost autarkic existence and by so doing contribute to the success of the struggle for national independence.

The alternative project, which also came to the fore during the 1920s and 1930s, claimed a socialist heritage. Jawaharlal Nehru, Subhas Bose, Jayaprakash Narayan, and Acharya Narendra Dev, among others, called for an end to privilege through drastic tenancy and land ownership reforms. They foresaw the rise of India as an industrial power. Impressed by the example of the young Soviet Union, they favored state ownership of large sectors of the economy and came to envisage central planning as essential for national progress.

The influence of the Socialist strand within the nationalist movement was at its height in 1937 when the Congress adopted a number of left-wing demands in its manifesto for the provincial elections. A year later Subhas Bose, as Congress President, set up a National Planning Committee with Jawaharlal Nehru as Chairman and Professor K. T. Shah as Secretary. Another member was Sir M. Visvesvarayya, not himself a socialist but an engineer with long practical experience in Mysore State who had set out in 1934 a program of rapid industrialization entitled *Planned Economy for India*. Although 29 separate subcommittees prepared reports, no comprehensive document was produced before the 1940 wartime arrest of Nehru brought the work to a virtual halt.

A group of leading industrialists—Sir Purshotamdas Thakurdas, J. D. Tata, Sir Ardeshir Dalal, A. D. Shroff, Dr. John Matthai, D. G. Birla, Sir Shri Ram, and Kasturbhai Lalbhai—stepped into the field with the publication in 1944 of

A Plan of Economic Development for India, drafted in large part by Professor K. T. Shah of the Congress Committee. Emphasis was placed on the need for industrialization, conceived in terms of large-scale modern enterprises. But to avoid inflation and create employment the planners also provided for the encouragement of village crafts and cottage industries. Cooperative farming was held out as the most promising prospect for the reorganization of agriculture. In effect, the "Bombay Plan", as it came to be known, foreshadowed the five-year plans later to be drawn up and implemented, and mirrored some of the ambiguities that were to continue to bedevil Indian development programs.

Two other plans which had much less impact also appeared in 1944, S. N. Agarwal's *The Gandhian Plan* and the *People's Plan for Economic Development of India* by B. N. Banerjee and others, issued under the auspices of M. N. Roy and the Indian Federation of Labor. The Government of India, for its part, set up in June of that year a Planning and Development Department. This organism issued a surprisingly radical report, proposing state ownership of new and necessary capital goods producing enterprises, for which private capital might not be forthcoming; the Department was subsequently disbanded.

There existed at that time no real forum for the discussion of these plans. As the second World War drew to its end, and immediately thereafter, the imminence of India's independence in one form or another brought into sharpest focus the question of to whom power would pass. Economic problems, although exacerbated by the cumulative effect of wartime shortages and the demobilization of soldiers, inevitably took a second place. The number of Indian economists was limited—the students Daniel[1] and I came to know in London in 1939–40 had been reading law, history, mathematics, medicine, even social anthropology—and there were only a handful of respected senior members of the profession. They served primarily in universities and colleges (e.g., the late Professors D. R. Gadgil, B. Ganguly, Gyan Chand, and the happily perennial Drs. V. K. R. V. Rao and A. K. Das Gupta). A few others held staff positions in government offices, banks, the largest industrial houses, or, like Da Costa and Lokanathan, in journalism. Their academic interests, which included industrialization, urbanization, population growth and national income trends, were highly relevant to India's future, but, except for Professor K. T. Shah, their expertise was scarcely exploited for national purposes.

Partition in a context of murderous communal strife confronted the nascent government of independent India with a massive influx of refugees as well as an armed conflict with Pakistan over Kashmir. Immediate crises had to be faced before public debate over long-term economic issues could resume. Nonetheless, the concept of governmental responsibility for improvement of the livelihood of the common people, and for fairness in the distribution of material goods was enshrined in the Directive Principles of State Policy embodied in the 1950 Constitution.

Although the Congress, as we have seen, had committed itself to planning in 1938, Nehru had to overcome the redoubtable resistance of the conservatives well-entrenched in the party, typified by Vallabhbhai Patel, before he could formally establish a Planning Commission in 1950. As Gunnar Myrdal has written: "Planning becomes the intellectual matrix of the entire modernization ideology", (*Asian Drama* (New York: 1968), Vol. **II,** p. 711). Nehru served as the principal protagonist of this ideology. He argued eloquently and incessantly for wider use of science and technology. So that the benefits of increased productivity should accrue to the masses of the people rather than to a handful of exploiters, he held out a vision of a socialist commonwealth, a socialist pattern of society. Private enterprise was to continue and even to be encouraged, but the commanding heights of the economy were to be occupied by state-owned undertakings.

Central planning was needed, Nehru explained, to coordinate the activities of the various sectors so as to avoid wasteful duplication. Imports were to be authorized primarily for capital goods, while local industry would produce the required consumers' goods. Controls would have to be organized to assure that factories were set up in accordance with these guidelines, that imports did not stifle domestic production, that India's resources in foreign exchange were not frittered away on non-essentials.

Despite his nontraditional tone, and even his occasional attacks upon religious or social traditions, Nehru aroused a large degree of popular enthusiasm for his program. He was able to attract a corps of competent and motivated planners and implementers. By the early 1950s, a new generation of economists had returned from postwar studies in the the London School of Economics, at Cambridge University and other leading institutions. Some of them had worked with Joan Robinson, Nicholas Kaldor or Arthur Lewis in England, Tinbergen in the Netherlands, Simon Kuznets in the United States, to mention only a few. Men of the calibre of K. N. Raj, I. G. Patel, K. S. Krishnaswamy and Iqbal Gulati took up posts in the Planning Commission, the Research Division of the Reserve Bank, and other governmental agencies. Others of equal distinction undertook the training of the next batch of economists in the rapidly expanding universities. Both senior and junior scholars contributed articles to the established press, and to new periodicals, most notably Sachin Choudhuri's Bombay-based *Economic Weekly,* which from its inception published the writings of sociologists, ethnologists and political scientists as well. Indian authors also expounded their views in prestigious British and American periodicals, such as the new *Economic Development and Cultural Change,* edited by Bert Hoselitz at the University of Chicago. In fact, the Indian experience furnished the subject matter for foreign as well as Indian researchers.

As his Scientific Adviser Nehru recruited the eminent statistician, Professor P. C. Mahalanobis. Starting with the mechanics of gathering and processing data on production, consumption and distribution, "Professor" transformed himself into an economic planner whose proposals evoked international consideration. With his broad array of interest in all fields of science and technology, "Professor" could keep Nehru up to date on the latest discoveries and controversies.

Gandhi had opposed the idea of national economic planning since, for him, the individual and the village were the basic units to be reformed. His followers, notably Vinobha Bhave, accordingly kept apart from the mainstream of the government's developmental activities and concentrated their efforts on other issues such as *Bhoodan,* persuading better-off peasants to donate part of their holdings to the landless. Nonetheless, the Gandhian heritage of concern for the lowly informed the thinking of the planners.

During the period of the first three plans (1951–65), the ways and means of development were discussed in great detail, and often at a very high level. The basic concept, however, was usually taken for granted. It involved the strengthening of India's infrastructure: development of hydroelectric power, expansion of irrigation, acquisition of new diesel and electric locomotives and introduction of automatic signaling systems for the railways, creation of new seaports and reconstruction of old ones, setting up of internal and international air lines, and improvement of roads. In terms of manufacture, development was seen to encompass the establishment of new fertilizer plants and pharmaceutical units, and the beginnings of a machine-tool industry. Later on the range expanded to include steelmaking, foundries, forges, alloys, heavy electricals, heavy machinery, petro-chemistry, atomic energy (for the most part in the public sector); and dye-stuffs, aluminium, textile machinery, automobiles, trucks, bicycles as well as the extension of already existing production of steel, cement, paper, soap and edible fats in the private sector.

Not that the governmental policies for promoting this or that industry, sector, geographical location, or level of investment went unchallenged. Far from it. But there was, or at least appeared to be, in the understanding of the politically interested public as well as of the professional economists a certain consensus that all of these additions to capacity could be counted as development.

So far as the rural zone was concerned (which accounted in the 1950s and early 1960s for more than 80% of India's population and almost as large a percentage of her working force, although only slightly more than half of the net national product), the meaning of development was somewhat more controversial. We have already mentioned the various village betterment schemes of the pre-independence period with their characteristic differences of emphasis on economic and non-economic amelioration. In the immediate post-independence years, the main initiative was taken by the Central Government under the label of "Community Projects".

These ambitious programs proposed to better the lives and the livelihoods of the entire village population by improvements in housing, drainage, health care, and education as well as introducing modern methods of crop production, animal husbandry, credit and marketing. Originally designed to be implemented by trained and devoted personnel having access to all existing departmental services as well as to substantial financial assistance, these projects were to be inaugurated in a limited number of carefully selected areas. The idea was that they would serve as demonstration centers of what could be accomplished, and that their rapid progress would evoke a spontaneous emulation on an ever-widening scale.

In the event, it proved politically impossible to concentrate governmental resources so narrowly. A network of "agricultural development blocs", emphasizing selected seeds, improved implements, chemical fertilizers and irrigation pumps soon covered the countryside. Cooperative societies were revived or created to distribute the new "inputs", to provide the necessary loans, and, in some cases, to take the produce to market.

Agricultural development became tangible when the figures for the overall production of food grains, which had been stagnant for several decades, began to move upward, albeit in jerks according to good or poor monsoons. Yet the debate never slackened as to whether the chief aim of policy in the rural areas should be to increase the supply of cereals (and, by that token, the diet of the people), or rather to promote the welfare of *all* the villagers. The series of land reform measures that the various States had passed and implemented from 1950 onwards had social justice as their stated goals. Whether through abolition of landlordism, through lowering of rents, or through making excess land (i.e., holdings above a statutory ceiling) available to the landless or the land-poor, these laws were intended to promote the prospects of the independent peasant cultivator.

Yet, to the proponents of higher agricultural production as a requisite for feeding the growing urban industrial classes at a reasonable price and avoiding reliance on cereal imports, these considerations took second place. It was more important, their argument went, to ensure that credit, high-quality seeds, suitable amounts of fertilizer, and a sufficient supply of water be supplied to cultivators who had enough land to make the best use of this governmental aid.

In point of fact, researchers later discovered, it was precisely this category of peasants—usually a handful in each village—which had both enhanced its land-holding position as a result of the reform legislation and had been able to obtain the lion's share of the advantages offered through the development agencies. In the process, numbers of weak tenants ended up working as agricultural labourers for their more prosperous neighbors. Was development turning into a game of making the rich richer and the poor poorer?

Much ink has been spent on arguments as to whether the "green revolution" leads ineluctably to a "red" revolution. Actually, the introduction of high-yielding varieties, particularly of wheat, in accordance with the policy of betting on the strong outlined above, has succeeded only in maintaining the upward trend attained in the 1950s—an achievement, but scarcely a revolutionary one. Long-standing controversies about returns to scale and the extent to which the improved agricultural package is, or can be made to be, neutral to size, remain unsettled. What appears clear is that the situation varies significantly from region to region, within the separate states, and that patterns of village power may explain part of the difference. So far as "red" revolution is concerned, there can be no doubt of the existence of sharp tensions in the rural area. To some degree, these stem from the demands of the lower castes for greater recognition of their human dignity, demands stimulated by their right to vote and their access to education. Violent incidents have been frequently reported in recent years. Upon investigation the cases turn out to be more often attacks by the haves upon the have-nots (or upon their womenfolk) than the reverse. Organized politically conscious rebellion has never again reached the level of the Naxalite movement of the late 1960s.

Even during the early plan period, development disappointed hopes of the creation of employment; particularly that the new industrial jobs would be sufficient to mop up the rural surplus labour. Economists and statisticians made heroic attempts to measure not only open joblessness but "underemployment". The latter term has gone out of fashion since it has proved impossible to obtain even an approximate figure in a predominantly agricultural economy. The amount of work available for any villager depends on the size of the family's landholding, if any; the technical level of cultivation; the household composition; the crop pattern; the season; the strength of the monsoon; the possibilities of non-agricultural employment or handicrafts. Recent governmental programs to provide paid work in the rural areas by the classic method of building roads, bridges and embankments have proved only moderately successful, in part because of poor administration and corruption.

The gloomy picture often drawn of multitudes of desperate jobless villagers pouring into the cities where they become a dangerous marginal mass of urban jobless does not fit the facts. Although India's towns have absorbed a very large absolute number of rural immigrants in the past 30 years, the rate of urbanization is low, as compared with Third World countries outside of South Asia. Newcomers to Indian cities typically belong to families with some land, but would like to better their financial prospects. Often if they do not succeed in finding a job or a source of revenue, they return to the countryside. Only the sons and daughters of the well-off can afford to remain totally without work.

On the industrial side, the machinery of planning produced a paradoxical result. The very system of permits and licenses, designed to promote a rational use of scarce resources and to prevent the concentration of economic power, constituted a formidable barrier that only the largest business houses with officers detailed for the task could hope to hurdle. Thousands of small enterprises and tens of thousands of little workshops did indeed spring up to produce simple consumer goods, machine parts and the like, and to service bicycles, tractors, buses and trucks. Yet, a tiny tightly-held group of huge conglomerates was able to strengthen its grasp on the national productive apparatus. It was these same big houses who managed to conclude profitable collaboration agreements with foreign companies. Meanwhile, a substantial number of the large-scale public sector enterprises ran into difficulties, which could be variously ascribed to unsuitability of the technical choices, rigidity of ministerial control, bottlenecks in the supply of basic machinery or of raw materials, and undisciplined labor. Research and development activities tended to be weak in both the private and public sectors; accordingly, new technology continued to be imported from abroad.

A period of disillusion ensued, during which economists and statisticians vied with one another to provide evidence that the great majority of the population were no better off, if not actually worse off, in the late 1960s than they had been at the beginning of the plan period. The discussion with regard to development narrowed down to a quarrel over the calculation of calories.

About the same time, new slogans were coined to give development a populist connotation. Elections were won under the banner of *Garibi Hatao* (Out with poverty!). Plan documents emphasized the importance of meeting human needs for proper nutrition, clean water, cooking fuel, shelter and accessible health care. Special highly publicized programs were devised to assist small and marginal farmers, tribals, arid-zone dwellers and the urban unemployed. International assistance was accepted for slum rehabilitation projects and urban transport amelioration.

By the 1980s a new source of concern came to the fore. India by now had taken giant steps toward modernity in the western sense of large-scale production, substitution of machines for men, and greatly enhanced use of non-animal energy. Insistent young environmentalists began to ask questions about the cost of this progress in denuding of forests, pollution of water and air and creation of seismic danger zones. Women complained that they had to walk further every year to collect twigs for cooking fuel. In parts of the country where agriculture has been intensified, this has sometimes involved longer hours of field work for peasant wives and female laborers, already heavily burdened with domestic duties.

From a quite different angle, Indian development is unfavorably compared with that of the four "little tigers". South Korea, Taiwan, Singapore and Hong Kong are cited as proving the possibility of rapid industrialization with full approval and cooperation on the part of the developed countries. Yet this type of modernization through expert processing zones (EPZs) with which India has experimented to a slight degree, clashes diametrically with the many times reiterated goal of national self-reliance.

There is no longer a consensus in India on the meaning of development, nor even on the proper assessment of the changes that have been brought about, willingly or unwittingly, since 1947. A decade ago Surendra Patel was bold enough to set out his views in *The India We Want*. What is required today is a reopening of the basic discussion as to where India is heading and by what means.

Note

1. When Surendra Patel was a student at the University of Pennsylvania in the late 1940s, Daniel Thorner (1915–74), the husband of the author, advised and encouraged him to write his doctoral thesis on agricultural laborers in India and Pakistan. This work had an important influence on Surendra's future intellectual development.

India: What's Missing?

NIKHIL CHAKRAVARTHY

A sustained mass movement by an unarmed nation—unprecedented in history—made India independent in 1947. But there were doubts in the minds of many of its friends and adversaries as to whether the country would be able to hold its own and build a structure that could be viable, politically and economically.

For one thing, the national movement had thrown up thousands of activists and supporters who had grown up in the culture of mass struggles but had not imbibed the rules of governance. For such a movement, it is not easy at the best of times to switch over from the role of the agitator and propagandist to that of the administrator.

Secondly, the newly independent country started with a backbreaking liability: the old India under the British Raj had been partitioned into two dominions—India and Pakistan. This partition brought in its train not only mutual anger and bitter hatred leading to bloody pogroms, but also complex processes like the division of assets and the allocation of administrative personnel, both civil and military, that could cripple any government machinery.

It was in this context that the national leadership decided to retain the old civil and military services of the Raj—the so-called steel frame—much to the confusion and distaste of many of the rank-and-file nationalists. But the national leadership in its wisdom realized that the art of administration could hardly be neglected, and even at the risk of being misunderstood in some quarters, it absorbed the old machinery into the structure of the new regime. The incongruity did not go unnoticed, both at home and abroad; but this did not deter the leadership from going ahead with it, as it felt that without an administrative machinery in hand there could be no way of ensuring that its writ would run over this far-flung country.

While this decision to retain the old services—civil and military—even after independence served its immediate purpose of keeping the country together, it brought in its train certain inescapable disadvantages. The most serious of these was the lack of a new outlook enjoined by the fact that the country which was the biggest colonial set-up within the British Empire, had become free and was destined to chart out its own independent course in the world of the second half of the 20th century. The old colonial administrative machinery had very little to do with any development program: at best, it maintained a railway structure mainly for the strategic and commercial requirements of the imperial power, while its welfare measures centered on ways and means of combating such natural calamities as famine, flood or pestilence. Its education system was meant primarily to provide the lower-level personnel for running the huge colonial administrative machinery.

Whatever might have been the justification for carrying on with such an administrative structure at the beginning, the failure to replace or restructure it in keeping with the needs and problems facing an independent country has been a serious drawback on the part of the national leadership. Many a time the need for such a change could be seen at different levels of political life, but nothing tangible was effected. For quite some time, the Government of India toyed with what came to be known as the Appleby Report. Then came the Administrative Reforms Commission. But no basic change was effected in the steel frame. Many new appurtenances appeared; various sections and departments were opened, but the red tape left behind by the colonial Raj not only persisted but was stretched longer and longer. Spread over three decades since independence, the Indian administrative machinery has grown in bulk and complexity only to make it almost unmanageable. The decision-making process has, if anything, become cumbersome, leaving little scope for monitoring and implementation. This could not but have a deleterious effect on any nation-building program. For a government commited to economic development and social justice, this amazing display of *laissez faire* in the matter of administration, could not but bring about serious distortions in the socioeconomic life of the nation. Inevitably, development planning did not yield the desired result of social justice; if anything, economic disparities have widened.

While the balance of internal class forces at the time of independence could not warrant any thrust towards a radically new social order—despite all the rhetoric on the subject—but could only give birth to the concept of the mixed economy, the anarchic drift that afflicted the management of the economy and the administration reflected the tenuous hold that the Government could, or chose to, exercise over these spheres. Hence, despite the Planning Commission the disorientation of the economy itself could not be checked; rather, it became commonplace to talk about this disorientation almost in a fatalistic manner. No serious effort was made over the years to find out what went wrong, why the objectives of growth and distributive justice could not be combined. This lopsidedness itself hampered the growth of purchasing power for the vast masses and thereby constricted the internal market.

It is not merely the administrative infrastructure that has been neglected, thereby leaving the country ill-equipped to take up the task of building a new India. On the political front, the erosion has been no less spectacular. If one were to compare the calibre of the members of Parliament and State Assemblies after the first General Election in 1952 with that of their counterparts of today, one cannot miss the decline in quality. The first generation of Indian legislators was really the product of the national struggle for independence, while the present lot reflects in large measure the vapidity and lack of direction in national life in recent decades.

The political purposefulness that marked the decade immediately after independence—epitomized in the personality of Jawaharlal Nehru—was largely dissipated in the years following his death. The 1969–72 interlude showed promise of checking this drift as issues of significance dominated the political stage for some time. But after the excitement of the electoral victory of 1971–72 subsided, drift became the order of the day. To a large extent, this contributed to the sense of insecurity on the part of the Government, which resulted in the thoughtless imposition of the Emergency, followed by the inevitable backlash in the 1977 General Election when the Congress under Indira Gandhi lost to a conglomerate composed of disparate elements, all brought together under the signboard of the Janata Party by the authority of Jayaprakash Narayan. By its very nature, such a coalition could not last long. It broke from within, as its different components, bereft of a single-minded national purpose, pulled in different directions, and finally pulled it apart. In such a state of disarray, Indira Gandhi found little difficulty in staging a spectacular comeback. Through these ups and downs in the nation's political life, most of the political parties and their leaders have emerged poorly. Those who have frequently changed sides or have been making one bid after another to form new coalitions without bothering about principles and programs, have grievously brought down their own credibility as well as that of political life as a whole.

Here, too, there is a perceptible lack of effort in setting up political instruments suited to our conditions and requirements. The Westminister type of parliamentary democracy was held up as a model. At every stage in the drafting of the Constitution, the British precedent got primacy and reference is often made in India today to the precedents and practice of Westminster.

Up to a point, this could work reasonably smoothly, though the results of the very first General Election in 1952 brought out the flaw of the British model—that a party can win an overwhelmingly large number of seats in Parliament or in a State Assembly without polling the majority of votes. In the subsequent General Elections, the shortcomings of the system have become glaring; and today it is common talk that the present type of election has become so expensive that money power with its accompanying corruption has become an important element in the process of choosing new legislature and the new Government. Here is one more example showing the lack of serious examination of the existing system in the last three decades. What the British Raj had left behind was neither restructured nor overhauled, with the result that many of its flaws have come in the way of ensuring maximum democratic content in the election machinery.

Apart from adopting the Westminster model for the election and functioning of our Parliament and legislatures, there has been very little input in forging new instruments of participatory democracy. Boards and Municipalities function mainly by the tenets left behind by the British Raj. At the grassroots level, the panchayat system has been introduced but its flaws, widely identified, have not been rectified.

The present scenario makes it abundantly clear that though there has been considerable mass urge to build India into a stable and powerful nation, there has been a dismal record in planning and implementing the efforts needed to reach that goal. There is no dearth of resources, material and human, in our country and there is no scarcity of talents, ready to be harnessed for any nation-building endeavor.

And yet there is a drawback somewhere. Our political life is no longer marked by the care and wisdom with which Nehru put forward the vision of a new India, striving for the greatest good of the greatest number. Nehru not only held out such a vision, but tirelessly engaged himself in educating the millions about it. The pride and confidence that exuded from his vision conditioned the entire nation with a sturdy sense of patriotism, fearless and vigilant. Since then, no political leader belonging to any political party has added a single inch to this ideology that he left behind. Rather surreptitious attempts are periodically made to undermine that legacy. It is amazing that in the two decades since Nehru's passing away, no political leader has come up with the confidence and capacity to inspire the millions in an enduring manner with a perspective that can galvanize the entire nation. The direction of national endeavor has not been set and hence the absence of the political will to mobilize the vast multitude that constitutes the Indian nation.

In the sphere of national economy, no strategy of growth-cum-social justice has emerged in the three decades since the First Five-Year Plan. There is no dearth of economists in India, many of them brilliant names in their specific spheres. But not a single one of them has contributed to the working out of a strategy for development as Mahalanobis did. It is significant that the only economic strategy for development formulated in India since independence was largely the work not of an economist, but of a physicist turned statistician who was deeply engrossed in finding ways and means to make this country great and strong while striving for social justice for its underprivileged millions. With this objective in view, Mahalanobis spared no effort to bring together for purposeful discussion a distinguished galaxy of specialists from the East and the West before drafting the Second Plan frame.

Understandably, the Mahalanobis Plan Frame came under severe attack from the *laissez faire* school, those who wanted India to become a free market economy. When it was put through mainly by the political will of Nehru, it was revealing that its subversion was accomplished by its opponents within the Government in collusion with big business interests. Consequently, growth has been registered, but social justice neglected. What is important to note is that while the opponents of the Mahalanobis strategy were active undermining it from within, the left claiming to be radical only paid lip service to it but cared neither to defend it effectively in a sustained manner against its saboteurs nor to mobilize the masses to forge the instruments of its implementation. The result has been that growth has become so bereft of social justice that today growth itself is retarded largely because of a stagnant internal market brought about by the neglect of social justice. Our economists and our political pundits have done precious little over the years to correct this monstrous distortion of the Mahalanobis Plan Frame. The impermissible coexistence of high-rise affluence and widespread impoverishment that has become the grotesque feature of the Indian scene has come about primarily because the economic strategy of the mid-1950s as embodied in the Mahalanobis Plan Frame has not been pursued steadfastly in the last three decades. This failure can certainly be ascribed to the bankruptcy of political will—a bankruptcy that has to be shared in equal measure by the ruling establishment and the radical left.

No doubt there has been a significant economic advance in India. The infrastructure of economic development has come up, the foundation of industrialization has been laid, agricultural productivity has risen. But the vulnerability of the entire structure has to be seen in the slow progress in the basic task of eliminating poverty. In fact, all the problems that beset the country today—slow rate of growth, increasing unemployment, youth unrest, caste and communal tension, regional discontent—can be traced to this one weakness. If famine has gone with the British Raj, starvation persists as the economists measure the large percentage of those below the poverty line. Here is the potential of internal instability that stares the political leadership in the face. Here is the fragile foundation of a widely acclaimed foreign policy.

In the age of the Technological Revolution that is about to set in, India is confronted with a grave challenge and a great opportunity. If the present drift continues, then the new technology will only help to widen the present disparities and thereby weaken the very fiber of nationhood: instead of progress, the danger of regression is likely to be intensified, and the technological revolution can very well threaten the very independence of the country. On the other hand, if the revolutionary implications of the new technology are absorbed, then it is legitimate to expect that with correct strategy India can achieve the breakthrough that may well place this country in the forefront of the coming age, the age of the Third World, as Surendra Patel has mapped it out.

This can be possible only if in the pursuit of the new technology, the objective of social justice is identified as the Number One priority. In the map of the Technological Revolution, India is uniquely placed. Geographically, it is endowed along with the rest of the Third World with perennial sunlight, while its industrial and technological development has reached a stage where it can harness the new renewable source of solar energy, particularly solar electricity—a stage which is yet to be reached by the rest of the Third World.

The great potentialities of the Technological Revolution beckon India to generate that political will which alone can develop and enforce a strategy that will ensure social justice and technological advance. It is this which is as yet missing in India's national life. Once that missing link is found, it will not take the country long to reach the destination of *The India We Want* about which Surendra Patel wrote 18 years ago.

Transnational Conglomerates: Reflections on Global Power

FREDERICK F. CLAIRMONTE

Introduction

In opposition to a mythologized New International Economic Order of the UN General Assembly, wedded to the voluminous ballyhoo that it has engendered, the staggering aggregation of economic power of the transnational corporation represents an authentic Corporate International Economic Order that dominates the world market.

The global megacorporation in its oligopolistic and conglomerate variants is not an historical aberration, but an emanation of global capital accumulation, i.e., savings and investment over several decades through a multiplicity of business cycles. The sheer magnitude of this unrelenting and almost wholly unaccountable aggregation of economic—and by extension political—power is manifested in the growth of world trade; and the grip of internationalization and transnationalization that partnered it.[1] Between 1948 and 1980, global trade escalated from $52 billion to over $2.4 trillion.

Globally this involved gigantic strides in manufacturing, mining, agriculture and, to a lesser measure, services that were sucked into the global market via the TNCs' outputs, marketing and promotional networks. Internationalization—that is, the share exported of a given sector—leaped from 16 to 27% over the decade of the 1970s.

Standing at the epicentre of this drive is the TNC that has already gouged out 80–90% of world trade (excluding the centrally planned economies). No less significant is that around two-fifths of global trade is transacted through TNC intra-firm transfers. The accelerated tempo of internationalization has been matched by the no less remarkable strides of transnationalization or the degree to which the output and marketing of a given sector is controlled by the global corporation.

TNCs have been defined as corporations with production facilities in more than one country, whereas trading, banking and other service TNCs may be categorized as business entities that control assets in two or more countries. Worldwide, the three TNC categories number 18,000. But, of the essence is that the locus of TNC power is embodied in a much smaller number. In the colorful

TABLE 1. *The internationalization of the world economy* (1970, 1980)*

	1970			1980		
	World GDP ($ billion)	World exports ($ billion)	Exports as per cent of GDP	World GDP ($ billion)	World exports ($ billion)	Exports as per cent of GDP
Agriculture	181.9	57.8	31.8	602.8	273.2	45.3
Mining and manufactures	726.7	217.5	29.9	2,769.5	1,522.1	55.0
Services	1,457.6	99.9	6.9	5,644.0	610.4	10.8
World total	2,366.2	375.2	15.9	9,016.3	2,405.7	26.7

Source: Calculated from data in UNCTAD, *Handbook of International Trade and Development Statistics*, TD/STAT/11 (New York: 1983); and from tapes of United Nations Department of International Economic and Social Affairs.
*Excluding socialist countries.

prose of the *Bangkok Bank Monthly Review*, "approximately 200 of them are the fearsome giants clutching world business in their grip".[2]

The extent of their grip, or better their stranglehold, on the global economy of the world's top 200 corporations, is seen in their combined sales that outpaced (1982) the astronomic sum of $3 trillion, the equivalent of one-third of global GNP—a concentration in numbers matched by a no less marked concentration in a handful of countries. Of the top 200 "fearsome giants", 166 are headquartered in just five countries: the USA (80), Japan (35), the UK (18), the FRG (17), and France (16). But these 166 have annexed over 85% of the 200s aggregate sales.[3]

A similarly devastating image is seen in financial services, with the top 100 giant commercial banks as their nucleus. In 1982, their joint assets of $4.5 trillion were almost half of global GDP, and more than one and a half times the combined sales of the top 200 corporations. Alone, the mega-Japanese and US banks control over two-fifths of the top 100's total assets.

Noteworthy is the prodigious transnationalization of the big US banks whose foreign profits soared from 22 to 60% in 1982. Joined to the state apparatus and the transnational banking system to which it is now organically connected, the TNCs are above all an instrument of power. Even as the Fabian J. K. Galbraith recognizes:

> When the modern corporation acquired power over markets, power in the community, power over belief it's a political instrument different in form and degree, but not in kind from the state itself.[4]

But that proposition, while having a surface lucidity, is incomplete since it gives the appearance that the TNC is a "state unto itself", which would be erroneous inasmuch as its real, and at times, unrealized power (the Japanese Sogo Shoshas and their predecessors, the Zaibatsus, provide the classic examples) is an outgrowth not only of its output, sales and marketing prowess, but of its inextricable connections to every facet of the state apparatus.

TABLE 2. *Profile of top 100 banks 1982*

Country*	Number of banks	Assets ($ billion)	Per cent of total assets
Japan	24	1,161.0	25.8
USA	15	743.9	16.5
France	8	514.3	11.4
FRG	11	466.1	10.4
UK	5	355.6	7.9
Italy	8	263.3	5.9
Canada	5	247.5	5.5
Netherlands	4	154.5	3.4
Switzerland	3	138.6	3.1
Belgium	4	91.7	2.0
Brazil	1	61.7	1.4
Hong Kong	1	57.1	1.3
Australia	2	49.1	1.1
Israel	2	42.6	1.0
Spain	2	41.6	0.9
Iran	1	27.1	0.6
India	1	23.6	0.5
Austria	1	19.6	0.4
Sweden	1	19.1	0.4
Iraq	1	18.6	0.4
Total	100	4,496.6	100.0

Source: Computed from *The Banker* (June 1983).
*Countries ranked by banks' assets.

TABLE 3. *Growth in foreign profits of leading US banks, 1970, 1981, and 1982*

	Foreign profits ($ mn)			Per cent of total profits		
Bank*	1970	1981	1982	1970	1981	1982
Citicorp	58	287	448	40	54	62
Bank America	25	245	253	15	55	65
Chase Manhattan	31	247	215	22	60	70
Manufacturers Hanover	11	120	147	13	48	50
J. P. Morgan	26	234	283	25	67	72
Chemical New York	8	74	104	10	34	39
Bankers Trust New York	8	116	113	15	62	51
Total	167	1,323	1,563	22	55	60

Source: Calculated from data from Salomon Bros. in *The Economist* (14 January 1978) and *Forbes* (5 July 1982 and 4 July 1983).
*Ranked on basis of 1982 assets.

The Political Economy of Power

Despite its pretentions to economic objectivity, bourgeois (alias mainstream) economic theory masks the morphology of power and its attendant inequalities in what it incongruously labels the "international free trade system". This

TABLE 4. *Top 200 transnational corporations: Services vs. manufacturers, 1982**

Country†	Service TNCs			Manufacturers TNCs			Total		
	Number	Revenues ($ million)	Per cent of top services 82	Number	Revenues ($ million)	Per cent of top manufacturers 118	Number	Revenues ($ million)	Per cent of top 200
United States	30	392,187	32.9	50	910,268	49.1	80	1,302,455	42.8
Japan	21	487,567	40.9	14	169,742	9.2	35	657,309	21.5
United Kingdom	9	99,548	8.4	9	165,174	8.9	18	264,722	8.7
FRG	4	45,034	3.8	13	162,507	8.8	17	207,541	6.8
France	9	92,904	7.8	7	89.733	4.8	16	182,637	6.0
Netherlands	—	—	—	4	86,377	4.6	4	86,377	2.8
Italy	1	10,150	0.8	4	74,331	4.0	5	84,481	2.8
Canada	5	38,772	3.2	2	16,309	0.9	7	55,081	1.8
Brazil	1	8,442	0.7	1	18,937	1.0	2	27,379	0.9
Spain	—	—	—	2	21,574	1.2	2	21,574	0.7
Switzerland	—	—	—	2	20,427	1.1	2	20,427	0.7
Israel	2	17,463	1.5	—	—	—	2	17,463	0.6
Others‡	—	—	—	10	118,233	6.4	10	118,233	3.9
Total	82	1,192,067	100.0	118	1,853,612	100.0	200	3,045,679	100.0
World GDP§								9,421,452	
Top 200 as per cent of GDP								32.3	

Source: Computed from data in *Forbes* (9 May 1983 and 4 July 1983).

*A service TNC is any TNC with over half of revenues from services; a manufactures TNC is any TNC with over half of revenues derived from manufactures.

†Countries ranked by combined sales of top service and manufactures TNCs.

‡Others includes ten countries with one TNC in the top 200: Mexico, Venezuela, Iran, Kuwait, Sweden, Austria, Belgium, Republic of Korea, India and South Africa.

§Excluding socialist countries.

theoretical cover-up has been made possible by the bogus construction over the last two centuries of an immutable economic model rooted in a competitive market on the basis of comparative advantage allegedly ensuring the optimum allocation of resources.

Translated into international trade theory, the neoclassical tradition purports to describe international economic relations in the context of market transactions between national firms and consumers ostensibly endowed with equal bargaining power. But the nostrum of comparative advantage embedded in economic liberalism has been given a special twist. Since the TNC conglomerate owns the modern equivalent of both Portuguese wineries and British cotton textile plants, to use David Ricardo's illustrious example, the TNC conglomerate in effect derives a double comparative advantage producing and arranging to trade with itself.[5]

The pivotal role of the political economy of power is seen in the global corporation which obliterated, in practice, the spindly theoretical constructs of neoclassical general equilibrium theory. Indeed, studies in commodities inspired by the neoclassical paradigm must therefore inevitably be shrouded in the price and quantitative analysis of econometric techniques in which social relations of power, dependence and marginalization have been ideologically abstracted by the creation of models that give wrong answers to trivial questions.[6]

But if bourgeois economics in its neoclassical garb and its practitioners, within and outside the United Nations system, are ignorant of the locus and motivations of the global corporations, its TNC spokesmen are clearly not.

> A corporation is a business structure whose sole reason for existence is the earning of profits by manufacturing products for as little as possible and selling them for as much as possible. It does not matter whether the product does good or evil; what counts is that it be consumed—in ever-increasing quantities. Since everything the corporation does has as its ultimate goal the creation of profit, it offers its workers no deep personal satisfaction, no feeling of contributing anything worthwhile to society, no true meaning to their activities. Go to work for a corporation and you are, through good salaries and various fringe benefits, installed as a faceless link in the lengthening chain—completing the circle by becoming one more consumer of all that junk.[7]

"History is junk", declared Henry Ford, another architect of the global corporation. Yet the origin and course of the global corporation is the culmination of an historical process that has not reached a terminal point. TNC corporation practice thus demarcates itself from Alfred Marshall's "representative firm".[8]

Historical antecedents

The shaping of the global corporation can be grasped schematically in three successive periodizations: 1895–1945; 1945–65; 1965 to the present.

Phase 1: 1895–1945

This 50-year span saw profound convulsions in global capitalism marked by the transition from relatively competitive conditions to monopolistic and oligopolistic output and market structures reflected in pooling arrangements, the formation of trusts, the financial holding company and a complex legal framework congenial to the concentration of capital. In the United States, the period 1895–1907 saw the advent of the corporate revolution with an annual average absorption of 266 firms; in 60% of these consolidations a single giant corporation acquired control of at least 62.5% of the industry's market as measured by its capitalization.

Nor was this tempo of economic aggrandizement confined to a single country highlighted by the merger (1903) of the American Tobacco Company (ATC) and the Imperial Tobacco Company into the British American Tobacco Company (BAT). James Buchanan Duke (1875–1925), architect of ATC and BATs first chairman of the board, grasped the crux of this annexationist thrust epitomized by Standard Oil. "If John D. Rockefeller can do what he is doing in oil, why should I not do it in tobacco?"[9]

"In brief", writes Professor Faulkener, "mechanical improvements made big business not only possible but imperative".[10] The holding company, fundamentally an American institution, traces its genesis to the last two decades of the 19th century thanks to the enactment of new corporate laws, notably in New Jersey: it laid the legal foundations for the global corporation. The revolutionary importance of the holding company for national and global capital accumulation was that it became the ideal instrument for speeding up capital concentration necessitated by mature capitalism beyond the ambit of partnerships and individual capitalists.

The techniques of what came to be known as "the new competition" enshrined in the holding company[11] included such practices as 'cutting prices in one locality to meet local competition while recouping in another, the operation of bogus 'independent' concerns to maintain the fiction of competition, the insistence upon tying or exclusive [marketing and output] arrangements, the securing of rebates and preferential agreements and use of blacklists, boycotts, espionage, coercion and intimidation".[12]

The annexationist momentum of the modern corporation had scaled such intoxicating heights that J. B. and J. M. Clark commented sardonically in 1912, that "the mere size of the consolidations which have recently appeared is enough to startle those who saw them in the making. If the carboniferous age had returned and the earth had re-peopled itself with dinosaurs, the change made in animal life would have scarcely seemed greater than that which has been made in the business world."[13]

While certain global corporations in North America and Western Europe were extending their operations, none approached the muscle of transnational power wielded by the Japanese Zaibatsus ("financial cliques") operating in conjunction with the state. Such symbiotic relationships underscored the political, cultural and social specificities of Japanese capitalism in general, and that of its TNCs in particular. "The usual means of welding the numerous companies in a Zaibatsu organization into a cohesive group", noted an official United States report, "has been the wide use of interlocking directorates, accompanied by the use of one individual as officer in several companies".[14]

By 1940, the dominance of Japanese finance, trading and industrial capital by the leading Zaibatsu had been completed. Despite the formal efforts by the United States occupation to dismantle them, a new species of Zaibatsu, immensely more flexible and penetrating in their reach than their precursors emerged. The centralizing force of the Ministry of Finance gave way to the Ministry of International Trade and Industry.

Phase 2: 1946–65

Arising from the technological explosion after the Second World War, the Japanese transnational corporate prototype became generalized despite the adoption of antitrust legislation in all countries—including Japan. In the 1960s, the operations of TNCs acquired entirely new dimensions. Whereas there were only two or three large (product-specialized) corporations with assets of $500 million in 1906, the present configuration is different.

> In contrast, by the first quarter of 1970, there were 111 industrial corporations with assets of $1 billion or more and nearly 200,000 partnerships and proprietorships engaged in manufacturing. These 111 corporations held at least 50% of the assets and 56% of the profits of all corporations engaged primarily in manufacturing; the 333 corporations with assets of $500 million or more accounted for fully 70% of all industrial assets, excluding their unconsolidated holdings. Indeed, by 1970, the two largest industrial corporations alone had sales of nearly $40 billion which is about as great (in constant dollars) as those of the over 200,000 manufacturing establishments operating in 1899.[15]

Likewise, in the United Kingdom, the 100 largest manufacturing firms by net output, which had acquired a relatively stable share of manufacturing (22–24%) between the early 1920s and the 1940s (with only a modest boost to 27% by 1953) had engulfed over 40% by 1970. Consequently the UK's industrial structure is now more concentrated than that of the USA.[16]

Phase 3: Mid-1960s to the present

From 1967 to 1969, concentration and centralization of capital was marked by an unprecedented spate of worldwide mergers. But, as Michael Pertschuk, Chairman of the Federal Trade Commission, explained, there was a qualitative difference between the earlier and more recent mergers. These mergers contrast with the largely "pure" conglomerate mergers of the 1970s, described as "firms with piles of cash buying up solid firms".[17] These involve firms acquiring functionally unrelated enterprises; the prototype was that of ITT which, in the colorful description of one of its annual reports, "is constantly at work around the clock, in 67 nations on six continents, in activities extending from the Arctic to the Antarctic and quite literally from the bottom of the sea to the moon". Any hope of curbing the power of such juggernauts through antitrust legislation is illusory as acknowledged by Congressman Neil Smith of Iowa. "Enforcement of this nation's existing antitrust laws rests with whoever may be in power at the Justice Department; and, if and when those laws are enforced, the process is so time-consuming that the injured firms may be long out of business before a decision is reached". Likewise, Canada's Anti-Combines Branch of the Department of Consumer and Corporate Affairs is virtually impotent. Notes one senior official: "We have a great concern over the size of these mergers, but the law is toothless. Mergers are impossible to stop. We must prove an 'undue' loss of competition to the public's detriment before that happens". To which the Director of the Anti-Combines Branch added: "What we will have if this march of increased capital concentration continues is a national oligarchy in which a few dozen people will interact to bargain about the economic future of millions".[18] In Canada's case, the future is already here.

Dr. Walter Adams testified to the overwhelming impact of conglomerate power even in the absence of oligopoly:

> A conglomerate giant is powerful, therefore, not because it has monopoly or oligopoly control over a particular market, but because its resources are diversified over many different markets. Its power, as Corwin Edwards says, derives from the fact that it can outbid, outspend, and outlose a smaller firm. It occupies a position much like the millionaire poker player who, in a game of unlimited stakes, can easily bankrupt his less opulent opponents, regardless of his comparative mastery over the dizzy virtues of probability theory.[19]

In fact, conglomerates derive prodigious gains from oligopolistic power; and conglomeration welded to oligopoly defines the power structure of the transnational corporation. While different markets expand and contract, the conglomerate which straddles several industrial and non-industrial oligopolies can ride with the tide shifting its resources from one profit centre to another.

Another technique deployed is that of "cross-subsidization", whereby the conglomerate operating over several fields shifts profits appropriated in its prospering corporate division to subsidize losses in any temporarily depressed lines of business, or better still to acquire a larger market share. Business reciprocity, the practice of making sales to and purchases from corporations that are clients of the conglomerate is, in addition to cross-subsidization, another powerful mechanism of conglomerate power.

The Major Sectors

The annexationist movement inherent in the operations of the transnational corporations is brought out more clearly at the sectoral level. The political economy of TNC aggrandizement has blurred or rendered far less useful for analytical purposes the conventional tripartite division of sectors into primary, secondary and tertiary. During the 1950s and 1960s, all primary commodity sectors experienced the disappearance (through mergers, takeovers and bankruptcies) of small- and medium-size trading companies so that by 1980, roughly 70–80% of global primary commodity trade ($980 billion) was accounted for by TNCs.

In particular, over the last two decades domination of primary commodity markets shifted from single commodity traders, of which the former United Fruit Company was the exemplar, to firms paramount in several global commodity markets. The trade in but three commodities, by no means exceptional, illustrates the dimensions of marketing leverage of the mega multicommodity transnational trading corporation: the trade in bananas, where three conglomerates dominate 70–75% of global markets; the cocoa trade, of which six corporations account for over 70%; and the trade in leaf tobacco, where 85–90% is under the direct control of six transnational buyers.

The annexationist power of the mega multicommodity traders stems from their self-reinforcing modes of conduct that contribute to enhance their bargaining stance with underdeveloped countries. In addition, most multicommodity traders are private and largely non-accountable. Many are integrated backwards into plantations and forward into processing, and hence are in an even stronger bargaining position *vis-a-vis* national marketing institutions. Another important trait of TNC-dominated primary commodity trade is intrafirm transfers carried out by trading affiliates. In consequence of progressive vertical integration, TNC intra-subsidiary trading has become a conspicuous trait of international trade.[20]

By contrast with primary commodity trade, global trade in manufactures is largely dominated by a different ensemble of protagonists (with the exception of the ubiquitous presence of the Japanese Sogo Shoshas). Essentially, these are trading subsidiaries of industrial TNCs which, jointly with the Sogo Shoshas, accounted in 1980 for roughly $810 billion of the $1015 billion worth of global manufactured exports.

While a voluminous literature has been devoted to the strides of the underdeveloped countries which doubled their share of manufactured goods exports in the last 25 years, what is often lost sight of is that such patterns of export-oriented development are still overwhelmingly under TNC ownership and marketing control. Contributing to this direction and volume of trade flows are the hundreds of free trade zones (FTZs) and other special enclaves (e.g., industrial estates) which offer a vast spectrum of incentives to attract TNCs, including among others, extensive tax holidays, subsidized infrastructure, and, of course, the elimination of militant trade unionism. Politically, such international operations have been massively bankrolled by the state and local governments in the poor countries.

What is novel about much of the export business carried on by the underdeveloped countries is that it involves a new TNC strategic approach. The latter have succeeded in fragmenting the production process by contracting out the most labor-intensive phases to the poor countries where wage costs are low.

In the textile and apparel sectors, two sectors that are by no means

TABLE 5. *Corporate control of global commodity trade, 1980*

Commodity	Total exports ($ million)	Percentage marketed by 15 largest transnationals*
Food		
Wheat	16,556	85–90
Sugar	14,367	60
Coffee	12,585	85–90
Corn	11,852	85–90
Rice	4,978	70
Cocoa	3,004	85
Tea	1,905	80
Bananas	1,260	70–75
Pineapples	440†	90
Agricultural raw materials		
Forest products	54,477	90
Cotton	7,886	85–90
Natural rubber	4,393	70–75
Tobacco	3,859	85–90
Hides and skins	2,743	25
Jute	203	85–90
Ores, minerals, and metals		
Crude petroleum	306,000	75
Copper	10,650	80–85
Iron ore	6,930	90–95
Tin	3,588	75–80
Phosphates	1,585	50–60
Bauxite	991	80–85

Source: UNCTAD secretariat estimates, based on extensive research and interviews with traders and marketing specialists. The figures represent orders of magnitude only.
*In most cases only 3 to 6 transnational traders account for the bulk of the market.
†Four-fifths consists of canned pineapples and one-fifth of fresh pineapple.

exceptional, outward processing involves domestic fabric design and cutting by the TNCs, followed by shipment to an underdeveloped country for yet another operational phase—sewing. In an ultimate phase, the partially processed goods are then reimported for finishing and packaging. In the advanced capitalist countries, this practice is promoted by special tariff concessions in which the TNCs concerned pay duty only on the value added to the product abroad.

While the growth of manufactured exports of certain poor countries has been numerically impressive, its foundations linked to the annexationist policies of TNC investment and marketing strategies are by no means uniformly solid. To be sure, with the same alacrity with which they located plants in the poor countries, the global corporation is able to uproot—and in several cases has already uprooted—such plants as advances in technology cancel the principle advantage of these countries—their cheap undifferentiated labor power.

Services, the third sector, have not escaped the encroachment of the TNCs now accounting for almost two-thirds of world GNP and growing at a faster clip than the primary and manufacturing sectors. As with the primary and manufacturing sectors, the drivewheel behind this internationalization and transnationalization has been a handful of TNCs that have speeded up the liquidation of the medium and small-scale firms that traditionally were the paramount economic agents. No less stark has been the disintegration of barriers between individual service sectors with the rapid upsurge of transnational service conglomerates (TSCs)[21] and transnational integral conglomerates (TICs).[22]

By the end of the current decade, the distinction between financial and non-financial institutions (assuming the present trend, intensity and direction of finance capital) will have largely vanished. A point in this direction has been the annexation of Salomon Brothers, one of the world's biggest investment bankers, by Phibro, one of the world's largest global traders (excluding the Sogo Shosha). But Phibro itself also overlaps with Englehardt's mining empire which, in turn, is a part of the wider galaxy of South Africa's Sir Henry Oppenheimer's Anglo-American mining corporation.

The evolution and systemic interaction of service corporations and non-service corporations is intensifying the thrusts of internationalization and transnationalization. The birth of the financial supermarket and its rapid growth in recent years is descriptive of the present configuration of forces. It would be difficult to improve on the portrayal of the mutations now revolutionizing services (and by extension non-services) than that of the American insurance periodical *Best's Review:*

> The financial services world is an intricate web of competing conglomerates, interlocking directorates, foreign and domestic subsidiaries and spin-off companies, all interwoven with provincial and federal legislative policies than can present a daunting prospect for newcomers to the field.[23]

What all this adds up to is the eradication of distinct categories or definitions and the emergence of new conceptions of a bank and even of money. The driving force behind these changes is what has been labelled the financial conglomerate or the financial supermarket, an institution spanning, among other services, commercial banking, securities broking, insurance services, travellers checks, credit cards, money market funds, point of sales debit systems and financial futures operations.

Corporate financial conglomeration is being fuelled by the interactive propellants of electronic technology, government deregulation and the corporate striving for the dismantlement of these barriers. The ongoing technological revolution opens up the possibility for financial transactions to be swung around the planet at the speed of light, conferring on individual corporations the ability to operate effectively in dozens of financial markets simultaneously.[24]

With technology flinging the door open for total conglomeration, what remains to consummate the truly global financial conglomerate is the dismantling of state regulation, a course already well underway globally, with the USA being the most conspicuous example. The bulk of existing US regulatory measures—institutional and geographic—will either be wiped out or better still neutralized, strikingly so the McFadden Act (1927), the Glass-Steagall Act (1933), and the Bank Holding Company Act of 1956.

Traditionally, in the USA, a rigid separation between banking and other financial dealings has been a basic principle of banking law since the early 1930s and part of the fundamental reforms of the New Deal administration. The walls, however, are crumbling, or in the summation of *The Financial Times:*

> Hardly a month goes by without another major development—another brick knocked out of the wall. Banks, particularly the major ones, are buying brokerage and insurance companies. Insurance companies are acquiring brokers and most recently banks. Big non-bank commercial and industrial companies are buying everything.[25]

This process is being propelled, certainly and conspicuously in the United States by two forces: first, the State itself, which condones and financially underpins the takeover of ailing banks by other banks; second, the changing attitudes of the US Government and state governments toward antitrust activities. As United States Assistant Attorney General William Baxter said in 1982:

> There is no question [the new antitrust guidelines] are somewhat more permissive. Their underlying philosophy is that mergers are a very, very healthy phenomenon of the capital markets and should not be interfered with except under exceptional circumstances.[26]

Case Histories

Certain recent takeovers underline the significance of the new directions of finance capital and the changes within global capitalism glimpsed in the service sectors. Intrusion of the newer breed of financial conglomerates means that supermarket style banking chains will become generalized, and initially more competition will inevitably lead to less competition.

Sears Roebuck, the number one US retailer (1982 sales: $30 billion), illustrates the dynamics of the financial conglomerate. "Our goal is to become the largest consumer oriented financial service entity", declares Sears Chairman of the Board. Sears's four major operating divisions comprise merchandizing, insurance, real estate and financial services. Annexation of Dean Witter Reynolds, the fifth largest stock brokerage firm in the US, was followed swiftly by the takeover of Coldwell Banker, the largest real estate broker.

These impressive annexations approaching a billion dollars, merely in 1981, are to be added to Sears' existing financial services which already include:

Table 6. *Selected financial mergers, early 1980s*

Acquirer	$ mn.	Acquired
Bank		
Citibank (USA) & Lloyds (UK)	n.a.	Grindlays Bank (UK)
Midland Bank (UK)	820	Crocker National (USA)
Mitsubishi Bank (Japan)	282	BanCal Tri-State (USA)
Bank of America	53	*Investment bank/brokerage house* Charles Schwab (USA)
Lloyds, National Westminster, Barclays, Midland, William & Glyn's (UK)	n.a.	*Futures company* International Commodity Clearing House (UK)
Fuji Bank (Japan)	425	*Financial conglomerate* Walter E. Heller, two subsidiaries (USA)
Financial conglomerate		
American Express (USA)	550	*Bank* Trade Development Bank (Luxembourg)
American Express (USA)	930	*Investment bank/brokerage house* Shearson Loeb Rhodes (USA)
Alexander and Alexander (USA)	300	*Insurance* Alexander Howdon (UK)
American Express (USA)	175	*Media firm* Warner Communications Cable subsidiary (USA)
Investment bank/brokerage house		
Lehman Brothers (1977, USA)	n.a.	*Investment bank/brokerage house* Kuhn Loeb (USA)
Goldman Sachs (USA)	n.a.	*Trading company* J. Aron (USA)

Acquirer	$ mn.	Acquired
Insurance		
Marsh and McLennan (USA)	n.a.	C. T. Bowring (UK)
Allianz (FRG)	n.a.	Eagle Star (UK)
Winterthur (Switzerland)	n.a.	Republic Financial Svcs. (USA)
Winterthur (Switzerland)	n.a.	Provident Insurance (UK)
Connecticut General (USA)	n.a.	INA (USA)
Prudential Insurance (UK)	385	*Investment bank/brokerage house* Bacho Group (UK)
Prudential Insurance Co. (USA)	n.a.	Hambro Bank (UK)
Aetna Life & Casualty (USA)	n.a.	Samuel Montague (UK)
Mutual fund Dreyfus Corporation (USA)	n.a.	*Bank* n.a. Lincoln State Bank (USA)
Retail company Sears Roebuck (USA)	600	*Investment bank/brokerage house* Dean Witter Reynolds (USA)
Sears Roebuck (USA)	n.a.	*Real estate brokerage house* Coldwell, Banker & Co. (USA)
Trading company Philbro (USA)	550	*Investment bank/brokerage house* Salomon Brothers (USA)
Jardine Matheson (HK)	30	*Insurance* Bache Insurance Services (USA)
Industrial corporation Dow Chemical Co (USA)	n.a.	*Investment bank/brokerage house* Arbuthnot Lutham (UK)
Texaco (USA)	n.a.	Charles Fulton Holdings (UK)
Bochtel Engineering (USA)	n.a.	Dillon Read & Co. (USA)
Société Générale de Belgique (B)	n.a.	Tanks Consolidated (UK)
Englehard Minerals (USA)	278	*Insurance* Northwestern National Insurance (USA)
ITT (early 1970s USA)	n.a.	Hartford Insurance

Source: Compiled from trade sources.

insurance; savings and loans operations; interest-bearing credit for some 25 million users of Sears' credit cards. Sears has already begun its own money market fund; its Allstate Insurance subsidiary already has 20 million policy holders, offering insurance from autos to annuities.

Texaco's link-up with one of London's large Eurocurrency brokerage houses is a strategy that displays certain surface similarities with that used by Sears. By the setting up of Ful-Tex Euro Services Texaco, one of the seven petroleum sisters challenges the exclusivism of commercial banks' monopoly on foreign exchange and time deposit services. Such a plunge gives Texaco access to the foreign exchange broker network, even though it is not involved in Ful-Tex's daily operations. Moreover, Texaco is in a position to save money, since it will be in a position to place parts of a large order discretely with several banks, rather than risk increasing the cost by using a single institution, or letting the market become aware that a large amount of money is being transacted, i.e., bought or sold. In addition, Texaco could tap directly Eurocurrency funds, bypassing commercial banks and thus obtaining their funds at the same price, or cheaper, than that obtained by commercial banks.

Other service sectors

These changes leading to extensive conglomeration are seen in other selected service sectors. The global explosion of insurance services since 1950 was stimulated by, and marched parallel with, the massive implantation of TNC subsidiaries overseas and the stunning growth of world trade. Corporations based in five countries dominate the international insurance market: the US, the UK, the FRG, Switzerland and France. Firms headquartered in these countries have about 75% of all overseas insurance offices.

This massive concentration is all the more crucial since the world insurance market in 1982 revealed an estimated global premium income of $260 billion (with reinsurance premiums added, this rises to over $300 billion) and a total insured sum of around $5 trillion, the equivalent of a little over half of global GDP.

The first major breach in the phalanx of specialized insurance companies was opened up in the 1950s with numerous mergers between life and nonlife insurance companies, a form of vertical integration which embraced annexations of insurance brokers and reinsurance companies. This process will continue and is likely to accelerate in the 1980s.

The same force is at work in the advertising industry controlled by a handful of giants. Worldwide mass advertising, now primarily controlled by transnational service conglomerates, is at once a major component of the services sector. An idea of the pervasive ideological impact of the TSCs advertising clout is conveyed by global advertising billings, topping more than $118 billion (measured media outlets) in 1982. The USA alone accounts for fully one-half of the advertising universe, with roughly $61 billion in advertising outlays in 1982.

Five other centers of corporate power—Japan, the FRG, the UK, France and Canada—added to the USA, have appropriated four-fifths of the total.

A comparison of the world's 50 biggest advertising agencies with the world's 50 biggest TNCs in 1982, illustrated the proportionately greater US dominance of advertising than of other sectors: whereas only 20 of the 50 biggest TNCs are based in the US, 35 of the top 50 advertising corporations are American.

Perhaps no worldwide advertising firm better exemplifies all of these phases than Dentsu—the world's largest advertising corporation—corralling a quarter of Japanese advertising. Today, as *Fortune* notes: "Its media properties include several newspapers and TV stations. In words which might apply to the vision of the other advertising majors, Dentsu's president Hideharu Tamaru proclaimed the agency's role in the 21st century to be 'organizer, producer, and consultant in the field of culture by maintaining closer ties with the government, public enterprises, academic bodies, international organizations and private corporations'." In short, big advertising has moved well beyond consumer manipulation into the epicentre of political manipulation.

Several of the largest companies have actively entered the political arena on a national level by representing political candidates/parties, which entails the creation of intelligence gathering divisions to monitor not only consumer behavior but political formations as well.[27]

Towards the future

The preceding highly summarized discussion on the nature of power of the leading economic agents of global capitalism—the transnational corporation—highlights the nature of power with all its far-reaching political implications. Such sedulous avoidance of the complexities and the perils inherent in the global economy has limited the capacity of many United Nations negotiators—and in their myopic vision they are by no means alone—to throw light on a swiftly changing reality.

The internationalization and transnationalization of goods, services and capital has been of crucial importance in deepening the economic crisis that has attained unprecedented heights and is still climbing. Merely in the OECD group of countries, the legions of unemployed already exceed the 30 million mark. The misery and tragedy, which these bare figures can never convey, is far more tragic in the poor countries whose level of mass unemployment is a multiple of that in the mature capitalist economies. What, it may be asked, has this suffering to do with the transnationalization of capital? In their annexationists momentum (for to arrest or even to attempt to arrest this process of expansionism would be tantamount to suicide), corporations are impelled to seize upon the newest technical innovations and adapt them to the entire range of their output, services and marketing operations. In this grand design, labour force liquidation becomes an overriding concern, as even a perfunctory reading of the annual reports of TNCs would indicate.

Inflation remains another phenomenon of the global capitalist generated TNC crisis. These inflationary pressures stem from several factors, of which concentration and oligopolistic corporate structures and pricing strategies are contributory. While obviously there are several sectoral variations, corporate concentration and centralization of capital has fuelled inflation through several techniques: administered and political prices in industrial sectors: speculation and manipulation of futures markets by the mega multicommodity traders; and the mark up practices of large-scale retailers. To this crisis ridden catalogue could be added the activities of those TNCs which are the major beneficiaries of the more than $700 billion global arms industry which, being nonproductive, remains one of inflation's classic breeding grounds.

Mandatory for any analysis of global capitalism—finance, industrial, trading, services—is that the emergence and subsequent evolution of the TNC marks the end of an earlier epoch of capitalism that was, in a certain way, related to the advent of a pluralist society. The rising tide of the TNC power, that is the increasing homogenization of capital, spells an end to pluralism in all forms.

Such massive fusions and aggregations of unrestrained economic power are necessarily part of a totalitarian political process which deliberately and erroneously identifies "a market economy", "private property", and "democracy" with the giant corporation. Rather, the transnational corporation is a mass liquidator of all three. This homogenization of capital, of which the breaking down of the three major sectors is but its surface expression, will multiply its degree of nonaccountability, already well beyond the threshold of democratic acceptability. It opposes (nationally and internationally) even the most innocuous species of nationalization which does not even touch its central nervous system.

In view of its political and sectoral imbrications, the totalitarian tendencies within global capital can only be countered by far-reaching nationalizations and democratic control that recognizes its scope, velocity and direction. As such its finalization is embedded in a political process that breaks with capitalism's propertied relationships of which the TNC is its supreme economic protagonist. But that is another story.

Notes

1. For an elaboration of these themes see Frederick F. Clairmonte and John H. Cavanagh, "Transnational corporations and services: The final frontier," *Trade and Development: An UNCTAD Review,* No. 5 (1984), pp. 215–273.
2. *Bangkok Bank Monthly Review* (November 1981).
3. The US corporations, however, account for over two-fifths of the top 200s revenues, trailed by Japan (22%) and the UK (9).
4. *American Economic Review* (March 1983).
5. Richard J. Barnet, *The Crisis of the Corporation* (Washington, D.C.: 1978), p. 6.
6. Referring to the "other-wordly mathematical models" of an important segment of neoliberal theorists, Wassily Leontief commented: "Page after page of professional economic journals are

filled with mathematical formulas leading the reader from sets of more or less plausible but entirely arbitrary assumptions to precisely stated but irrelevant conclusions". Quoted in *The Economist* (26 June 1982).

7. The judgement of Mr. Maisonrouge, Senior executive of IBM. Quoted in Barnet, *op. cit.*, p. 4.
8. This was conceptualized within a competitive structure where market power is diffused among atomistic economic agents, where all producers sell at an externally and objectively determined price, and price equilibrium is reached when marginal cost and marginal revenue are equal. In sum, an idealized competitive model which purported to explain what should have been happening theoretically but was not in fact happening historically (UNCTAD, *Marketing and Distribution of Tobacco* (Geneva: 1978)). Also, as has been noted, neoclassical theory fails to provide a satisfactory explanation of the growth of the firm. Its assumptions of atomistic firms operating in undifferentiated markets, with identical knowledge and technology and no economies of scale, leave no basis for distinguishing individual firms or for predicting which will grow faster than the others. See N. S. Siddhartan and S. Lall, "The recent growth of the largest multinationals", *Oxford Bulletin of Economics and Statistics* (February 1982).
9. *New York Times* (11 October 1925).
10. H. U. Faulkner, *The Decline of Laissez-Faire, 1897–1917* (New York: 1951), p. 171.
11. The holding company has been defined as "any company, incorporated or unincorporated, which is in a position to control, or materially to influence, the management of one or more other companies by virtue, in part at least, of its ownership of securities in the other company or companies". See T. C. Bonbright and G. C. Means, *The Holding Company: Its Public Significance and its Regulation* (Reprint) (New York: 1969).
12. H. U. Faulkner, *op. cit.*, p. 172.
13. J. B. Clark and J. M. Clark, *The Control of the Trusts* (New York: 1912), pp. 14–15.
14. *Report of the Mission Japanese Combines. A Report to the Department of State and the War Department* (Washington, D.C.: March 1946).
15. United States Congress, Hearings before the Senate, Sub-Committee on Monopoly of the Select Committee on Small Business, *The Role of Giant Corporations in the American and World Economies, Corporate Secrecy: Overviews.* 92nd Congress, 1st session (9 and 12 November 1971), pt. 2:1111. See also Federal Trade Commission, *Economic Report: Conglomerate Merger Performance. An Empirical Analysis of Nine Corporations* (Washington, D.C.: 1972). By the middle of the 1950s, four firms controlled over two-thirds of the United States market in many industries: automobiles, aluminium, copper, rubber, cigarettes, soap and detergents, whiskey, heavy electrical gear, computers, sugar, iron. Cf., John K. Galbraith, *The New Industrial State* (Toronto: Signet Books, 1968), p. 190.
16. S. J. Prais, *The Evolution of Giant Firms in Britain: A Study of the Growth of Concentration in Manufacturing Industry in Britain: 1909–1970,* (Cambridge University Press, 1976).
17. Testimony to United States Senate, Subcommittee on Antitrust and Monopoly of the Senate Committee of the Judiciary, *Mergers and Industrial Concentration,* Hearings (September 1978), p. 152.
18. Quoted in *The New York Times* (25 March 1980). An almost comic twist to this level of impotence is that Canada recorded only one monopoly conviction since 1945 and that on the basis that the defendant pleaded guilty.
19. United States Senate, 95th Congress, *Mergers and Industrial Concentration, Hearings before the Subcommittee on Antitrust and Monopoly. Acquisitions and Mergers by Conglomerates of Unrelated Businesses* (Washington, D.C.: 12 May, 27 and 28 July and 21 September 1978), p. 172.
20. This now familiar strategem is one in which the TNC arranges "buying and selling" transactions between their own subsidiaries and has given rise to the widespread technique of transfer pricing, designed to boost corporate profits and circumvent national economic policies.
21. A TSC is one which operates in two or more service sectors.
22. A TIC is one which spans a large spectrum of service and industrial sectors and whose highest expression to date is the giant Japanese conglomerate groups (e.g. Mitsubishi, Mitsui, Sumitomo etc.).
23. *Best's Review* (July 1983).
24. F. F. Clairmonte, "Dynamics of finance capital", *Journal of Contemporary Asia*, Vol. **12,** No. 2 (1982).
25. *Financial Times* (24 May 1983).

26. *International Herald Tribune* (6 June 1982). Indubitably, antitrust legislation in the US, not to speak of Japan to which it was an alien injection, was never a coherent body of doctrine, since it was countered by other legislation that favored capital concentration and the penetration of foreign markets. This was the case of the Webb-Pomerene Act of 1918, that promoted corporate collusion through the formation of export cartels, notably joint business ventures, supported by the state. An illustration of such a joint venture (which later proved important in the setting up of Aramco in the 1930s and 1940s) was the activism of the US government in the mid-1920s.
27. These agencies have not been content to confine their political promotions to a national level, but have pushed on to assault United Nations bodies as well. By no means atypical is the language of the Interpublic Group of Companies Annual Report. "UNESCO and other bodies are encouraging government control and ownership of the media—a concept that is anathema to freedom-loving people and to the communications business, of which advertising is an integral part".

3

PERCEPTIONS OF THE SOUTH

Is an Endogenous Development Strategy Possible in Africa?

SAMIR AMIN

The deterioration over the last decade of the economic and social situation in the Third World as a whole, and notably in the large majority of African countries, is increasingly making demands on authorities, at national and international levels, to take up anew the concepts that underlie the theory and practice of conventional development policies. The alternative choice of an autocentred development, as opposed to extraverted conventional policies, has thereby suddenly benefited from an unexpected popularity. It is this success and the use of this expression in different contexts (or even in the vaguest of speeches) which impose the utmost prudence in ascertaining first the history of the elaboration of this concept and its content, before considering the question of its operationality in Africa today.

The first formulation of this concept goes back to 1957: "Whereas in the autocentered accumulation model, external relations abide in logic and imperatives of international accumulation, in the extraverted development model, on the contrary, it is relations with foreign countries which determine almost entirely the rhythm and character of internal accumulation". In this concise formula remains the definition of the question, *perforce* requiring clarification.

The conflict between the autocentered model versus the extraverted model was not deduced from an abstract, *a priori* formulated theory, or from an ideological theme. It was the conclusion of a comparative historical analysis, and as in all interpretation of facts, in the framework of a theory (that of the accumulation on a world scale).

Three sets of historical experience were considered: (1) that of the development of countries and regions of developed capitalism (Europe, Japan, North America), taken in its overall historical background from the mercantilist era up until the present; (2) that of the development of the regions of underdeveloped capitalism (Asia, Africa and Latin America), taken in the same historical setting; and (3) that of the contemporary socialist development (USSR, Eastern Europe, China).

The theory of accumulation on a world scale explicitly considers that the history of (capitalist) development from the very outset (around the 16th century) until today, is not that of juxtaposed "national" developments, "backward countries" making headway like their predecessors, but rather that of a world system dividing itself into driving centers, and dominating peripheries shaped by those same centers. The world economic system, therefore, is logically superior to—not equal to the sum of—its parts ("local economies"). This theory, progressively elaborated through a collective effort of closely associated names of the author, Giovanni Arrighi, André Gunder Frank and Immanuel Wallerstein, calls for four conclusions that are outlined below:

(1) It was opposed to the concept of the "successive stages" of growth, both in its conventional version (according to Rostow) and in a version drawing from Marxism and the language of its concepts (modes of production).

(2) Therefore, it considered that "underdevelopment", far from being synonymous with "backwardness", is the product of "development". Development of some (centers) and underdevelopment of others (peripheries) are the concomitant products of the same development (world accumulation).

(3) It recognized, between "centers" and "peripheries", some situations of "semi-peripheries" oscillating at one time or another, either to attain the stage of "new centers" or to become "peripheries". But the space of indetermination that enabled the passage from the stage of semi-peripheries to the stage of center, ceased to exist from the end of the 19th century. Thereafter, the world system was such that universalization of the control of the productive process, through the capital of the centers' monopolies (today the multinationals), made a qualitative leap reducing this margin to nil; there has been no new Japan (the existing NICs are not that of a Japan in gestation).[1]

(4) The deduction from the preceding conclusions was that the development of the "South" (other than the dominated peripheral growth) implied thereafter a "rupture" with the logic of world accumulation.

The theory outlined above, which it is unnecessary to evaluate here, enabled the regrouping of all the main aspects of the phenomenon of the historical development of the modern world.

In the centers, development has, from the outset, always been autocentered, but has never been autarkical (no external relations). On the contrary, the centers that have progressed at the most accelerated pace are those that have benefited from a more extensive or deeper domination over the peripheries, which have submitted to the exigence of their autocentred accumulation. During certain periods, a center which had become hegemonical on a world scale (Great Britain from 1763 to 1870; the United States from 1944 to 1970) preached the ideology of economic freedom in forms adequate to the system at that time: freedom of trade for the Pax Britannica, and freedom of enterprise for the Pax Americana. The non-hegemonical centers accepted the consequences of this ideology. They preached, and above all practiced, various forms of protectionism (in the broad sense of the term), which was necessary to avoid a peripherization for them; and have thus often succeeded in accelerating their autocentered development, and even in calling the hegemonies themselves into question. This protectionism (towards hegemonical centers) was not synonymous with autarky, but instead, with an aggressive opening in the direction of the peripheries. "Liberal" hegemonical or "protected" non-hegemonical centers have all registered their autocentered accumulation in an international division of labor that they modulated themselves.

On the other hand, the socialist countries (in a restrictive sense: USSR, Eastern Europe, China, North Korea, Vietnam, . . .) which are "less advanced" countries (to varying degrees indeed, and with the exception of Czechoslovakia and East Germany), have not only accepted the principle of an autocentered development (submitting the external relation to their internal accumulation), but have practically passed through an almost autarkical phase. This autarky was not only imposed by the world system rather than wanted by themselves during two decades or more; it was also accepted, and disseminated even to small East European countries during the Stalinist plans of the 1950s. The USSR is attempting, with Eastern Europe, to restructure this system through COMECON (and to include in it some non-European newcomers) and to open up extensively to the international division of labor. China, likewise, is opening up to the international division of labor. This shift does not mean that these countries have renounced the principle of autocentered development.

The Third World forms part of the system of the capitalist international division of labor in a peripheral position. It has never practiced an autocentered strategy, but in certain circumstances, it has begun to question certain aspects of the extraverted peripheral strategy. It has at times also indirectly benefited from the loosening central control during wars and crises; and it is interesting to note that often in such periods an autocentred development has been initiated and then destroyed in the subsequent phase. In the extreme, during the liberation wars, the resistance movements, forced *de facto* into autarky, have pushed changes towards a basic autocentered development further than anywhere else in the Third World. Some countersigns like "self-reliance", etc., which obviously imply a political sense—to which we acknowledge our sympathy—are born from the peoples' national liberation movement. This is not merely accidental.

To try to schematize the opposing autocentered model/extraverted model, a four-sector plan is proposed: (1) production of the means of production; (2) production of goods for mass consumption; (3) luxury production/consumption; (4) exports. The autocentred model is defined as that which is mainly governed by the articulation of sectors 1 and 2, and the extraverted model as that which is mainly determined by the articulation of sectors 4 and 3.

The proposed blueprint leads to a major conclusion. In the autocentered model labor remuneration (wages and peasants' incomes) must necessarily increase according to the rhythm of the progress of productivity. On the other hand, in the extraverted model the labor remuneration can be disconnected from the productivity growth.[2]

This conclusion now takes on a political form which is the following:

(1) The development of a country from the contemporary Third World cannot be achieved through the adjustment of its economy to the requirements of the international division of labor; on the contrary, what is required is the delinking of this economy as opposed to the international division of labor;

(2) This delinking is a necessary (but not sufficient) condition for an autocentered development that remains impossible if it is not intended for the people (that is, if the benefits of the productivity rise do not go straight to the greater majority);

(3) On the other hand, growth, the benefits of which are chiefly assigned to a minority, not only is possible on the basis of extraverted development (not always and everywhere possible), but also calls for such a development, more effective for this objective than an autocentered model.

Therefore, for the Third World, autocentered development can be synonymous with national and peoples' development.

The previous developments have, it is hoped, clarified some ambiguities of language. The concept of autocentered, "national" and "peoples" development should not be vulgarized, compared to a series of "protectionist measures", or simply to autarky. Neither should it be associated with that of "dependence". Of course, the peripheral economies are dependent, in the sense that the rhythms and structures of their growth are governed by those of the centers (whereas the contrary is not the case).[3]

If we can speak of interdependence, be it even an unequal one, between the centers (hegemonical and other powers, big and small peasants of the center . . .), we cannot in the same sense speak of interdependence between centers and peripheries, but of dependence of the peripheries. The difference between the two concepts (periphery or extraverted model on the one hand, dependence on the other) emerges when we consider the cases of non-peripheral dependent economies: for instance, Canada, whose national capital only holds a subordinate position, hence dominated by the United States capital, but whose growth in labor incomes is parallel to that of productivity. Canada is therefore more of a United States' province (and, as it is not one, is thus a dependent state in a certain sense) than anything else. The dynamics of its accumulation (through articulation of sectors 1 and 2) is similar to that of California or Alabama, but not comparable to that of Haiti or Brazil.

One should not associate periphery and backwardness ("underdeveloped" in the banal sense). Spain has known a growth during the 1950s and 1960s characterized by the parallel growth of labor incomes and productivity; a growth characterized by an increasing inequality in the income distribution (drop in actual salaries and peasants' incomes; increase in the incomes of the middle classes, which have been the main source of demand growth). From there, it is concluded that Spain was a backward center and Brazil a periphery.

It is now possible to see that the policies implemented in Africa during the 1960s and 1970s (and those proposed by fora like the World Bank, up to the present) are mainly extraverted. It is not possible here to review all the diverse experiences of Africa. Neither is it our intention to reduce them all to a simple, unique model, deprived of nuances. But bearing in mind the contrasts and differences, even giving them more importance than what constitutes their common denominator, one could at least recall that in this vast continent, which has known different social regimes and numerous political uprisings, there have been four sets of experience:

(1) The "stagnation" cases (in terms of growth), associated with a lack of natural resources which, without necessarily being absolute, is stagnant in terms of world system demand.

(2) The "stagnation" cases, in spite of the existence of such resources either potential (but well-known), or even exploited (and at times on a large scale).

(3) The cases of "relatively marked growth", at times even high, associated with the exploitation of these resources, either by the multinationals or by the nation state.

(4) The cases of "marked growth", despite the fact that the exploited resources (then often agricultural rather than mining resources) are "moderate", due in general to an extensive opening to the exterior; this "marked growth" is associated with a distribution of benefits that is more or less uneven.

One could also reorganize these experiences according to the following political criteria:

(1) Assertion of a target of national independence, at times (and even quite often) of a "socialist" aim, and effective measures accompanying this assertion at least at the level of state intervention (nationalization, at times agrarian reforms, systems of cooperatives, formal controls over external relations, and so on); in liaison (not accidentally), with general speeches (on the world situation, etc.) and obvious international alliances; or

(2) Assertion of an aim of neutral appearance of "development first", call upon capitals (predominantly Western), refusal to "condemn" the principles of capitalism, private initiative, of the worldwide strategy of the multinationals and Western states, etc.

The reader who knows Africa, even superficially, will have to confront the aforementioned criteria, and will have to identify the names of the countries and heads of states for each subgroup considered.

In the framework of this formal classification, pursuing conventional analysis of the "economic performances" (as is being done in the World Bank reports, or those of administrations, and decision centers of the national, international, public and private system), we shall easily distinguish:

(1) Some driving activities of the effective growth when it has existed: (a) oil and mines; (b) export agriculture (relatively rich—coffee, cocoa—or poor—groundnut); (c) light consumption industries managed in an acceptable way, established by multinationals or the state, modern in their techniques, responding to the internal market (import substitution); (d) a lively building sector (linked to the accelerated urbanization and to "prosperity"); (e) administrative expenditures conceived in classical terms, mimicking the West in its form and to one degree or another, so-called "social" (education first), growing at a sustained rhythm; (f) tertiary activities (such as trade and finance), nearly always more lively in their growth than the other sectors.

When the global growth has been slight—or zero, or negative—it seems attributable to an insufficient dynamism of (a) and (b) and/or to the doubtful character of (c). If, in addition, (d) and (e) have been realized, then the linked double crisis of public finances and balance of payments, ineluctably worsens the situation.

The lack of dynamism of (a), (b) and (c) is attributed either to the evident shortcomings of the country, or to its lofty "nationalism" that refuses foreign capital, a rare factor. It is, or can be, aggravated by the indifference of the élite, its "corruption", its "demagogy", etc. Is the diagnosis of the World Bank reports recognized here, and through an easy deduction, its recommendations?

(2) A continuously backward agriculture, underdeveloped, almost always stagnant (except in the products of the export sector) and incapable therefore of releasing a marketed food surplus up to the standard of the urban effective demand. In the most extreme cases, it becomes increasingly difficult for the rural world to feed itself. In the most favorable cases, statistics show a positive growth of food production, even *per capita* (but rarely more than 2 to 3% per annum in long periods for the global production of the country); a marketed surplus may rise, but it is practically always insufficient to respond to the urban effective demand which, it is true, may easily increase from 4 to 5% per annum.

These disasters or shortages are easily attributed to the climate (drought) or to the careless administrative bureaucracy of the rural world. We rarely get into the analysis of those policies which drain the resources of the rural economy. Are the World Bank analyses recognized here again?

(3) An industrial aggregate that is hardly ever the driving force of the growth, but widely the product of a response of adjustment to the latter, whose chain effects are limited: upstream—through the basic industries shortage and the weak inter-industrial integration—and downstream—through the limited quantitative character of the incomes it distributes. Three nuances can be added here.

First, if the industry restricts itself to a definite number of production units in a position of quasi-monopoly in a small market and these units provide consumption goods, this industry, efficiently managed (that is, without need for subsidies and with prices competitive with those of imports) is derivative rather than a driving force.

Second, if the state wishes to control this kind of industry too tightly, it does so badly (the African experience), and therefore in addition, must call for subsidization.

Third, some countries try to go further, and proclaim their will to make industry the chief driving force sector of the economy. Some "industrializing industries", national and integrated on the basis of a priority to basic industry, must facilitate "catching up" with the developed world. Therefore, the question is to know whether this last important nuance qualitatively distinguishes the African experience from the others. This fundamental question is rarely touched upon.

Therefore and beyond the apparent and short-term economic performances, integrating the political factor, two sets of experience will need to be recognized in Africa:

(1) *Those who have not questioned the basic external driving force* and have satisfied themselves by exploiting their international, agricultural and mining "comparative advantage," grafting on the growth of these sectors that of light industries: import substitution, services, and administration. "National" or "nationalist" nuances—the state being called upon to control, through public property, more or less important sectors of this growth; "social", not to say "socialist" nuance—the fruits of this growth being appropriated by 1 or 15% of the population do not question the extraverted character of development. The latter, qualified as "neocolonial", deserves its name, in the sense that it perpetuates the colonial development, satisfying itself in associating the local élite to it. That this type of development has been possible here, no doubt. But it has been objectively impossible for the neighboring country, that this one was, for example, a supplier of cheap labor to the first, or deprived of leading export activities less because of the nature than of the structure of world demand.

(2) *Those who have wanted to question this external driving force.* These experiences, the outcome of a political and social history of the national liberation movement, of the association of the popular masses to this movement, have endeavored to be national. But if we draw the balance sheet from them, beyond intentions, it hardly seems that these experiences (about 20 over the two decades of independence) were capable of starting the setting up of a national integrated economic system, relatively capable autonomy. The only experience that has succeeded in going far enough in this direction (that of Nasserian Egypt) has been overthrown by the conjuncture of its insufficiencies and the hostility of the West. In spite of the financial factors that enhance it, does not the second serious experience (that of Algeria) meet the same internal and external obstacles? Are not the intentions of some others (Nigeria among others and in the first instance) only verbal rather than actual? For example, if we consider the steel industry, one criterion for autonomous development, we shall be

forced to acknowledge that there is no steel industry anywhere in Africa, except in Egypt and Algeria, whose autonomy, not only through formal ownership but particularly by the outlet, has been organized. Elsewhere, the steel industry, most often an extension of the colonial trade of building materials, remains linked to the demand of largely luxury construction, and not of that of industries to be equipped. The attempt of Tunisia to imitate Egypt and Algeria has been limited by the restricted market of the country, its political choices and its illusions (tourism!). That of Nigeria, still in the project stage, would deserve to be examined more closely because of its de-industrialization through the implantation of multinationals (which destroy the small and medium-size undertakings of the country and direct national capital towards trade and speculation) comes into objective conflict with the national options as presented in official pronouncements.

This second set of experiences, the only interesting one from the point of view of the autocentered option, if it deserves to be qualified as "national" (reality considering intention) is scarcely "popular". The envisaged national industrialization was not in fact meant in the first instance to serve as an improvement of the standard of living of the poorest masses (and therefore to center itself on their actual, non-"effective" demând), but to orient itself towards the expanding demand of the middle classes. The "backwardness" of the agriculture is therefore not an accident, as the basic food demand interésts more the poor masses than the middle classes. Then the agriculture of rich speculations is preferred (fruit, vegetables, meat for the privileged) to the basic productions (cereals): one good example is Egypt, encouraged in that direction by Arab and international capital. The consumption model of the middle classes, which is the Western model, then absorbs all the said "scarce" resources: capital (local and foreign) and qualified manpower. Our estimate of its use in the Arab world leads us to the conclusion that three-quarters of this "scarce" factor (in the circumstances of the labor stock having gone through a secondary, technical or higher education), were directly or indirectly assigned to the production intended for the satisfaction of the overconsumption of the privileged. Of course, the productive system that corresponds to this option implies a total technological dependence. The theory of increasing first the size of the cake before sharing it is absurd. Highways are not built (to satisfy the needs of those who possess cars) to be distributed tomorrow by the meter to beggars. The distribution controls the nature of the cake, even more than its size.

If, therefore, the experiences in question remain extraverted that is because they are not "popular" in the sense defined here ("development for whom?"). This last characteristic common to one and the other, expresses itself through the respect of the sacrosanct principle of the "profitability". If in fact this criterion is retained as decisive (even with sporadic attenuations), if it is based (as it is) on the world price system (which reflects the international division of labor, the market sharing between monopolies, etc.), the criterion cannot but lead only to one option of extraverted strategy. Neither the political language nor the diplomatic alliances modify the meaning of this choice.

The alternative option of a national and popular autocentered development is not only possible in Africa as elsewhere—there is no absolute "technical" obstacle that would make it impossible (the obstacle is always social). But it is even objectively necessary to persist in rejecting this option to find oneself in a deadlock.

The national and popular autocentered strategy rests first on the principle of distribution of income as evenly as possible, especially between rural and urban sectors, between the modern sectors with higher productivity and the backward ones. The surplus of production over the remuneration of labor thus equalized constitutes an excess which, if it is national and retained for accumulation, permits a marked growth and a parallel and even progression of popular consumption. The structure of demand, constituted in this way, would give priority to basic needs and orient the productive system towards the satisfaction of the latter.

To illustrate this principle, let us assume through the starting situation (year 0) characterized by the predominance of the rural population (80%) and its inferiority in terms of productivity (the production of agriculture represents only 60% of the GDP). If all distributed incomes were equal to the average rural incomes (that is 60/80=0.75), the surplus of the GDP over the remuneration thus determined (60 for the rural population, 15 for the urban population) would allow, if it were not exported and wasted through the consumption of the owners, a considerable accumulation (25% of GDP), ensuring, for example, a growth on the order of 7% per annum (doubling in 10 years). At the end of a decade, the GDP having doubled and the population increased for instance by 25% (2% annual growth), the average national productivity (per capita production) would have increased by 60% (200/125), the urban population having increased from 20 to 30% of the total population.

Let us admit then the principle of a national and popular autocentered development: over the entire decade, the planning of prices and incomes will ensure an even growth of incomes for all, parallel to that of productivity. If, for instance, rural productivity increases by 40% (less than that of the entire economy), the per capita income in the urban and rural sectors will equally increase by 40% (going then from 0.75 to 1.05). The surplus assigned to the accumulation will by itself increase very heavily from 25 to 35% of the GDP.[4]

The blueprint in Table 1 illustrates this model of autocentered national and popular growth.

This type of growth pattern cannot result from the operation of market laws based on the world pricing system. The decision to regulate the remuneration of labor on an egalitarian basis, determined in various ways by the average rural productivity (equality of the cereal "ration" in urban and rural areas, hierarchy of urban salaries consolidated and not copied from that of the West), by the decision to nationalize the surplus and by ensuring its centralization and

TABLE 1.

		(1) Year 0	(2) Year 10	(2)/(1) Index
Production:	GDP	100	200	200
	Agricultural	60	92	153
	Urban	40	108	270
Population:	Total	100	125	125
	Rural	80	87	109
	Urban	20	38	190
	% of Urban Population	20%	30%	—
Per capita production				
	All*	1.00	1.60	160
	Rural	0.75	1.05	140
	Urban	2.00	2.85	143
Incomes:	Distributed income (per capita)	0.75	1.05	140
	Rural	60	92	153
	Urban	15	40	268
	Surplus	25	68	270

*The apparent paradox (the average per capita production increases faster for the national economy than for each of its rural and urban components) is easily explained: the urban absolute productivity is much higher than that of the rural sector and the urban proportion rises from 20 to 30% of the global workforce.

redistribution on a national scale. These are all political decisions which imply the separation of the system of economic choice from that which would derive from "project analysis" and from "profitability", sacrosanct principles such as exist at the World Bank amongst others.

Without entering into an illusory description of concrete details of measures necessary to implement a development pattern of this type, one can assume that:

(a) It not only implies declaration of the "agricultural priority", but also its effective implementation. This priority requires that other activities with higher productivity should not be an opportunity for the distribution of incomes that exceed those distributed in agriculture. The reason for this is that the demand is structured in such a way that the satisfaction of the needs expressed by the privileged would of necessity absorb an essential portion of the means allocated to accumulation—at the sacrifice of agriculture. It is recognized that priority given to agriculture, especially food cropping agriculture, should not be restricted to the choice of "food cropping projects". To prefer millet to groundnuts, with obvious common sense, is not at all adequate and not even practicable, if the entire income distribution policy and those related to industrialization, prices, wages, and financing (return of the surplus by the state and its reallocation to satisfy the priority in question) is not revised. Failing this, the food cropping projects will not avoid deterioration of producers' incomes, through taxation, credit, deterioration of their terms of trade, etc. Then the peasants, through experience, will resist the so-called "progress" thus proposed to them, because they know that they will not profit from these benefits. At most, priority to "food crops", if the global policies are not consequently reconsidered, is synonymous to that of wanting to produce cheap food products (at the expense of the peasants) to ensure the perpetuation of conditions favorable to the maintenance of cheap manual labor (to the detriment of urban workers and for the sole benefit of the capital, more especially through specialization controlled by the international division of labor of foreign capital). The choices are: satisfaction of basic needs, or raising of the benefit rate?

(b) It implies that industrialization first be conceived to maintain the progress of productivity in agriculture: production of adequate inputs (fertilizers, equipment, etc.), infrastructural works (irrigation, transport), preserving and processing of the produce, etc. It then follows that this industry satisfies the non-food consumption needs of the rural and urban population, on a basis as egalitarian as possible and, from this demand, again one goes up the chain of intermediate and equipment industries that condition the efficiency of production of manufactured consumption products. It is understood that we cannot give up this national industry in substituting for it through imports because imports have to be paid for through exports, and the comparative advantages are those which result from the price and income system of the world order, in conflict with the political coherence outlined above. Imports must therefore, at each step, be reduced to the minimum.

(c) It implies national and popular forms of social organization of production: the control of peasants over agricultural projects, real cooperatives (which should not be a way of draining the peasants by providing industries, depriving the peasants of their hold on production), institutions of collective bargaining to set agricultural prices, national control of industries, a national wage policy, and the redistribution of the financing means on a country scale. It is difficult to imagine how the multinationals would find a place in this organizational pattern, except to provide, in time and under strict national control, some models as to production or organization.

(d) It implies that a relation to technology be not reduced to its "transfer". It is, in fact, a question of creating an inventive capacity, not for reasons of cultural nationalism, but simply because the available techniques, especially the more advanced ones, are not neutral with regard to the range of products, the structure of demand to be satisfied (Western patterns), and the price and income structures that control the profitability of these techniques.

(e) It implies limited external relations radically different from those derived from the various industrialization strategies, import substitution or export promotion. The import substitution is based on already existing demand, within a structure of income distribution characterized by its very strong inequality; and on this basis, it respects the principles of the profitability (with at most some arguments of "moderate protection of infant industries" during a brief transition). Therefore, it cannot but displace imports towards the intermediate goods (the industrial apparatus remaining non-integrated), and sophisticated capital goods (as the demand to be satisfied, in competition with the imports, reproduces the Western capital consumption model). It thus remains extraverted. On the other hand, national and popular autocentered industry is not built in response to a pre-existing demand; but is created on the basis of the satisfaction of peoples' needs (incomes policy) and intermediate and derived capital needs. Those continuing imports aim at filling the holes, in the range of these derived needs, reducing progressively their relative importance (but not necessarily their absolute bulk). It binds therefore the external relations to the logic of internal accumulation and, in this way, merits its name. As for export industry, it is, by its very definition, extraverted. Forced to compete with the industry of advanced countries on their own territory, it must import advanced technology extensively. This explains why the NICs, which are the most advanced in that direction, are also the most indebted: the export industry does not alleviate balance of payments situation (contrary to the argument put forward to that effect)—it aggravates it.

(f) It implies the building of a national structure of interdependent pricing and financial instruments which conflicts with the very principles of the criterion of micro-economic profitability. In fact, the autocentred industry, if it must comply with the peoples' needs, must accept the juxtaposition of the very uneven productive units—modern industries, semi-mechanized manufactures, handicrafts manufactures. The unit of labor remuneration and that of prices thus entail unequal surplus. It must be redistributed so as to avoid the polarization of progress in the modern units; and, on the contrary, to finance the progressive modernization of the backward sectors with the surplus of the modern sectors. This is hardly possible on a vast scale without large public property: the private national enterprise and *a fortiori* the multinational subcompany are not prepared to give up this level of profitability. As we know, they act in a directly opposite sense: in destroying the non-competitive cottage industry, they have contributed to increasing unemployment and at the same time depriving the population of useful products.

The obstacles to the implementation of these principles are evident. But are they insurmountable? What is the nature of the most serious of these obstacles?

The obstacle of the size of each African country is evident. But it operates particularly in the field of industry, much less in that of agriculture. In fact, many of the immediate problems of agriculture can be solved through direct action at the root level, and from the root level. The solution of these problems probably requires some material means which will have to be imported if they cannot be supplied by the local industry. The obstacle, here, is less the impossibility of supplying these means, than the social framework imposed by the global strategy. If the function of agriculture is to supply a surplus for—at best—an industrialization to which it is not related or—at worst—for parasitic consumption, reliance on the mobilization of the rural energies may be in vain, as we are seeking here to break the peasant's autarky not to improve his condition, but to exploit him. The methods considered move towards this choice: large barriers and perimeters of enrolled settler, agricultural investments by multinationals, and the Green Revolution associated with large and medium-size modernized property (driving the poor peasants out of their lands), state and cooperative farms imposing on the peasants means of production and techniques which produce a growing surplus, will be preferred. These methods adjust well to the input supplying agribusiness; but they are at the same time, extremely costly and with a doubtful efficacy, because of peasant resistance. Diffused progress, based on thousands of little improvements—water management over small areas, intensive agriculture/cattle raising association, etc.—requires quite a different social framework and a global strategy which do not aim at extracting a growing surplus from the peasants. The obstacle of the size of the country operates little at this point. It is evident that some of the large African countries do not do any better than the others.

The obstacle of the size itself should not prevent the initiation of an autocentered industrialization, at least for the countries with over five million inhabitants. Undoubtedly, this industrialization could not be "complete" for these countries, and therefore it will be difficult for them to avoid a strong relative independence with respect to imports. In addition, the obstacle of the size could be reduced by an intra-African cooperation, based on the identification of planned complementarities. But in turn, this cooperation is not compatible with the formulae of "common markets" which would accentuate the inequalities instead of absorbing them.

The actual obstacle is therefore primarily social: the autocentered development sacrifices the privileged expansion of the middle classes. Thereafter, it is international in a sense; where the system of multinational interests and of developed states anxious, above all, to assure themselves the security of their raw material supply, are opposed to the autocentered option. Is it not symptomatic that the efforts of the IMF aim at participating in the destabilization of experience which will initiate this development? Is it not symptomatic that the conditions imposed by this organization always systematically sacrifice peoples' interests (wage freeze, liberalization of prices, savings on social expenditure), in spite of the declarations on basic needs? Thus, this policy, increasing the social unevenness, strengthens the deadlock of the extraverted strategy.

An external obstacle might undoubtedly be attenuated if the Third World requests for NIEO were fulfilled. Indeed NIEO is not an autocentered strategy, since it basically calls for a revision of the international division of labor in a direction favorable to the South. Meanwhile, this improvement of the means at the disposal of the South might be exploited by an autocentered development made less difficult. As it might equally be exploited by a non-popular national project, though it would be a deadlock!. This is on condition that we do not misinterpret the NIEO in the sense in which the Group of 77 understand it in their notion of redeployment of multinationals. We have expressed ourselves elsewhere on these questions. But, whatever the circumstances, the North has rejected the NIEO!

Finally, it very much appears that autocentered development is synonymous with "delinking". The latter concerns a long historical period (at least several decades). Only later, in a world modified through the strengthening of nations and peoples of the South, will we be able to speak anew of "relinking", recognizing the advantages of the international division of labor.

Notes

1. This opinion is, of course, very controversial. It would be too long to explain here the reasons why we do not believe that *now*, starting from a largely extraverted growth, through the massive import of capital and technologies, imposing an active participation of the international division of labor, it is possible to think of national development indeed opened to foreign countries, but capable of dictating itself, that is of finding itself at dominating positions in the international division of labor.
2. The diagram (see "The Law of Value and Historical Materialism", part V, the end of a debate, *Monthly Press,* 1978) links, in fact, the demand (and the incomes which determine it) to the progress of labor productivity. In a closed system, this linkage imposes an increase in labor incomes, in the absence of which insufficiency of demand blocks growth. In an open system, this linkage can disappear to emerge again at the global level of all the countries interconnected in the international division of labor. That is precisely what characterizes the present world system. The peoples' insufficient demand in the periphery does not block the system because it is compensated by the local demand of the privileged and that of dominating countries which benefit from the transfer of value associated with unequal exchange, a consequence of the unequal international division of labour. Even here, our definition qualitatively separates the countries of the center from those of the periphery.
3. Can we make here an exception concerning the oil economies? It is obvious that, if we take the word "dependence" in the ordinary sense of common language, the Northern economies "depend" on imported energy, as in fact they largely "depend" on other vital raw materials from the Third World for their development. But, this is not the point. It is located in the nature of the international division of labor which organizes the system. Really, the oil-exporting countries do not hold the dominating positions in the international division of labor.
4. This blueprint shows that a sufficient surplus to finance development is possible even in a "poor" country; the Chinese experience demonstrates this. Obviously, there are other aspects of the problem, linked to the size of the countries, to their endowment of potential natural resources, etc., which are not considered here. The argument in favor of the "opening" rests on three considerations: (1) the need for a massive appeal to foreign capital and to imported technology; (2) the comparative advantage of the specialization; and (3) the question of the size of the countries (natural resources, market outlets, etc.). The first consideration neglects the price to be paid for the import of capital and technology, and favors indolence. The second consideration is based precisely on a thesis of comparative advantages which is blind to the concept of the unequal international division of labor and therefore to the transfers of values inherent in the system of world prices. Only the third consideration is valid.

Africa in the 1980s: Some Key Political Economic Issues, Problems and Prospects

REGINALD HERBOLD GREEN

The prospect of impending disaster is not just a figment of the imagination. The facts are there to be seen, pointing to the past and to the present with an accusing finger and to the future with scornful laughter.

—OAU/ECA Symposium "Africa Towards the Year 2000", 1979

Our own reality—however fine and attractive the reality of others may be—can only be transformed by detailed knowledge of it, by our own effort, by our own sacrifice.

—Amilcar Cabral, late President, PAIGC

Output per capita rose more slowly in Sub-Saharan Africa than in any other part of the world, particularly in the 1970s, and it rose more slowly in the 1970s than in the 1960s . . . The tragedy of this slow growth in the African setting is that incomes are so low and access to basic services so limited . . . Now, against a backdrop of global economic recession, the outlook for all less-developed nations—but especially for the Sub-Saharan region—is grim.

—World Bank, *Accelerated Development in Sub-Saharan Africa*, 1981

Mistakes are mistakes.

—President Julius K. Nyerere

Shape without form, shade without colour,
Paralysed force, gesture without motion; . . .
What are the roots that clutch, what branches grow
Out of this stony rubbish?
—T. S. Eliot, "The Hollow Men"/"Gerontion"

Crises, Consolidation and Choice

Viewed as a continent or from the perspective of the majority of the citizens of most of its over 50 polities, Africa is in crises—economic, social, political and cultural. The general optimism—and fairly general progress—of the 1960s was shaken by crises in the early 1970s, only partially reaffirmed in short-lived recoveries and dashed by the continuing post-1978 crisis. Looking at the early 1980s balance-of-payment deficit, output (and in particular agricultural) growth rates, government deficits, inflation rates, deteriorating capital stock and falling standards of living and of access to basic services, an observer could well think an appropriate title for a preview of much of the continent in the 1980s was "Africa: Into Oblivion".[1]

Further, the post-1978 crises are clearly dominated by external events. The 1979–80 oil price increases, the 1980–82 primary commodity price collapses, the sudden worsening of the post-1969 industrial economy malaise (including in recent years industrial socialist economies)—all have buffetted African economies and politics as have a Pandora's box of droughts and wars. So different are the policies followed by different states and so general the worsening of the economic results that the dominance of a worsening external economic environment emerges clearly as the central cause of the present generalization and intensification of crises in Africa.[2]

Under these circumstances, it might be argued that picking out key political economic issues, problems and prospects was rather pointless—all depended on the evolution of the central industrial economies and the global economy they dominate. Equally in the context of central banks juggling—and not always succeeding—to avoid defaults in debt and gaps in petroleum supplies, governments scrabbling—also with mixed results—for food aid and funds to cover wage bills while workers and peasants first seek to find basic goods and then to find the means to pay their rapidly rising (legal or "parallel") prices—with diminishing success, one may wonder whether anything beyond short-term survival is worth projecting or possible to plan.

To the first challenge, there are two answers. Even if the causes of the present crises are external, domestic (and regional) responses are needed. Africa can hardly afford to wait for global economic recovery and the severity of the 1979–82 depression in Africa underlines the cost of continued dependent externally oriented, open economies specializing in vulnerable, low-growth prospect exports.[3] Further, all was not well in Africa before 1973. Neither in terms of output growth nor meeting basic needs, earned import capacity to finance transformations to national economic integration[4] nor availability of food, government revenue to cover recurrent spending nor patterns of income distribution, political stability nor participatory state and society building were the trends satisfactory in all, or even a majority, of countries. The ECA/OAU Symposium looking to 2000 was in 1979, before the force of the present external economic gale was apparent and its combination of sharp criticism and grim warning by Africans was just as sharp as the OECD "Interfutures" Group conclusion that barring major changes one is:[5]

> . . . Forced to envisage a continuation of the present situation of virtual economic stagnation particularly with respect to agriculture, with escalating local conflicts and external interference.

The case for devoting all attention to crisis management and consolidation is somewhat stronger. To reorganize to renew attempts to win development—however defined—does require halting the disintegration[6] that grips many and threatens more African economies and politics. However, to do that does require identifying key problems and alternative ways of grappling with them. Further, consolidation at present levels is an interim goal—its justification is to make possible more ambitious goals that require looking backward and forward to identify key issues and to highlight crucial choices.

Africa is not homogeneous. Different sets of issues, of priorities among problems, of prospects and of practicable scenarios apply to different states and to regions and communities within states. However, a series of problems apparently related to deep-seated problems that show few signs of early or general resolution are common to most African states and economies and to Africa as a region. Whether the struggles against these problems and contradictions lead to transcending syntheses (even if partial and temporary) or to continued blockages (even if some groups and states are partially exempted from the costs) will be central to the political economic evolution of Africa over the next decade at the continental, national, village and household levels.

Therefore—however unpromising the setting and however pressing the immediate practical demands on time—it is appropriate to reflect and to look ahead in terms of broad limits, possibilities and choices. The reflection can in large part base itself on the body of analysis begun by Surendra Patel's path-breaking surveys of present and potential during the highly intellectual creative founding days of the United Nations Economic Commission for Africa. Long-term trends and their implications and the greater width of the productivity gap between Africa and the North in agriculture than in other sectors remain of practical as well as intellectual significance. However, because a decade and a half have passed, there is a need to review and to reassess the future as well as the past; though futurology as an art may have made progress, as a "science" it retains the fallibility of meteorology and some hints of the methodology of astrology.

What then are issues meeting the test of overriding importance and continental scope? There are a number of alternative lists,[7] but seven isues will be considered here:

1. External imbalance/import strangulation;
2. Food supply;
3. Income distribution and basic service access;
4. Participation;
5. Legitimacy of states;
6. Southern African liberation;
7. African development coordination.

External Imbalance/Import Strangulation

In the 1980s, almost all African states are facing unprecedented balance-of-payments crises.[8] In the 1960s and the first half of the 1970s, trends varied but, in retrospect, tended to worsen. The 1973–75 (food and oil price plus global inflation) crises were surmounted[9] at very considerable costs in a majority of

African states; and 1976–77 appeared to mark a soundly based recovery. That appearance was false; in many cases, the better export performance depended on the short-lived beverage price boom; the cost of adjustment in deferred maintenance (i.e., capital stock deterioration), in loss of momentum and in use of leeway (e.g., external reserves, borrowing capacity, internal credibility, people's ability to make sacrifices) was not fully recognized and left Africa ill placed for renewed external batterings; the second oil price shocks (1979–80) and the sustained industrial economy stagnation since placed greater pressure on Africa than the events of 1973–75.

Combined with very slow rises of net real resource transfers—whether on concessional or commercial terms—these shocks have led to reserve exhaustion; radical cuts in real imports and real public spending (on basic services and on capital formation); accumulation of an array of burdensome short-term debts, deferred payments and delayed asset maintenance and repair (all mortgaging future export and government receipts); and to what, for a growing majority of economies, can best be characterized as import strangulation (symbolized by manufacturing sectors operating at 25–35% of capacity, versus past achieved levels of 65–80%, because imported raw materials and spares to complement local labor and inputs plus existing plant are simply not available).

Import strangulation is not a shortage of imported consumer goods, a limited variety of staple foods and/or limits on capital goods for new projects. Those are the necessary consequences of serious national economic integration strategies (e.g., in Tanzania), as well as of overly ambitious spending programs (e.g., in Nigeria); of socioeconomic reconstruction efforts (e.g., in Mozambique) as well as of long years of economic drift (e.g., in Ghana); of serious attempts at dependent development (e.g., in Kenya) as well as of attempts to protect selected subclasses amid a disintegrating polity-economy-society (e.g., Zaire). Rather, acute import strangulation (as exemplified in Tanzania, Zambia, Ghana, Kenya, Guinea-Bissau and Mozambique as well as Zaire and Uganda) comprises:

(a) earned import capacity (export purchasing power) 35 to 60% of actual and 25 to 40% of the minimum necessary (to operate the existing economy near capacity) imports;

(b) fuel and food import bills totaling over 50% of earned import capacity;

(c) inability to import adequate inputs (e.g., processing plant and transport spares as much as fertilizers, pesticides, implements) to sustain production, procurement and movement of agricultural exports;

(d) parallel inability to import inputs (including transport, spares, fuel) to operate existing—let alone newly commissioned and pipeline—manufacturing capacity at or near normal 1960s and mid-1970s utilization levels;

(e) inability to import equipment and materials (e.g., bitumen for road sealing) to halt deterioration of infrastructure or enough basic service inputs (e.g., paper for textbooks, spares for rural water systems, basic drugs for clinics) to maintain plausible quality and availability levels in education, health, water and transport;

(f) inability to import fuel, vehicles, construction materials to provide physical capacity for either the public or private sector to procure, transport and store staple foods—a constriction greatly exacerbating the food supply problem (and the "need" for food imports) by deterring production, blocking stocks in isolated areas, raising levels of wastage and spoilage;

(g) consequential on loss of exports, imports and domestic manufactured output, dramatic shortfalls in government revenue (especially where fiscal reform has centered revenue on sales and company taxes on domestic industrial goods and enterprises) leading to unprecedented recurrent budget deficits (often above 5% of Gross Domestic Product) with massive inflationary impact (e.g., early 1980s annual cost of living increases running at over 30% annually versus an average of around 10% in the 1970s).

These constraints do illustrate how externally dependent and open African economies are. But they are not limited to states which have failed to seek domestic economic integration based on producing a broad array of products with significant inputs from most parts of the country. Indeed precisely those states that began such a process in the late 1960s and 1970s are now most severely affected. For them, imports (beyond present achievable earned import capacity) are still needed to operate their economies near installed physical capital, human skills and institutional capacity. Thus cutbacks are necessary. But because imports are now largely inputs into domestic production, not final consumer goods, the negative multiplier impact on domestic production is far worse than in the pure unprocessed primary export/manufactured good import economy. For example, in Tanzania a \$1 cut in imported spares and materials for the manufacturing sector reduces domestic value added by \$4 (wages, local inputs, use of capital stock), tax revenue by \$2 (sales and company tax), investible surplus by \$2 (as operating profits turn to losses when capacity utilization falls) and availability of manufactured goods at retail by over \$8 (ex-factory \$5, indirect tax \$1.50, distribution \$1.50). Similarly, large states that have achieved agricultural growth in widely spread areas are most affected by the inability to maintain and operate transport infrastructure and vehicle fleets while, within them, the isolated area peasants who have lost effective access to the market achieved in the 1960s and 1970s are the hardest hit.

The largely external causation of this crisis—at 1972 or 1977 terms of trade, it would not exist for a majority of African states in the form outlined above—does not alter the need for national action. As Tanzania's Finance Minister Amir Jamal, among others, saw clearly in 1970, the reality was that any serious development strategy in Africa requires an increase in absolute import levels, even if it reduces their ratio to total domestic production. The only import capacity fully under African control is earned import capacity—that derived from exports. A country unable to keep earned import capacity growth in line with basic import requirements plus normal sustainable net loan and grant levels delivers itself bound hand and foot into the hands of its creditors and donors. As President Nyerere put it over 1980–81 (reversing two decades of absence, or merely verbal formulation, of an export strategy), adequate export earnings are a precondition for self-reliance and in the present crisis the choice is "export or die". "Delinking" from exports first is not a strategy for self-reliance but for penury and mendicancy.

This is not a case for returning to an unprocessed primary export-led growth strategy—as even the World Bank is now ready to admit.[10] That approach is at best short term and provides possible conditions for internal production and use pattern reconstruction necessary to sustain development (or even growth). The record of Malawi—now entering a nil export growth, import strangulation phase—illustrates that.[11] But exports—old or new primary (e.g., cocoa rehabilitation in Ghana, perhaps soy beans in Tanzania), raw or processed (e.g., leather for hides, cocoa powder-butter-paste for beans), promising natural resource based (e.g., coal, fertilizer from natural gas, pulp and paper from softwood and/or sisal)—must be raised significantly and rapidly; or overall national and regional economic integration concepts like the Lagos Plan of Action[12] will remain nonimplementable, mocking visions of what might have been.

In the short run, export growth and import capacity allocation to minimize damage (such as shifting from new projects which cannot be operated to maintenance, rehabilitation, operation of existing capacity) will not be enough to halt the slide into economic oblivion in a number of African states or to rescue those—such as Ghana and Uganda—already up to their necks in political economic quicksands. More effective and better used resource transfers are needed. The chances for rapid rises in total real concessional transfers are poor[13]; most African states neither have substantial access to commercial finance nor can they afford it at present interest rates and repayment periods. Therefore, better allocation—e.g., to fuel and bitumen and road maintenance equipment and away from new highways; to spares and inputs for existing core manufacturing capacity and away from building new factories to swell unutilizable capacity; to basic drugs and dressings and vaccines for rural clinics and preventive medicine and away from monumental central hospitals—must have priority. Of course, selected new investments are vital to raise exports, to develop domestic power sources (including seeking to locate oil where plausible), to clear bottlenecks inhibiting domestic food production, procurement and storage. But ending import strangulation and halting falls in output first means concentrating on capacity utilization and maintenance; where non-vital consumer goods imports are already gone, the only way to achieve that is to reduce imports for new fixed capital creation until an at least temporarily sustainable overall external and internal balance position has been won back.

Food Supply

Most Africans do not have enough to eat. The average calorie deficiency is at least 10%; for the poorest 40% of the population, it approaches 25%. Most African states import substantial amounts of staple foodstuffs.[14] On the first count, the situation is at best improving very slowly in most states and worsening in some, albeit the 1960–80 record varies widely, e.g., +16½% per capita in Tanzania, +6% in Upper Volta, −17½%in Ethiopia.[15] On the second, deterioration, not progress, is the general pattern.

Production failure has been attributed to many causes. The Sahel and Ethiopia, Somalia and Tanzania, the Mahgreb and Zambia have been affected by severe droughts. Funds for seed or fertilizer or extension services have often been lacking. Land needed for staple foods has been under export crops—for example, in the Senegal groundnut country. Prices to farmers have been too low.[16]

Arguably, the main present constraints are both not on the farm and not centred on overall prices received by farmers.[17] They turn on:

(a) failure of procurement, transport, storage and distribution to adapt to rapid urbanization, longer weather cycles (requiring larger reserves) and higher transport costs;

(b) inability of farmers to reach buyers (public or private) who can and do pay promptly, in cash and/or to reach shops with goods to sell;

(c) worsening water and fuel (firewood) availability increasing non-crop production work for women and interacting with weak and slowly growing (or worsening) health and education access to create a strong negative incentive toward urban drift as well as weakening (often literally) peasant ability to produce;

(d) failure to understand or interact with peasant perceptions of what needs to be and can be done, leading to even viable improvements (e.g., animal-drawn implements, improved seed) not being adequately field tested, extended or taken up.

But, valid as they are as parts of the explanation, these cannot be the whole. Nigeria's per capita, perhaps even absolute, food production has been on a downward trend for 15 years. Kenya's per capita domestic food production has been drifting downward for over a decade. In the Fleuve and Casamance, Senegal has plenty of land to raise food production dramatically. Nor are the cases of trend growth of food production above population growth—for example, in Malawi and Tanzania—marked by uniform absence of drought or presence of very effective extension services or supplies of inputs.

Low availability of food to the hungry also has varied particular causes. Money for commercial imports or state-financed drought relief was lacking, e.g., in Senegal in the early and mid-1970s. Physical and organizational ability to distribute food to the worst hit rural areas has not existed; Ethiopia both before and after the revolution is one example of this.

Again, true enough, but two 1970s "success" stories on distribution of food to drought-affected rural areas—Somalia and Tanzania—were marked neither by high levels of foreign exchange nor high rural incomes nor by strong general transport systems and still less by general ability to organize and operate all state activities quickly and efficiently. The difference appears to have been in the priority given to averting starvation which resulted in prompt response, concentration of organizing capacity, allocation of resources (all of Tanzania's opening 1974 foreign exchange reserves) and rapid distribution.

One general and growing weakness is urban food supply. Peasant systems cannot handle transport or storage for urban markets. National (public or private) systems have transported and stored increasingly badly as urban populations have grown. A glaring case is Tanzania which lost (rotted, turned into poultry feed, exported at a huge loss) 500,000 to 600,000 tons of its 1976–78 food crop surpluses because of totally inadequate storage strategy, policy and facilities. That figure is roughly equal to 1979/80 through 1981/82 maize imports for the three bad weather crop years following the 1976–78 run of good weather crop years.

If the 1980s are a rerun of the 1970s, then for most of Africa SWAPO's comment[18] will be the best possible result:

> A nation which is dependent on other countries for the food consumption of its population cannot but be a dependent hostage of the particular country which feeds its population.

Best, because for many Africans the result will in fact be death from the diseases of extreme malnutrition or literal starvation even if they are virtually within sight of ports where food could be landed, as in the Bas Congo today.

The bottom line of the challenge of production reads: productivity, rural/urban terms of trade, income distribution. That of availability reads: income distribution, decision-taker priority to basic human needs, effective storage and transportation systems.

For production, adequate prices—to the grower (on time, in cash, at a point he can reach), not just paid by the consumer—and a chance to earn a decent income and to receive basic services (education, health, water, enough law and order to be able to go about his concerns in safety) are vital. Better marketing and distribution (including something to be bought by the rural crop seller) are critical. Often knowledge and inputs must be made available (and first discovered) for really large, sustained productivity gains. All of this is in the end rather boring, except during a food crisis, and requires sustained hard work—not just during a food crisis.

This partly depends on participation and legitimacy. If peasants win a say in states, they do demand attention to agriculture and rural services; if a state in Africa is seriously seeking legitimacy, it can hardly be totally unconcerned with rural welfare or food availability.

The uses of adversity may also help. Shortages of foreign exchange—even in major oil exporters like Algeria and Nigeria—do concentrate the political and official mind on agricultural production and, less certainly, on those who do the producing. There will be more and more prolonged foreign exchange crises in the 1980s.

Income Distribution and Access to Basic Services

While the formal data are at best fragmentary,[19] both they and observation reveal very wide disparities of income within all African states. More critical for future prospects is that, in a majority of cases, these are widening. In terms of the ratios between the top 1 or 2% of salary earners, minimum wage earners and peasants, Africa has probably already "achieved" the greatest relative disparities of any continent.

This is not a set of phenomena characterizing only the successful capitalist road states such as Kenya and the Ivory Coast, the newly oil rich without coherent strategies to raise the incomes of the masses, such as Gabon, Nigeria, or the states *en voie de désintegration*, such as Zaire.[20] It is also true of those deeply concerned with peasant welfare, e.g., Botswana; those seeking to build non-capitalist mixed economies, e.g., Zambia; and those engaged in serious efforts to achieve transitions to socialism, e.g., Algeria.

Accompanying and interlocking with the growing inequality are two failures: failure to make significant progress in raising the average incomes and purchasing power of the poor majorities of peasants and "informal" (residual?) sector non-agricultural workers. Failure even to develop a really broad and growing middle class of peasant proprietors, small businessmen, professionals and officials with a real stake in the system is all too common. Neither the Maoist principle of "all boats float higher" broad front advance nor the "equality of access to the elite" principle of the creative variant of capitalism can credibly be asserted to operate in more than a handful of states. The result is the grim realities of *Petals of Blood*[21] and of the stark, wasteland dump for people that is Pekine—Senegal's second largest city.

There is nothing inevitable about this. Tanzania has reduced income inequality and has made systematic advances toward basic human needs fulfilment,[22] even in a context of very low levels of atttained productive forces and relatively modest (even if well above the African average) rates of growth. However, since 1977, terms of trade shifts have reduced real national command over resources by 15% and war costs by 3–4% more than offsetting physical output per capita rises so the result is "all boats float lower". The same can be said of Mozambique. While Algeria has had growing inequality—and a very open dialogue about the threat it poses—it has also made progress toward meeting the basic service access needs of the majority of its peoples, albeit not their capacity to produce and to earn. It also moved toward creating a broad middle (petty bourgeois) class, even though this was in fact an undesired byproduct of national solidarity and key sector surplus generation promotion and now poses very severe tensions in terms of resource allocation and incomes policy.

The conditions for reversing these trends are stringent. Ceilings are needed as well as floors—given the resources available in Africa real luxury for some necessarily means real deprivation for many. Rural strategies designed to promote production and raise incomes, not extract surplus and control peasants, are equally essential. (And vital if the food "problem" is to be surmounted.) So is breaking those coalitions of elites that amount to transnational alliances (TNCs and local political/bureaucratic/managerial partners) against the workers and peasants.

Provision of services—health, education, water, housing—is, in principle, easier. But in fact it will not be done without the more basic changes just cited. Priority allocations of funds and personnel will not happen—it is Tanzania not Kenya which has achieved at least 60% adult literacy and has millions of adult continuing education students.

These requirements interact with the crisis resolution requirement of halting decline. Redistribution out of falling national purchasing power imposes massive social and political as well as economic strains and cannot be pursued indefinitely: in 1982, Tanzania had to abandon its 1980–81 policy of raising minimum wages parallel to inflation; Mauritius had to reverse (ironically under its first socialist government) a two-decade-old policy of real wage enhancement. Basic services are among the main victims of import starvation. Precisely because elites have also been hit badly in a majority of African states, positive redistribution in the 1980s will have to be largely out of renewed growth (starting with food production growth and the peasant incomes it generates), not out of existing, let alone falling output.

Two forces may lead to change in the 1980s. The dictum of Anton Rupert, the Afrikaaner magnate, "If they do not eat, we shall not sleep", may become an item in the political economic strategy of the less poor, capitalist road states, such as Gabon, Ivory Coast, Kenya, Cameroon, Nigeria. The uses of adversity may enforce such strategies on states with poor growth prospects or results, such as e.g., Zambia, Upper Volta, Mali, Mauritius, as a condition for the survival of legitimate, functioning polities.

Participation

The OAU/ECA symposium grimly warned:[23]

> the issues of freedom and justice can no longer be left in abeyance. Only yesterday, the birth of a State that respected basic freedoms was one of the most important demands in the struggle for independence. Has this erstwhile dream now turned into a nightmare in which repression and censorship condemn whole generations of Africans to silence?

and again:[24]

> The problem of underemployment and unemployment is more and more serious . . . to place employment, in other words the social usefulness of men and women, at the centre . . .

The no-party state, increased repression, growing slums and *bidonvilles*, peasant retreat from the State and the market—these are the common coin of African political realities. Without participation in two senses, they will continue to be: first, participation in production at high enough levels of output and fair enough levels of remuneration (whether wages or agricultural prices) to be both economically self-reliant and economically important to the state; second, direct participation in decisions affecting peasants and workers in their villages and neighborhoods, their workplaces and primary markets and indirect participation through leaders chosen by and responsible to them in national policy-making.

Without such participation access to food and to services and to income distribution cannot be tackled. There will be no effective discipline imposed either on leaders or on TNCs and no real priority given to decent rural incomes or to universal access to basic services.

This is not an area in which great success stories can be cited, in Africa or anywhere else. The problems of elitism, bureaucratism, isolation of leaders, arrogance of officials, greed of the rich are deep rooted, as is only too evident in the states in which they have been struggled against, held at bay, perhaps turned back in part, such as Mozambique, Cabo Verde and Tanzania; and those in which the struggle has been made and has the backing of sincere and dedicated leaders but may well, in the opinion of some not unsympathetic observers, be being lost, such as Angola, Zambia, Botswana and Algeria; and those in which, despite efforts, it seems to have been lost, such as Somalia, Ethiopia and Guinea-Bissau.

The necessity for participation (like that for legitimacy) does not turn on particular forms. Both multi-party and one-party states can (and do) exclude and can (and in some cases do) include. What is essential is that all citizens can be involved, and all leaders can be held accountable (including by challenge and removal through elections).

The question for the 1980s is whether necessity will enforce the freedom to participate beyond a handful of states. Perhaps. The end of military rule may force the attention of the returning political figures to be less exclusive than before, e.g., in Nigeria. The 1980–82 Ghanaian experience is hardly a happy omen, and that of Uganda over 1979–82 is at best mixed and problematic. The fact that bourgeois democracy cannot live without a broader participatory base (on production and on decision-taking) than now exists may be acknowledged by present leadership in a few states, such as Nigeria, Tunisia and Senegal. The liberation movements in power may be able to hold to their genuine commitment to and practice of participation; Mozambique points that way as, despite adversity, does Angola; but the same cannot be said of Ethiopia and the Guinea Bissau experience is, on any reading, highly problematic. States seeking to reduce dependence or foreign exchange crisis through food and basic services strategies may see that production and decision participation are instrumental as well as moral necessities. An example is Algeria's return towards *autogestion*, decentralization and rural development, triggered by 1972–74 explosions in the volume and cost of agricultural imports.

Three parallel cases can be made for participation. The first is that without it sustaining legitimacy becomes even more difficult, and sustaining stability requires increasing levels of coercion that (quite apart from normative considerations) African states cannot afford or sustain for very long. The second is that people who participate in making decisions and choosing goals work harder and more productively toward carrying them out. The third is that, contrary to "expert" and bureaucratic opinion, African workers and peasants usually do understand many elements of their human and economic condition and what needs to be done to alter them better than "outsiders" (including at least some African elite and bureaucratic or commercial structure members). Participation is no guarantee of absence of errors; peasant and worker perceptions may need broadening or deepening (though not dismissal as stubborn ignorance or "false consciousness" as right and left authoritarians advocate and, in power, practice); but state and private enterprise decisions are also hardly free from error or more attuned to the low-income African household's interests.

President Nyerere summed up the challenge in his 1979 keynote speech to the World Conference on Agrarian Reform:

> If the people are to develop they must have power. They must be able to control their own activities within the framework of their village communities . . . At present the best intentioned governments—my own included—too readily move from a conviction of the need for rural development into acting as if the people had no ideas of their own. This is quite wrong . . . people do know what their basic needs are . . . if they have sufficient freedom they can be relied upon to determine their own priorities of development.

Legitimacy of States

There are not two types of polities, economies, societies in Africa—*en voie de développement* and *en voie de développement dépendant (ou de développement de sous développement)*. There is a third—*en voie de désintegration.*[25] These are states whose basic legitimacy is either in question or is no longer in question because it is a fading memory from the past.

This category is not limited to the vicious personalized dictatorial systems that earned fairly universal opprobrium: Nguema, Amin, Micombero, Bokassa. It also extends beyond cases in which there is no functioning state—oppressive or otherwise—for example, Chad. And it includes cases in which a capital and a tiny elite are propped up by a revenue earning enclave and its foreign *de jure* or *de facto* external proprietors and friends, while the national economy, society and polity approach deliquescence—for example, Zaire.

Worse, there are dangers that more polities will enter *en voie de désintegration.* The reasons may be external—copper prices and the Rhodesian rebels drove Zambia to the brink; the heritage of Secretary Kissinger's attempts to make Angola "SideshowII" to follow Kampuchea and the continued South African intervention create the same danger for Angola. They may be internal; in 1965, the Nigerian elite had so thoroughly made the misuse of public office for private gain the governing ideology that the Sardauna's regime had lost legitimacy. Mauritania abandoned its claims to the southern zone of the Sahara Democratic Republic because it stood on the brink of losing legitimacy at home. The present Moroccan state's legitimacy is now so intertwined with its annexation of the Saharan Democratic Republic that it is hard to see how it can extricate itself from a war of colonization it can neither win nor carry on indefinitely without, in either event, facing grave peril to its ability to sustain the burden of legitimacy at home. The degree of participation and the degree of material advance of the Ethiopian revolution in its central zone are placed in peril by the fact that it has not been able to achieve legitimacy in the eyes of the people of the Haud, the Ogaden, the Awash and Eritrea. Until it does, it is forced to live under siege and with a siege mentality. In Ghana, the cumulative errors of three civil and four military governments have both brought the economic and social fabric to the brink of collapse and eroded the legitimacy of the state. Whether Lieutenant Rawlings's second populist military coup is likely to be part of the solution or a further deepening of the problem remains to be seen. Liberia's "progress" from politically and morally bankrupt elitism through a populist military coup with a radical intellectual fringe toward a right-wing authoritarian neo-elitism suggests that populist coups in conditions of economic chaos do win a temporary "suspension of disbelief" in government legitimacy, but one which is very difficult to use constructively to reestablish positive legitimacy.

Legitimacy does not require full participation or socialism or even full bourgeois democracy. The Ivory Coast, Senegal, Swaziland, Tunisia and Togo are legitimate; until 1982, Kenya was as well, though more doubts arise now. The test is a state that has, acts on, produces some results from its policies and that is responsive and responsible to substantial domestic class coalitions, not to a handful of power holders, entrepreneurs of adversity, and foreign backers or creditors. One element is the ability of ordinary people to go about their occupations and daily lives in some security (from state officials and bandits alike). Law and order is not everything: its total absence usually guarantees there will be nothing else.

The slide of a significant number of states to the edge—or beyond the edge—of illegitimacy, as perceived by their people, is chilling. If it is not reversed the "scornful laughter" warned against by the OAU/ECA symposium will triumph and only the vultures (winged and otherwise, domestic and foreign) will have cause to exult. It is very hard to claw back. The post-Amin course of Uganda is arduous and perilous, the desperate efforts to recreate the Chad look all too like trying to water its deserts with a teaspoon.

Two strands can be picked out. The first is solidarity with those in danger: helping Zambia break the downward spiral triggered by Rhodesian and copper price assaults; exploring ways in which Morocco can extricate itself from the

Sahara and revolutionary Ethiopia can overcome the heritage of hatred it has been bequeathed in what were the Imperial Outlands. The second is more effective waging of the liberation struggle against South Africa and its outriders; only that can avert a series of mortal challenges to Zimbabwe, Mozambique and Angola, Swaziland and Lesotho, Botswana and Zambia.

Southern African Liberation

The course of the Southern African liberation struggle is critical for Africa in the 1980s (and perhaps for the nature of great power relations in the 1980s too). That view is shared by the trilateralist intelligentsia,[26] by the Front Line States, by the Liberation Movements.

This is evident enough for Namibians, Zimbabweans and South Africans. It is also clear enough for the Front Line States which face armed raids, security costs, economic dislocations that can run as high as 25% of fixed investment and 5% of national output for "backup zone" ones like Tanzania and Zambia, and much more for those with recurrent forays and regular destabilization on their soil like Angola and Mozambique. For Africa as a whole it may seem less clear; not every state speaks or acts with the force of Algeria and Nigeria.

It is critical to African self respect. The continued existence of compulsory, legislated racism at the core of the ideology of states in Africa diminishes all Africans. Its ending would demonstrate that African states could act and achieve something substantial in a cause beyond personal or national material gain.

Liberation is equally critical for external respect. It is a brutal fact that the "accusing finger" the OAU/ECA Symposium sees is also seen by many who were the friends of Africa and Africans. It has cooled their desire to stand in solidarity, placed weapons in the hands of Africa's enemies, turned perceptions and phrases of hope into hollow mockeries. Just as Amin's perverted use of the slogans of self-reliance and struggle demeaned and degraded the image of all Africa, so the first steps toward the liberation of Uganda and Ugandans by Africans has helped restore dignity. Southern Africa is a similar struggle and a continuing one despite the landmark victory of Zimbabweans in 1980.

Finally, many African states and Africans deeply need major successes to restore their own hope and faith and will to work for victories at home as well as externally. The years 1974–82 have been grim for most; the dogged battle against calamity is less daunting to maintain when there are clear battles won.

Battles—not the campaign. Namibia must be genuinely independent long before 1990, or the cost to Africa does not bear estimation. But South Africa will not be "Azania" by then; the process may have taken fitful first steps, but the last laager will not have been broken. Nor will the independent states of Southern Africa be fully free from dependence on the Republic despite the dynamic toward economic liberation begun by the Southern African Development Coordination Conference (SADCC).[27]

What is to be done? As President Nyerere has warned, "Killing people is not a sacrament". But it is an inescapable fact that only the dynamic counter violence of liberation will bring the static violence of repression to serious negotiations for surrender of power when the power to be surrendered is vital to the repressor. Algeria, Mozambique, Angola, Guinea-Bissau, and the Sahara Democratic Republic underline that.

International pressure is not a substitute but a consequential complement. Until Africa has proved it can and will win by force if it cannot win by negotiation, there will not be effective international pressure—only sinister charades like the "Tarbaby neutrality" strategy of Secretary Kissinger and the Reagan administration's quest for closer relations with South Africa; brutal sellouts like the Vorster-Duncan-Benn arrangements for Rossing uranium in Namibia; half-serious, half-maimed ventures like the "Gang of Five" proximity talks on Namibia (and even those are possible only because the Peoples Liberation Army of Namibia has at last been striking effectively and frequently).

This is not to say political consciousness is not the key—without it there will be no sustained guerilla wars. Nor is it to imply that the tactics used in Angola, Mozambique, Zimbabwe and Namibia are appropriate to South Africa. A more urban and less transborder struggle appears inevitable there. Still less is it to advise lowering the pressure for international action and negotiation—quite the contrary. The goal is not Armageddon or war to the last wrecked house, burnt field and dead combattant; it is a negotiated transfer, as one had in Mozambique and Algeria. It is to warn that those who hold the triggers of power in Windhoek and Pretoria—like those in Lisbon and Salisbury before them—will not negotiate unloading them until they see the barrels of African guns much nearer and more precisely zeroed in on themselves than they do today.

African Development Coordination

"Small nations are like indecently clad women, they tempt the evil minded," said President Nuyerere some years ago. "Consequently, African unity is not just a slogan, a pious dream or an irresponsible ambition. It is a necessity," concluded the OAU/ECA Symposium.[28]

The questions are why has the past been marked by debacles? How can African states move forward? The answers are probably interlocked: the debacles have largely been in respect to partially colonial inheritance schemes with only slightly modified *laissez faire* cores operating in a continent of independent states which—would be capitalist and would be socialist alike—were committed to national economic intervention.[29] If that is the right reading, then ECA's continued promotion of preferential and free trade areas and ECOWAS's attempt to float what remains in the last analysis a complex, imported, European model integration scheme[30] may be steps in the wrong direction.

Coordination must rest on durable common interests perceived by participants as accomplished better together than separately. These need not—one is tempted to say under African conditions usually do not—include generalized free trade. They may encompass coordinated production of specific goods which will lead to trade, e.g., Nigeria-Benin, or coordinated use of specifically identified otherwise idle productive capacity through trade, e.g., the Mozambique-Tanzania agreement; but both of these approaches are very different from, and likely to be hampered by, accepting standard customs union theory.

Further, the concerns may not center on trade at all, at least not at first. The initiatives toward a Southern African Development Coordination dynamic launched by the Front Line States, and broadened to all nine independent Southern African States in the 1980 Lusaka Declaration "Toward Economic Liberation", are an example.[31] The key goal is reduction of dependence on the Republic of South Africa. The first priority is delinking transport and communication by creating a regional network. Second is enhancement of national and regional food security—to end both hunger and 500,000 tons a year grain imports from RSA, and coordination of industrial production to broaden the range of goods available within the region rather than only from outside it. A subsidiary goal is mobilizing finance to do it. Then come coordinating exercises of "permanent sovereignty" *vis-à-vis* mineral companies, balanced trade agreements, coordination of production, no one of which is seen as necessarily having more urgency or priority than the others.

This is not an accidental incoherence of approach; it is a deliberate choice of unconventional wisdom. Chairman Quett Masire underlined this in his concluding remarks to the initial Arusha SADCC in 1979:[32]

> Until our economies are free from the domination of the Republic of South Africa and linked regionally with one another we cannot go forward with assurance or in safety. We must ensure that the efforts of our people to achieve development, to meet their basic human needs are in a setting which gives them the greatest measure of success. That setting is Southern African regional development coordination . . . we have affirmed our determination to set out on the long journey regional level . . . we believe that we have taken that first step and have taken it in the right direction.

Clearly the key goals and routes in West Africa, Central Africa and the Maghreb[33] are not the same as in Southern Africa. The lack of any single overriding need like that confronting the FLS to loosen ties that bind to RSA may, ironically, prove a disadvantage for establishing effective coordination. But more serious reflection on what precise common interests were and how they could be met—even if this led to unconventional models more like the Andean Pact than CARICOM[34] or SADCC than EAC—would serve other regions and regional organizations better than slightly tarted up repetitions of past models. The 1980 Lagos Summit of the OAU may have been a step toward recognizing the need to act as well as talk about economic regionalism and to do so in ways more attuned to African needs and realities and more liberated from "off the shelf" import models.[35]

Paths Not Taken

A number of standard "problems" are noticeable by their absence from this short list. This is largely because they seem to be unsatisfactory ways either of posing or of tackling the problem. Five "old reliables" of this type are: ethnicity;[36] production;[37] neocolonialism;[38] efficiency;[39] socialism.

Ethnicity covers a multitude of evils—and goods. A border war, a clan skirmish, a national "objection" to marginalization in a multinational state, regional loyalties . . . It is very often a manipulated—even if quite real—consciousness. Legitimacy and participation are better ways of posing the challenge—and especially of meeting it. So is Southern African Liberation if the "problem tribe" in mind is the Boers.

Production in the abstract begs too many questions. Assuming food, income distribution and participation can be achieved, then production of the goods and services most needed is likely to follow. Resources will remain scarce, but that is precisely the reason it is not helpful to call for more production without specifying what? for whom? why? how? where?

Neocolonialism—or its more advanced form, dependent development—is very real. But to talk of tackling it requires posing questions about states, decision-takers, participation, income distribution. Otherwise one at most chips away rough corners—Kenya is quite good at tough micro negotiation—or punches a feather pillow—the internal logic of the Zairean state precludes tough external economic negotiation. The context of change and who benefits are critical to any meaningful attempt to alter concentrated, unilateral external dependence.

Efficiency is a goal. But it is a dependent or instrumental goal—efficiency in the pursuit of some basic, substantive goal. It is not copying European management, despite McKinsey and Co. and many other consultants and technical assistance experts[40]—nor European social democracy's smooth running state corporate bureaucratism, despite Gunnar Myrdal and Jan Tinbergen. It cannot be discussed outside the objective correlative of a time and place and of a set of decision-takers with time, resource and accomplishment constraints.

Capitalism in Africa is most unlikely to be African capitalism. Adjectival socialisms are usually variants of fascism. There is little to suggest either President Nyerere or President Touré judged wrong in these assertions, even if this suggests that what now exists in Guinea deserves the appellation Guinean Socialism. But that is not the same thing as arguing that the form of the present challenge in most African contexts is planning socialist revolutions. In the first place, the six challenges cited would—do for that matter—confront African states in transition to socialism just as starkly (if perhaps somewhat more manageably) as they do those seeking a transition to domestic capitalism. In the second place, if socialism is necessary for development, then facing food, income distribution and participation squarely will demonstrate the limits of dependent capitalist development and create the conditions to launch a coherent, mass-based transition to socialism. Especially in much of West Africa those conditions simply do not exist today and will not be created by intellectuals' affirmations. Nor in many of the *voie de dèsintegration* cases is it easy to envisage a direct transition to a socialist road path without at least an interim national reaffirmation/reconstruction stage. In this respect at least, President Obote's 1980 shelving of his own pre-Amin "move to the left" was almost certainly a correct reading of necessity.

Roads Almost Walked

Reflection on priority lists always raises doubts: Should not more points have been added? The problem is that to decide yes in too many cases defeats the purpose of selecting priorities. "Candidates" in this category of "almost included" comprise personpower development, technology transfer and development, transnational corporations and the state of the world economy—all of which are perhaps surprising omissions both in a volume reflecting from the work of Surendra Patel and in terms of much of the present author's own writing.

Personpower development is no longer a crisis area in Africa. Certainly further progress—especially in respect to middle-level personnel who are relatively much scarcer than high level and therefore greatly limit the achieved efficiency of the high-level personnel—is needed. But much has been achieved and in most cases continued effort and gradual rethinking of programs, approaches, institutions—not total structural and directional changes—would appear appropriate.

Technology transfer and development are important. But in most African states and economies it is a short list priority for the 1990s not 1980s. Until a larger mass of experience and personnel—and a greater use of even moderately advanced technology—are built up, a massive concentration on this area, as opposed to sustained, low key build up of assessment, operation, negotiation and research/development capacity and experience, would probably not be appropriate. Further, until there are clearer goals with respect to production and service patterns, it is hard to work out appropriate technology strategies to support them.

TNCs (Transnational Corporations) are certainly present and influential in Africa and in its external economic relations. They are certainly not benign, selfless agents of progress. Equally certainly in some sectors and activities they are necessary (a necessary evil perhaps, but one with no better practicable alternative). Therefore, the questions are ones of negotiation, control and selection/rejection. These ultimately turn on the political and economic strength of the African states facing TNCs. Therefore, crisis resolution, food supply, income distribution, participation and legitimacy are preconditions for making major progress in African/TNC relations (at least progress from the viewpoint of African peasants and workers). Technical expertise is needed but without firmer political and economic bases the best informed and skilled African negotiators will still be quite unable to secure equitable bargains because their options are so limited and needs so urgent contrasted to those of TNCs.

World Economy issues are important to Africa—as emphasized in surveying the present crises genesis. African strategy, however, must primarily aim at reducing external dependence and vulnerability by broadening options and domestic sources of strength (national and regional economic integration) and at securing crises bridging resource transfers. The depth of a global recession is not the most promising time to negotiate global economic restructuring, nor do African states (except in broader contexts such as the Group of 77 or Non-aligned) have either the economic power or expertise to play a leading role in formulating and pushing through such changes. This is not to argue against continuing to think about and press for such changes—especially in South-South and North-South groupings which can cumulate the strength of their participants—quite the contrary. It is to warn that the 1974 New International Economic Order proposals really proposed restructuring the 1945–69 world economic order 5 years after it had begun to disintegrate and are simply not relevant to the present new international economic disorder while Brandt-type mutual interest (global Keynesianism and fair business practices) proposals, though relevant to the present crises, do not appear to be politically viable in the North. The 1990s may be a decade in which major reforms can be won (as—and if—the global economy begins to recover or totally disintegrates); regrettably, there is no reason to expect that more than useful but secondary gains can be made on this front (especially by Africa) in the 1980s.

In Beginning

The purpose of projecting the future from the present and past is, or ought to be, to change it. A futurologist, therefore, should presumably end "in beginning" and not "in conclusion".

The prospects sketched may prove to be uneven and uncertain, not uniformly grim and foreboding. That does suggest that President Nyerere's dictum, "To plan is to choose, choose to go forward", is relevant. Bas Congo famines, the lunar landscapes of Pekines, the consignment of African majorities to the worlds of *Petals of Blood,* regimes of Amins and Nguemas, the lengthy survival of Mudges and Bothas, the grounded planes and closed borders of East African Non-Communities are not inevitable. The adverse external settings confronting Africa and most African states during the 1980s do not invalidate that. Challenges can bring forth responses, external penetration can give rise to national resurgence.

In this quest, two quotations underline critical points. The first is that progress from a bad and disintegrating starting point requires persistence and acceptance of costs or, as President Nyerere put it:

A man who has inherited a tumbledown cottage has to line in even worse conditions while he is rebuilding it and making a decent house for himself.

The second is that persistence in commitment must not exclude critical examination of actions and effects lest it become persistence in error. Mozambican Finance Minister Rui Baltazar Santos has summed this up very succinctly:

We must look at our mistakes lest we fall into repeating them.

Total avoidance of error is possible only by the greater error of never making any choices at all. The key goals must be to limit the number of errors, identify and correct them promptly, and learn from them to avoid their repetition.

One concerned with the course of events in Africa can best echo Gramsci's letter from prison counselling "pessimism of the intellect, optimism of the will". As a futurologist, one should perhaps conclude with the bemused comment of the first recorded practitioner—Pliny the Elder—when his predictions went wrong:

Out of Africa there is always something new.

Notes

1. "'Nought for your comfort': African Economies in the Mid 1980's", Scandinavian Institute of African Studies, Conference on "Africa: Which Way Out of Recession?" (September 1982) by the present author is an example of such an examination of the present crisis and possible immediate consolidation measures to halt decline.
2. For a more conservative presentation stressing internal African policy and management weaknesses see *Accelerated Development in Africa: An Agenda for Action* (Washington: The World Bank; 1981); and for a critique also advocating changes by Africans, but rather different ones, see C. Allison and R. H. Green (Eds.), *Accelerated Development in Africa: What Agendas for Action?*, Institute of Development Studies (Sussex) *Bulletin*, Vol. **14**, No. 1 (January 1983).
3. *Lagos Plan of Action* (Addis Ababa: OAU, 1980) (adopted by special summit of African Heads of State) presents this view of African economics more fully. *Accelerated Development*—somewhat against the empirical evidence—regards them as closed and internally oriented.
4. The term national economic integration is used in preference to delinking for two reasons. First it emphasizes the positive requirement and second delinking, in practice, has tended to start with cutting exports (by positive decision or inattention) which is a sure route to blocking the strategy since that "delinking" raises dependence on present and potential lenders and donors.
5. *Facing the Future: Mastering the Probable and Managing the Unpredictable* (Paris: OECD, 1979), p. 225.
6. See R. H. Green, "L'economie politique de désintegration, de développement du sous-développement et de développement: Quelques reflections", *Cahiers Economique et Sociaux* (Faculte des Sciences Economique, Universite Nationale du Zaire, Kinshasa), Vol. XVIII, No. 4 (December 1979).
7. For a rather different list see I. W. Zartman (scarcity of resources, external penetration, Southern Africa), L. K. Mytelka and Steven Langdon (neo-colonialism) and C. Legum (ethnicity) in *Africa in the 1980's—A Continent in Crisis* (New York: McGraw Hill for the Council on Foreign Relations, 1979). The OAU/ECA Symposium "Africa Towards the Year 2000" (see Report by A. Tevoedjre in *IFDA Dossier* 7 (Nyon: International Foundation for Development Alternatives, 1979)) selected a set of themes very similar to those of the present projection except that it treated Southern African Liberation as an achieved fact, which seems rather premature.
8. cf. sources cited at Notes 1 and 2 for a fuller presentation and dialogue.
9. See R. Leibenthal, *Adjustment in Low Income Africa 1974–78* (Washington, D.C.: The World

Bank, 1981); R. H. Green, B. Van Arkadie and D. G. Rwegasira, *Economic Shocks and National Policy Making: Tanzania in the 1970's* (The Hague: Institute of Social Studies, 1980).

10. At the September 1982 "Africa: Which Way Out of Recession" Conference in Uppsala, Stanley Please, then of the World Bank, expressed the view that primary export-led growth was unsound, accelerated expansion of coffee-tea-cocoa production counterproductive, and better designed internally oriented economic strategies often as sound as externally oriented ones. While presented as an interpretation of *Accelerated Development (op. cit.)*, these remarks seem to reflect a distinct rethinking of some of its premises as does the 1983 *World Development Report's* much more nuanced treatment of public policy intervention and public enterprises (Washington: The World Bank, 1983).
11. See C. Harvey, "The Case of Malawi" in Allison and Green, *op. cit.*
12. *Lagos Plan of Action* (OAU, 1980), *op. cit.*
13. For a fuller presentation, see R. H. Green, "'Things fall apart': The world economy in the 1980's", *Third World Quarterly*, Vol. **5**, No. 1 (January 1983).
14. See *Accelerated Development, op. cit.*, Statistical Annex and *World Development Report* 1982 (Washington: World Bank, 1982).
15. Derived from *Production Yearbook* (Rome: Food and Agriculture Organisation, 1981).
16. cf. *Accelerated Development, op. cit.*
17. See C. Allison and R. H. Green, "Stagnation and decay in Sub-Saharan Africa: Dialogues, dialectics and doubts"; and R. H. Green, "Incentives, policies, participation and response: Reflections on IBRD 'Policies and Priorities in Agriculture'", in Allison and Green, *op. cit.*
18. SWAPO, *Political Program of the South West African People's Organisation* (Lusaka: 1976), p. 12.
19. But see Botswana, *Rural Income Distribution Survey 1974/75* (Gaborone: Government Printer, 1976), which shows very marked intrarural inequality and (by comparison with wage data) between the majority of rural households and urban wage earners.
20. For fuller discussion, see reflections on legitimacy below and R. H. Green, *Magendo in the Political Economy of Uganda: Pathology, Parallel System or Dominant Sub-mode of Production?*, IDS (Sussex) Discussion Paper 164 (August 1981); and "L'Economic politique de désintegration . . .", *op. cit.*
21. Ngugi wa Thiongo (London: Heineman, 1977).
22. cf. B. Mwansasu, and C. Pratt, *Towards Socialism in Tanzania* (University of Toronto/Dar es Salaam, 1979); R. Van de Hoeven, "Meeting Basic Needs in a Socialist Framework: The Example of Tanzania" (Geneva: ILO, 1979); J. K. Nyerere, *The Arusha Declaration: Ten Years After* (Dar es Salaam: Government Printer, 1977).
23. *op. cit*, p. 3.
24. *ibid.*, pp. 2, 6.
25. See references at note 20.
26. See, e.g., I. W. Zartman, "Social and political trends in Africa in the 1980's" in *Africa in the 1980's, op. cit.*
27. See SADCC, *Southern African Development Coordination: From Dependence and Poverty Toward Economic Liberation* (Blantyre: SADCC, 1982); *SADCC-II, Maputo* (London: 1981); *Southern Africa: Toward Economic Liberation* (London: Rex Collings, 1981). R. H. Green, "Constellation, association, liberation: The struggle for Southern African development coordination", and "Toward economic coordination in Southern Africa: The struggle continues", in *Africa Contemporary Record*, 1979–80 and 1980–81, (New York: Africana, 1981 and 1982).
28. *op. cit.*, p. 4.
29. In fact, the demise of the EAC at the time and in the way it happened related primarily to the illegitimate state in Uganda, not to mere generalizable causal explanations. cf. R. H. Green, "The East African Community: The end of the road", and "The East African Community: Death, funeral, inheritance", in *Africa Contemporary Record* (London: Collins, 1976–77) and (New York: Africana, 1977–78).
30. For a detailed study of ECOWAS and its weaknesses beyond the free trade core, see J. P. Renninger, *Multinational Corporations for Development in West Africa.*
31. See sources cited at Note 27.
32. SADCC, Chairman's Concluding Remarks (Arusha: 4 July 1979).
33. With the EAC dead beyond resurrection and the Horn in violent conflict a viable East African body seems unlikely. The Eastern and Southern African Preferential Trade Area launched in 1981 faces an uncertain future as certain key potential members view it as inherently capitalist

in theoretical and institutional design and because an orderly agreement on division of labor is needed with the smaller but more cohesive and already operational SADCC to avoid damaging conflicts between PTA and SADCC member states.

34. cf. W. A. Axline, *Caribbean Integration: The Politics of Regionalism* (London/New York: Frances Pinter/Nichols, 1979).
35. The Lagos Plan of Action and the ECA's recent statements that production not trade is basic to regional cooperation mark a certain shift of emphasis away from imported European neoclassical regional economic models. But the Lagos Plan still makes a continental common market its symbolic goal for 2000 and ECA's latest venture in promoting cooperation is the Eastern and Southern African Preferential Trade Area whose charter very much places trade and payments in the center and production in the wings.
36. cf. C. Legum, "Communal conflict and international intervention in Africa", in *Africa in the 1980's, op. cit.*
37. cf. *Interfutures, op. cit.*, pp. 218–220, 223–226.
38. cf. S. Langdon and L. K. Mytelka, "Africa in the changing world economy", in *Africa in the 1980's, op. cit.*
39. cf. C. Pratt, "Tanzania's transition to socialism: Reflections of a democratic socialist", in Mwansasu and Pratt, *op. cit.*
40. The World Bank has learned from experience (some of it bitter) to take a different and less Eurocentric position. In the 1983 *World Development Report* (centering on government and public enterprise management), it argues that local values, practice, institutional innovations and personnel availability must be taken into account if functional, efficient and accountable management is to be achieved.

The Cultural Dimension in Asian Development

GODFREY GUNATILLEKE

Time present and time past
Are both perhaps present in time future
And time future contained in time past . . .
What might have been and what has been
Point to one end which is always present.

T. S. Eliot—"Burnt Norton"

Death said, "Choose sons and grandsons who shall
live a hundred years, herds of cattle, elephants, gold
and horses, choose the wide abode of Earth, and live
thyself as many harvests as thou desirest.
. . . But do not ask me about dying . . .."

Nachiketas said, "These things last until tomorrow,
O Death, they wear out this vigour of all the senses.
Even the whole of life is short, Keep thou thy horses,
Keep dance and song for thyself . . .

"No, that on which there is doubt, O Death, tell us
what there is in that great hereafter. Nachiketas
does not choose another boon but that which enters
into the hidden world. . . ."

Katha Upanishad

Introduction

The diffusion of technology has almost always been accompanied by far-reaching changes in ways of living, and man's perception of his role in society and place in the universe. It alters the value systems that organize human relationships and creates meaning for the individual and society. The "utilities" in the economic system are transformed into "meanings" in the value system. Development is as much concerned with the "utilities" that are produced as with the "meanings" that are generated or renewed with the changing pattern of utilities.

This paper deals principally with the changes in value systems which are part of the process of modernization, the pressures which they exert on Asian societies, and the capacity of these societies to make critical autonomous responses in coping with these pressures. Therefore it does not deal with a whole range of problems relating to economic and technological growth, fundamentally important though they are for Asian societies; consequently, it does not attempt to reconsider "modernization" in terms of an alternative strategy for overcoming the constraints to development in its economic and technological dimensions. The discussion in the paper focuses mainly on the cultural discontinuities that result from the process of modernization, and deals with selected issues that have special relevance for Asian societies.

The first section of the paper is a brief comment on the heterogeneity of Asia. In discussing the broad pattern of Asian diversity, it also draws attention to the strength of the historical presence—the continuity of tradition in most Asian societies. The cultural dimension in the development of Asia seems to assume a special character and importance. The paper then goes on to discuss two sets of issues. In Section 2 the set of issues relating to the pattern of structural change in Asian societies is described. It is argued that the large agrarian societies, given their demographic constraints and sociocultural conditions, will not replicate the structural transformation of industrialized societies. These Asian societies could produce a significantly different socioeconomic organization and spatial distribution of economic activity with a different mix of growth, equity and participation, and with alternative life styles and cultures. This discussion leads to Section 3 which proceeds to examine the conflict of value systems resulting from the modernization process in Asian societies. It explores some important clusters of values associated with modernization and compares them broadly with the value framework of the Asian cultures. It goes on to discuss the possible intellectual and cultural responses of Asian societies to the entry of modern structures of knowledge and modern value systems, and suggests that these societies have the capacity to achieve a unique equilibrium of material and spiritual well-being, which has relevance and meaning for the human community as a whole.

No Single Design in Asia

It would obviously be incorrect to talk of Asian development as though it were a single homogeneous phenomenon. Politically, culturally and economically, Asia at first presents a bewildering pattern of diversity and differentiation. Developing Asia is almost equally divided between the centrally planned politico-economic system, essentially based on Marxist ideology, and systems that operate principally through private capitalistic enterprise. Within this division itself are different adaptations that have provided a wide variety of socialistic, capitalistic and "mixed economy" models. Forms of political organization in Asia range from monarchies and autocracies that retain their semi-feudal despotic characteristics, to multi-party parliamentary democracies; here again numerous intermediate systems with different combinations of central control and popular participation have emerged as responses to the specific problems of the political transition.

Economic development in Asia spans a similarly wide disparity and differentiation from least developed Nepal and Afghanistan to newly industrializing South Korea and Singapore. Surveys of developing Asia usually define three broad economic configurations—the oil-rich affluent countries of West Asia, the fast developing relatively resource-rich countries of South-east and East Asia and the slow developing comparatively resource-poor countries of South Asia. On the other hand, there is little correspondence between the configuration of incomes and the growth of technological capacity in Asia. India with the largest share of the world's population which is in absolute poverty, has been able to advance to the frontiers of modern science and technology, and has a productive and technological capacity that can cover the entire range of modern industrial goods. The technological foundation in China and India is not very dissimilar. The coexistence of a high level of technological capability with mass poverty of a deep persistent character is a phenomenon which has not been witnessed in the development process as it has occurred in the past. In such contradictory conditions, the potential for rapid structural change and eradication of poverty is not easy to determine.

Asia also presents an unusual dichotomy of the big and small. It contains the world's two most populous nations, which between them contain nearly 38% of the world's population and 71% of the total population in developing Asia. Therefore, when we talk of heterogeneity of Asia we need to take account of these two large agglomerations of population, each of which has an extraordinary continuity of historical tradition and a deep-rooted cultural identity. Their economic power and cultural weight will have a strong gravitational pull on the countries of the neighboring regions. At the same time, the problems of relationships between small and big nations, the issue of dominance and dependence between mega-powers and their periphery are perhaps present in their most striking dimensions in developing Asia.

Cutting across this pattern of political and economic diversity are the main religio-cultural demarcations, representing the four great spiritual traditions that gave birth to major human civilizations—the Islamic tradition originating in West Asia and stretching to East Asia including the world's most populous Islamic nation, Indonesia; the Hindu-Buddhist tradition of South Asia; the value systems and ethos of the Confucian tradition which came out of the civilizations of East Asia and merged with the Buddhist-Hindu tradition of South Asia; and finally the Christian tradition which, mainly as a result of European expansion, has taken root in Asia and counts a minority with significant weight in the total cultural mix in Asia. The ASEAN grouping, for example, includes countries in which each of these traditions is dominant and represents a very interesting potential for cultural balance and synthesis.

Therefore, attempting to seek and discover a single design in this vastly variegated and rich tapestry of Asian development would indeed be a fruitless task. There is, however, one constant element in the heterogeneity we can use as a point of departure for our inquiry into the response of Asian societies to their modernization, to their penetration by the industrialization of the North. This constant element lies in the historical dimension of these societies. Asia contains societies in which history has a more profound presence than perhaps in any other part of the world. They are evolving into the modern age out of settled agrarian societies with a cultural accumulation that often extends into the past with remarkable continuity for more than two to three millennia. They are, therefore, societies with the longest human memory, with the greatest span of past experience, consciously carried forward and clearly articulated as they move into the future. If the weight of history is heavy on them, at times constraining growth and inhibiting change, it also provides the necessary spiritual ballast, the maturity and detachment of old civilizations that like Tiresias, have "foresuffered all". In seeking new forms of civilized living at the level of technology available to contemporary humans, these societies constantly tend to draw nourishment from the forms of civilized living they were able to create and the inquiry into the human condition they were able to undertake at lower levels of technology available to them in the past.

It may be argued that the way in which history and the cultural inheritance act on the ongoing processes of development in Asian societies is not fundamentally different from how they act in other developing societies or the way they have acted in the developed societies themselves. But the special quality of the historical presence in Asian societies, the interaction of "time past and time present on time future" lies perhaps in the degree of its continuity, its intensity and intellectual range. In a developing region such as Latin America, it would be correct to say that the migrant culture rooted in its European origins, has largely displaced indigenous history and tradition, resulting in sharp discontinuity and cultural loss. Sub-Saharan Africa with its younger tribal cultures does not appear to carry the same historical accretion as the settled agrarian societies of Asia. It is above all in the traditional civilizations of Asia that there exists, more than elsewhere, the highly developed alternative structures of philosophical reasoning, the complex systems of thought dealing with the human condition which are capable of questioning and challenging the forms of Western modernization on their own intellectual terms. Furthermore, the cultural and historical setting in these societies make the relationships of the elites and masses a more cohesive one, depending as they do on a framework of social relationships and network of reciprocity which have evolved over long periods. While these obviously have their negative features, they also help to create moral restraints and sanctions for social action, and provide for cultural continuity in the process of change.

Yet another aspect is relevant to our discussion. The developing societies are seeking to create the modern socioeconomic systems and enter the industrial civilization at a point of time when these have already reached a stage of profound crisis in the developed societies in which they have evolved. It is this juxtaposition that is crucially important and specific to the present crisis in civilization—the compelling attraction of the industrial society to those who have started on the path, counterpoised with growing disenchantment of those who have completed the journey. The developing countries need to evaluate and comprehend the changes they are undergoing, which are, at the same time, organically linked to the crisis of the industrial civilization. They need to perceive clearly the reflection of their future in that crisis and to seek the alternative future.

It is in this context that what comes out of Asian cultures in response to the processes of modernization could have a special value and meaning for mankind's future. The paradigms of man and society that have been clearly articulated in these cultures provide a mold in which political, social and economic changes acquire their own distinctive human values and ethical characteristics. This can be perceived in the ideological rationale of many Asian experiments and efforts—Nehru's vision of a mixed economy as a basis of a non-aligned society with its unique distribution of social and economic power, the Indonesian Panchasila, the mix of welfare and growth in Sri Lanka with its dominant Buddhist ideology, the Nepali search for a participatory political system free of elitist partisan politics, the effort to accommodate modernization within the transcendental Islamic vision. Each of these would require treatment in depth to illustrate the importance of the cultural mold in which development takes place, and the unique configuration of Asian values which each represents. However, within the limited scope of the present paper these themes can only be considered terms of broad generalizations.

The Weight of History

If the immense weight and presence of history acts as the center of gravity in Asian societies, then the process of modernization has to be regarded as its opposite. The discussion in this paper does not attempt to go into different definitions of the term "modernization" and the controversies around them. In the general discourse on development, "modernization" has signified the broad range of societal changes that accompany the process of industrialization and development. We have recognized that development requires growth, the accumulation of modern new capital stock and a steady increase in that stock enabling us to create an industrial system, and that industrialization in turn implies modernization which not only means new methods of producing new goods, but also a new set of values governing the entire set of relationships in society. In this paradigm, the transformation of values, the restructuring of the subjective world must precede, accompany, and follow upon the changes in the physical world. The imagery and thought forms implicit in such a concept of modernization signify discontinuity, a severance from history, an emancipation from the past, and a shifting of the centre of gravity to a future that has yet to be realized. Modernization demands the overthrow of inherited institutions and value systems perceived as an obstacle to change. The dialogue on development in the 1950s and 1960s singled out "tradition" as one of the chief obstructions to development, which was almost synonymous with modernization.

A clear and forceful statement of the modernist position in the introductory chapter of Myrdal's *Asian Drama,* serves as a useful point of departure for the discussion in this paper. Modernization is presented almost entirely in terms of its positive attributes. It is perceived essentially as the transition from autocratic and oligarchical structures of power to democratic forms of government; it replaces inegalitarian and rigidly stratified social systems with systems that have greater mobility, and a more equitable distribution of income and wealth. With modernization, the rational scientific structures of knowledge replace traditional, non-rational explanations of reality. It motivates societies towards higher productivity and better levels of living. It inculcates the virtues of disciplined and efficient work. It integrates loosely knit social and economic units with national economies and creates nation states. There is no mention here of any negative aspects of modernization—the deformities and excesses that are discussed in later sections of this paper. Modernization can also breed totalitarian ideologies. Massive structures of domination can establish themselves within democratic forms, making a mockery of civic participation and popular power. Secular structures of knowledge can destroy the base of man's religious experience. The impulses of growth and higher productivity can lead to wanton consumerism, which aggravates the problems of world poverty. They can promote highly inegalitarian patterns of growth. The ethic of modernization can produce a competitive, self-centered, individualistic man. Integration into national structures can lead to suppression of ethno-cultural identities. The problem of development is that of discriminating between those polarities, and making the right choices.

Myrdal goes on to make a sharp abrasive criticism of the popular stereotypes that are often presented regarding the contrast between Asian spirituality and Western materialism. It is argued that the Asian man does not act very differently from his Western counterpart. Despite the high value supposedly given by the Asian value system to nonviolence, to contemplative life styles and to spiritual pursuits, political violence persists; high rates of crime prevail; the behavior of elites in the pursuit of worldly gain is not dissimilar to that of elites in more ostensibly materialistic societies. Of course, such an analysis does not add up to much more than the common observation that people do not always live according to their beliefs and the values they profess. But then, what is the conclusion that is to be drawn from such an observation? The disparity between the social ideal and the social reality is true of all societies and all cultures. It also applies to the modernization ideal which Myrdal describes.

It is also possible to find a common core of human imperfection and virtue across cultural boundaries and across the broad history of human civilization. This, however, does not mean that the "ideal" component in a society, its image of perfection, the core values to be striven after but seldom achieved, have no deep societal significance. The way in which this disparity is perceived, how it is woven into the fabric of social and individual action will depend on the vision of man and reality within which this disparity exists. In the religious vision, the disparity is the inherent imperfection in the temporal order, to be continuously assailed but never to be overcome within the dimension of time. In the nonreligious vision, as in Myrdal's analysis, such ideals and values that are not manifested extensively in men's overt behavior and social action are of little significance. In the two visions, the understanding of the inner human reality, the tensions between values and action, the moral quality of life remain very different.

The argument developed in the paper suggests that these ideals and core values are important; they shape the visions and the future images of the good life in the process of development, and the conflicts that are generated beween the alternative visions as they grow out of Asian culture on the one hand and those that come out of the process of modernization on the other, have far-reaching implications for the human content of development. The intellectual revolution that ushered in the technological societies of the West was essentially a historical and positivistic in direction. History was a linear progression into the future, a forgetting and discarding of the past. While the past became a field of historical exploration and discovery, a field in which human curiosity ranged in all directions, the existential links with the past were increasingly severed. Although sometimes this exploration brought with it a new humility to contemporary man, the past ceased to be a real source of life-giving knowledge and experience. On the other hand, the traditional Asian system of thought has tended to perceive history and society as sharing the life cycle of other life forms, to see the death of the old and the birth of the new as an organically linked and reproductive process. They recognize that recollection and renewal is as necessary to the movement of history as the casting away and the forgetting. In such a framework, the relationship of "development" to history, the encounter of tradition and modernity, is a much more positive intersection than is depicted in a simple model of conflict and opposition, of modernization which liberates and tradition which enchains.

Furthermore, the majority of Asian societies have a compelling need to reestablish and renew their links with the past. After long periods of foreign domination, they need to rediscover and reaffirm their own cultural identity across the hiatus created by colonial rule. It is mainly through this rediscovery that they are able to develop the national self-consciousness, the cultural integrity and the collective will to organize their societies for development. Therefore, they are caught up in a process fraught with deep contradictions. On the one hand, these societies are restored to the spiritual centers which sustained them and gave meaning to their existence over their long history. They thereby acquire the capacity to respond discriminatingly and selectively to the forms of domination, cultural and economic, pressing on them from outside. On the other hand, this very process unquestionably tends to preserve and entrench the grave imperfections that are part of the cultural heritage and past social structures.

A great many elements in the process we describe as "modernization" will unquestionably form part of the changes that must inevitably occur in developing societies. But for each of these elements, the question arises whether it need replicate the experience of the developed countries both in quantity and in quality. One set of questions relates to the technology and structures of production which the Asian societies need to develop. These societies must industrialize, but should the pattern of industrialization be identical to that of developed countries in terms of scale, spatial distribution and urban growth? If it cannot be identical, what are the implications of the alternative for life styles, and social structures? The second set of questions relates to the value system which is implicit in the industrialization process. Is the technology and productive capacity of the industrial societies inseparable from the entirety of the value system and the organization of human life that have evolved in those societies in the process of their industrialization? Do the Asian societies have the choice of acquiring the former without the latter in its totality? And if so, what are the forces that will determine the character of this choice to create viable alternatives? Within the limits of this brief paper, it is not possible to examine these two sets of issues in any depth. What could be attempted is a broad intimation of the alternative futures by examining one or two key elements in each set of issues.

New Patterns of Change

Asian societies, like all developing societies, would need to modernize their systems of production, acquire industrial technology and vastly increase their productivity if they are to eradicate absolute poverty and achieve reasonable standards of living for their population as a whole. One accepted indicator in this process of development and growth of productivity is the change in the sectoral composition of the gross domestic product and the concomitant change in the sectoral mix of the workforce. We have seen that in the industrialized societies, development has meant a steady decline in the share of the workforce engaged in primary activities such as agriculture and mining and an increase in the share engaged in secondary activities arising from modern services. This, of course, has been possible as a result in the application of modern technology that has enabled developed societies to increase steadily the productivity of human labor and transfer labor from primary to secondary and tertiary activities. This basic change in the structure of human activity must take place in Asian societies as well if they are to create a system that can produce an abundance of goods and services that will appreciably raise their level of material well-being.

In Asian societies, this structural transformation has to take place in a demographic and sociocultural environment fundamentally different from that of industrialized societies. The transformation in industrial societies was accompanied by vast movements of population and a very high degree of mobility. It required a massive transfer of people not only in sectoral, but also in spatial terms. Major streams of migration from rural to urban and across national boundaries to new continents characterized the demographic changes that formed part of the process of industrial development in these societies. One basic feature of this process was the magnitude of scale both for the organization of economic activities and human communities in large factories and huge conurbations. Capital and return to capital dictated where human beings should be moved and assembled. Apart from this question as to whether a replication of this process is desirable, it would seem that the demographic and socioeconomic conditions in Asian societies rule out the feasibility of this model. It is difficult to conceive of a far-reaching spatial redistribution in which the majority of rural populations in Asian society are transferred to urban concentrations to be mobilized into industrial production in a large-scale factory system in a manner that would eventually correspond to the spatial distribution of population and activity in developed countries.

Most Asian societies are large, densely populated agrarian societies which, in themselves, are collectivities of a vast number of small rural communities. The major problem of development in Asian societies has been that of transforming these principally agricultural communities into economic units that are diversified and capable of dynamic growth, providing an increasing share of employment in non-agricultural activities. It is in this context that there has been an intensive search for a pattern of development in rural Asia that brings a process of diversified development to where the human beings are, and that delivers the development process to the small rural communities in which the people are organized. This search is reflected in the policy prescriptions of international agencies, the strategies of national governments and the activities of non-governmental voluntary movements. It is seen in various experiments in local-level development or development from below; in attempts to create systems and structures of decentralized planning and development; in the strategies that have been described as integrated rural development. In all these efforts, the pattern of development sought is one in which the existing human settlements will need to expand and diversify the base of their economic activities in their present habitat. Consequently, the process of urbanization grows organically out of the transformation of these economic units and the upgrading of their man-made environment. The pattern of urban growth could then be radically different from the centripetal concentrations of the Western model of industrialization with its high element of population transfer and mobility. It would be more widely dispersed and evenly distributed.

The alternative patterns of structural change would be far less disruptive; they do not require the dislocation and uprooting of existing communities, but would be more of a humane process that preserves cultural continuity and is better able to cope with the encounter of tradition and modernity. The future patterns also require that economic activities be on a smaller scale. The units of organization therefore provide greater scope for participation; the decision-making processes are more amenable to community involvement and control; technology itself being on a smaller scale shares some of these characteristics and is less of a monopoly by a technocratic elite. The distribution of power that consequently evolves is less elitist and more equitable.

Furthermore, this modular form of development promises to lead to more efficient and less wasteful use of resources than the conventional patterns of growth. The spatial distribution of agriculture, industry, residential settlements and services produces a pattern of exchange and transportation that makes a lower demand on energy. There is less movement of goods and people. The module of socioeconomic organization is such that the major part of its economic transactions take place within the module itself. It offers an alternative that is less energy-intensive than the conventional model for comparable levels of productivity. It provides scope for the development of new alternative energy systems based on renewable resources such as biomass, biogas, mini-hydro power, wind and solar energy. Both because of the scale of its activity and its agro-industrial mix it is better adapted for the use of recycling technologies and conservation. It is at the same time capable of achieving a better ecological balance.

Another vision of life styles

If we examine some of the global scenarios for the economic well-being of the developing Asian countries, these often project three components. In one such scenario, by the turn of the century, West Asia (including North Africa) unfolds a scenario in which per capita income raises to US$2450 at 1970 prices; in South-east Asia including China, the per capita income reaches a figure in the region of US$900. South Asia which includes India, however, stagnates at a level around US$200. Some of the underlying assumptions in these scenarios, taking as they do the economic relationships of investment and output as the principal determinants, might be disputed. This might apply particularly to the South Asian scenario where even the physical and technological potential may not have been taken fully into account. But these qualifications apart, there are more fundamental questions which arise. Given the different levels of development realized by different societies and different generations in the societies, what do these differences, this development "ladder" imply for a satisfying quality of life, for a meaningful equilibrium of material and spiritual satisfactions at each of these levels, for each of these generations? Do these different levels of per capita income imply that societies at lower levels have a less meaningful and satisfying equilibrium? Or, that they must wait until they reach a certain level before they can begin living a meaningful human life?

In asking these questions we are inquiring into the value system that guides the development process; we are getting involved in the persistent dilemmas which confront that process. There is in the concept of development a tacit unexpressed assumption that the satisfying level of well-being is yet to be achieved, that it is in the ever-receding future towards which effort has to be constantly directed. But the lived human experience is totally different; all communities at whatever stage they are, are concerned with the material and spiritual fulfilment in the life available to them, as much as with the effort to change that life in order to enjoy a better life in the future. They are as much concerned with "being" in the sense of experiencing life fully in the present as with "becoming" in the sense of developing into something better in the future. The human quality of development that is taking place depends very much on the balance that is achieved—this balance between "being" and "becoming", between the fulfilment in the life "here and now" in its spiritual and cultural totality within the limitations of the present, and the suspension of that fulfilment in the constant striving for a larger, more productive material base to enjoy life better in the future.

Depending on the extent to which this balance is achieved, the development process can be aggressively and impatiently directed towards higher growth, become destructive of human values and neglectful of equity, or it can be a more patient humane forward movement of the whole community. The future-centred vision, the intense concentration on enlarging the material base for a better life in the future can lead to a devaluation of the past and present. It can result in the breakdown of community and the rich network of primal bonds and human relationships based on non-economic reciprocity and personal interdependence. On the other hand, we know only too well that the emphasis on the spiritual and cultural fulfilment in the present, the effort to realize a meaningful life within the available material base can perpetuate existing structures of inequality and oppression, preserve modes of perception which cripple human relationships and give social sanction to unacceptable inhumanities. The former promotes a society that is restless in a constant state of high mobility both physically and in terms of values, and that is, at the same time, liberating. In the very rootlessness it produces, the individual is loosened from traditional ties which constrict human freedom; social rigidity and immobility are reduced. In the case of the latter, the tendency is to reach a state of greater rest as it were, there is a much lower degree of mobility; the space and time for concern with the non-material ends of life seem to be greater. But in the terminology of development the restful conditions, the slow movement, also denotes a stagnant society. In each of these approaches, the positive and negative elements are inextricably intertwined: seeking the balance between the two calls, therefore, for a vision of reality which encompasses both and which perceives their positive complementarities. Here we need to pick up the thread of the argument developed earlier in our discussion of the interaction of past and present in the future growth of Asian societies. The value patterns in Asian societies, their perception of life, has greater capacity to accommodate the opposites of rest and movement, of "being" and "becoming".

This line of argument leads to several other complex issues concerning the relationships of development to human values in Asian societies. Indeed, there is no question that the agenda for development in these societies requires structural changes that eliminate poverty and satisfy the basic needs of the population as a whole. But what is equally clear is that this objective could be achieved long before they reach states of material well-being approximate to those of developed societies. At the same time, the socioeconomic forms that promote equity, participation, self-reliance and ecological balance and the social and political substance of development that generates a satisfying quality of life need not themselves be specific to any particular level of material and technological development. They can exist at the different levels in each of the scenarios of Asian development to which we referred. It would then seem that it would not be always necessary to be achieving continuously higher levels of well-being to enrich the quality of life of a society and its population. That enrichment could be the product of other non-material resources of a society—its cultural and spiritual vitality. But given the capacity for technological advancement, the drive and the compulsion to reach successively higher levels of material well-being seems to become inescapable and to assert itself without limit. In this unlimited drive towards the improvement and enlargement of the material base, the distribution of time, energy and resources is skewed sharply in the direction of technological and economic goals. In the predominantly economic and technological goals and value system, the non-material human pursuits increasingly lose their collective social character; they decline as decisive and motivating forces in a society, and are pushed steadily into a peripheral and private world. It is in this drive and compulsion that the more profound dilemmas of development manifest themselves in Asian societies.

In the traditional framework, the disposition of time and human energy gave a central place to religio-cultural pursuits. This disposition depicted the configuration of values in these societies; it reflected their vision of reality and human life within it. This is most visibly represented in the way in which they directed a significant part of their capital accumulation to places of worship, to the tangible symbols of the holy, the "points of intersection of the timeless with time". They also made a clear separation of time between the sacred and the mundane, allocated days and hours during the day especially for worship and prayer and communication with the transcendental. It is this rationale, both spiritual and material, which sustained the distribution of time and energy in the traditional Asian societies, and it is this which is fundamentally disturbed by the new paradigm that comes to these societies with the industrial civilization. It is true that these societies urgently need the technological capability that the industrial societies have to offer. But, at the same time, they need to acquire and apply this capability within essentially Asian value systems, visions of life which have evolved from their own civilization and their own major spiritual traditions. They need to preserve the distribution of time and energy which in the final analysis was saner and more meaningful. Therefore, beneath the techno-economic problems of development, Asian societies have to grapple with this profound cultural dilemma—the effort to achieve material well-being within the value system that provided them with their spiritual well-being and inner equilibrium.

We can select a few key components of the modernization paradigm that will help to define the nature of this dilemma. There is first in this paradigm, the overriding emphasis on *economic rationality* as the basic organizing principle of social and individual activity and behavior. It provides the normative framework, as well as acting as the critical determinant. In such a frame of reference, other forms of rationality become non-rational. An increasing share of human experience and human relationships become transferred into the world of economic transactions. The world of primary voluntary relationships based on love, ethical values, moral determinants and interpersonal obligations arising out of these, steadily contracts; the interior personal reality becomes increasingly externalized and expressed and perceived in terms of their monetary equivalent. Everything has its "shadow price", and is substitutable for everything else when the right price relationships are determined.

The rationality of traditional societies is more complex and cannot be reduced to a single denominator. The thought forms underlying this rationality spring from a recognition of the paradox of human existence, the coexistence of opposite states of being in conflict as well as complementarity. This is seen in the categories of thought such as "ying" and "yang", Siva-Hakthi; the Divine Being of the Creator, Destroyer and Preserver, the non-reality as well as the reality of Maya or existence. There are several ways of life or paths (margas or yogas) of liberation, each with its own rationale and sanctity. They include the way of knowledge (Gnana), action (Karma) and faith (Bhakti). Economic well-being (Artha), sensual enjoyment (Kama), responsibility in action and every day life (Dharma) and the final release (Moksha each have their place and are all a unity. It is a rationality which is multi-dimensional, which is related to several clusters of human needs, material and spiritual, each of which makes autonomous demands on human energy, time and resources for its fulfilment.

Development would then have to take place within such a value framework. It should simultaneously contain its econo-political, cultural and spiritual rationale. These are all related to the developmental needs which we defined earlier in the alternative paradigm. But these needs do not stand in a hierarchical relationship to each other. They are not depicted adequately in the image of an inorganic columnar or pyramidal structure. They are best represented as a living system in which all parts have to grow organically together. The need for food, clothing and shelter must always exist together with the nourishment of the inner man, the foundation of interpersonal relationships, and the space for the spiritual inquiry. There is no genuine trade-off possible between the different clusters of developmental values; between equity and growth, between political freedom and the rest of development, between human rights and economic development, between spiritual inquiry and the pursuit of material well-being. As in the case of living organisms, the loss or neglect of any one of these clusters is a malformation, a mutilation. Any loss in each of these clusters has to be made good in that cluster itself.

Development implies the capacity of the system to correct itself in satisfying each cluster of needs when imbalances occur and when the harmonious process of growth is disturbed. It is a capacity which depends on the vision of man and society that informs the development process.

Another key element in the modernization paradigm that is at odds with Asian perspectives is its epistemology and its cognitive structure. The modernization paradigm bases itself on the *mode of knowing and the structure of knowledge which are positivistic*. The reality which is not capable of empirical verification and not capable of being identified and treated as an object of knowledge through the analytical methodology of the empirical sciences is not reality. In the final analysis, it is dismissed as a linguistic creation that has been produced by the structure of language itself. For the Asian societies, the metaphysical is real, and the modes of knowing which go with it, the intuitive, existential, holistic modes, illuminate reality more than the analytical empirical modes of knowing. The former is principally concerned with the intensification of knowledge of the parts which is analytical and dissective—knowledge which is capable of technological and practical uses. The latter is essentially concerned with knowledge that puts the parts together, which is configurative and visionary. In the former the unifying principles are conceptual instruments for the development of technology; they have limited validity in themselves and are successively replaced. In the latter, the search for the unifying principle is the goal of knowledge itself. It is a means of communion with reality. The Asian society of the future has to be one in which these different modes of knowing and these different structures of knowledge are not in irreconcilable conflict, and destructive of each other.

Another key attribute of the modernization paradigm is its essentially *anthropocentric and materialistic character*. At best, it is agnostic within a materialistic frame. In the final analysis, the frame of reference is man, and the uses to which man can apply the material world available to him. The touchstone of reality is ultimately man's experience, and this experience itself has a limited life span with no verifiable before or after. In contrast, the traditional vision sets the human drama in a transcendental scheme. Man's known life span and the total body of knowledge he can accumulate exists in a reality which transcends empirical verification; it is encompassed by the Unknown. This is the non-human reality to which man himself must relate, and for which purpose the life of the individual and society has to be consciously organized. On the other hand, the modern paradigm while being *anthropocentric* is not necessarily *humanistic*. It makes man the ultimate source of meaning; but in the process of limiting all meaning to man, the larger configuration which contains the human scheme becomes devoid of meaning and man himself is reduced to a meaningless entity. When the transcendental worth of man is denied, the value system which grows out of that denial becomes capable of creating collective ideologies, technocracies and totalitarian systems that destroy the spiritual rationale of human rights.

At the same time, it is necessary to recognize that the religio-cultural value system plays multiple conflicting roles in the process of modernization. While it affirms the transcendental worth of man, renews the perception of human dignity and provides the spiritual motivation for a just social order, it is at the same time capable of moving in the direction of extreme intolerance. It can rationalize grave social injustices and activate the fierce suppression of individual liberties. Throughout human civilization, the transcendental significance of the person and his relationship to the total religious system have been a source of profound tension. This has been manifested in the opposites and complementaries of the religious life in various traditions. The polarities of order and love, the law and the forgiving redeemer, the mystical way of personal communion with the Transcendental and the mediation through the religious organization and the ritualistic system are all part of this spiritual travail. One of the most moving explorations of this dilemma is in Dostoevsky's "Grand Inquisitor". The problem of protecting the spiritual and human core against the despiritualizing and dehumanizing system is therefore an inherent part of the religio-cultural value system itself and the vision of man on which it is based.

It is in this context that we have to perceive the religio-cultural responses of Asian societies and their effort to reinterpret the transcendental core of their cultures for contemporary living. What we have in these societies are as yet intimations, rudimentary ideological efforts. They exist at varying levels of effectiveness and authenticity—national policies and pronouncements such as Panchaseela (Indonesia), Dharmistha rule (Sri Lanka), Islamic revivalism (Pakistan), and popular movements as well as intellectual initiatives such as Sarvodaya or Gandhian ideology. But to have an impact, to offer a counter-model of development derived from Asian cultures, a much more far-reaching intellectual and ideological effort is required. The human concern which informs such an effort will have a different perception of human rights in the development process. A large part of the vulnerability and dependence of the individual exists in human situations and in areas which are not readily accessible for adjudication, not easily reached by the formal processes of the law. In these situations, violations of "human rights" are often suffered in silence and without redress, whether they be in the family, the group or the larger community, which have their own informal structures of oppression, violence and crime. The protection of the person, the safeguarding of the conditions for his or her full human growth require processes of a more voluntary informal character that are sustained in society by institutions which act with moral and spiritual authority and which internalize the norms that govern human behavior and relationships. The protection of human rights therefore requires not only the formal institutions of the state, but also the informal structure of the community; and it is best achieved in a social order in which the sources of moral and spiritual authority are powerful and active.

The three elements of the modernization paradigm which we observed—its use of economic rationality as the principal criterion of individual and social action, its positivistic structure of knowledge, and its anthropocentric vision of life—seem to lead logically to a way of life which is *hedonistic and consumerist*. The driving force in the industrializing societies is the continuous expansion of demand, the ceaseless multiplication of new material wants and the flow of new products and services to satisfy them. The capacity to sustain this process becomes the final measurement of a society's well-being. The way of life that inevitably evolves out of such a process is incompatible with the basic disciplines for living cultivated in the traditional societies. These disciplines have been in rhythm with the cycle of human life—childhood, adolescence, adulthood and old age. The Hindu vision of man and society for instance, distinguishes four important life stages, "asramas" or dwelling places—the stage of learning for the child and adolescent, the "sisya"; the stage of the householder with the pleasures, responsibilities and worldly successes, "grhasti"; the stage of contemplation and spiritual search when he withdraws from the world of sensual pleasures, the "vanaprasti'; and finally the stage when he travels along the path of deliverance or "moksha", the "sanyasi" or "bhikshu" who is the holy mendicant. The social order had to provide this total perception, preserve the dynamic tension between the four stages, and enable the individual to move harmoniously from one life stage to another, from one state of being to the succeeding state not only teaching man how to live, but also preparing him for death. "Mendicancy" and "renunciation" become the visible symbols of holiness in these societies, the constant reminder of the last stage of the journey, the monastic presence in society which helped it to achieve the inner equilibrium between material and spiritual well-being.

In terms of the values in the Asian religio-cultural traditions, the immoderate pursuit of material well-being in the modernization paradigm creates a world of inordinate desire and craving. It makes the enjoyment of material goods the main center of values for all life stages, it produces an inner drive of expectation and need that obstructs the natural movement from one state of being to the other in the progression of an individual's life. To be able to organize human experience to provide for this harmonious movement, the value clusters of each life-stage must be in balance with the clusters of other stages. In the traditional societies, each life-stage had its value-cluster and its discipline; and the individual could not be enchained or inordinately attached to any single value-cluster—he had to achieve the inner capacity for free movement from one to the other. This is not an equilibrium which a society can achieve through the interaction of individual preferences in the market place. The market's capacity is limited, its efficiency of little worth when it comes to the allocation of time, energy and resources between the short-term and the long-term, between pleasures which are intense but transient and joys which are permanent and enduring, between things which produce sensory excitation and those that bring inner serenity. The capacity for such an allocation can only come from a social order which through its ideology and total value system acts consciously to guide individual and collective choices and which, in achieving material well-being, yet responds to it with appropriate detachment from a spiritual centre of gravity. We need to remember that such an equilibrium did not prevent these Asian societies in the past from developing civilizations which were technologically among the most advanced in the era preceding the industrial civilization.

Conclusion

The main theme of the discussion in this paper is that an alternative vision of reality and an alternative value system will underpin the economic and technological transformation of Asian societies. These alternatives are immanent in the deep ideological contradictions, and the powerful polarities of opposing value systems which form part of the development processes in these societies. The discussion in this paper, however, limits itself to a conceptual analysis. It is not possible within the limited scope of this paper to relate this discussion to a more detailed factual analysis of specific ideological trends and movements in these societies. As might be expected in any of these societies, a multiplicity of ideological forces are at work. The technocratic modernizing elites in these societies who are the carriers of the modern value systems and ideologies are most often close to, if not in actual control of the centers of power. To them, the first priority is the socioeconomic transformation and the expansion of productive capacity. The ideological side-effects of this transformation are of secondary importance. On the other hand, the traditional elites who are protagonists of the indigenous value systems and ideologies while retaining a powerful hold on the popular consciousness, have not adequately developed and reinterpreted these ideologies within a conceptual frame which can handle the modern structures of knowledge and critically examine their ideological foundations. Nor have they been quite able to project the socio-economic transformation in a manner which has contemporary relevance.

The basic premise in the argument outlined in this paper is that ideology has an independent existence and is not an epiphenomenon of the material and economic reality; it is not a byproduct of the structures of production and the relations governing them, however much it may be affected by them. On the contrary, it is a fundamental need and a primary resource of the human mind and spirit, shaping the social forms, providing the rationale for the distribution of time, energy and resources in a society and giving the framework of meaning for organizing the totality of human experience and the personal world. There is in this approach a partial acceptance of the Hegelian premise. It is a premise that ideological contradictions emerging as a result of the processes of development in man have to be resolved in specifically ideological terms. The resolution has to evolve existentially from the ongoing cultural and spiritual encounter itself.

The main burden of seeking a satisfying ideological equilibrium falls on the intellectual communities in these societies. It falls on groups who are able to liberate themselves ideologically from the existing polarity, who are thus able to separate the modern technological capability and knowledge structure relevant to it from the value system and life goals it has engendered; and in doing so, to absorb and develop that capability within specifically Asian cultures, and the visions of life and value system generated within them. Such an equilibrium must be capable of creating societies which are technologically dynamic, which continuously expand the frontiers of man's empirical knowledge and are still able to master the technological drive and the structures of knowledge which support it within a larger cognitive structure related to a fuller vision of reality.

Reflections on Development Perspectives since the 1950s

NURUL ISLAM

The decades since the 1950s have witnessed significant evolution in the approach towards and perspectives on development processes, patterns and policies. This is particularly so with respect to growth and equity issues. There was a time when economic growth was considered to be a sufficient objective of development policy. But during the last two decades, it has been increasingly recognized that economic growth, while necessary, is not a sufficient objective, and that equity, stability and people's participation are equally important objectives to pursue.

Those who propounded economic growth as a sole objective of development argued that it was possible to attain other objectives such as equity through growth itself. Research, analysis and empirical experience, however, subsequently demonstrated that economic growth unassisted by supplementary policy measures may, in fact, aggravate inequality. Simon Kuznets offered the hypothesis that growth, unaccompanied by any corrective measures, would lead to an increasing inequality of income until a level of per capita income characterizing present day middle-income countries—say $800–1000 per capita in 1970 prices, is reached; afterwards, as growth continued, its cumulative effects would help reduce inequality. It must be noted here that this hypothesis of a changing relationship between growth and inequality over time did not comment on the subject of the impact of growth on absolute poverty.

In the early stages of growth, there are various factors that tend to increase inequality, both between groups within a country or within a region, as well as between regions within a country. It is most unlikely that the various regions within a country would enjoy the same rates of growth, or equally enjoy the benefits of growth; they have dissimilar endowments, and differential human and physical infrastructure. Moreover, as the process of growth acquires momentum, increasing demands are made on technical, professional and managerial skills; there is a mounting demand on entrepreneurship, innovative spirit and ability, capacity and willingness to take risks, to explore and exploit new opportunities for financial rewards for pride of achievement. The supply of such skills and abilities is scarce in the initial stages in development; they have to be acquired and learned through experience and training over time. Hence, in view of scarcity of such skills and abilities, their financial rewards are high compared to the average income level. Also, economic growth is associated with a transfer of population from the agricultural to the nonagricultural and from the rural to the urban sectors. The nonagricultural and urban sectors have higher per capita incomes so that the increasing weight of the high-income sector would tend to increase inequality.

The various factors, which cause increasing inequality in the early stages of growth, become less important in the later stages. The lagging regions develop infrastructure or receive an inflow of resources, human and material; simultaneously, migration out of them increases the income of those left behind. Their weight in the national economy declines. Similarly, urban and rural, agricultural and nonagricultural disparities decline; scarce resources (such as skilled labor or managerial capacity) increase in supply, partly through "learning by doing", and partly through institutional development, training and education.

This is what in general has happened in the developed countries of today in their earlier stages of growth. Over time, inequality of income has ceased to increase and, indeed, has started to decline. There is a political consensus in the developed countries that corrective measures to redress extreme income inequalities are not only socially desirable, but also politically necessary.

So much for the interrelationship between growth and inequality; what about the impact of growth on absolute poverty? Doubts have been raised as to whether absolute poverty declines with growth. However, a decline in absolute poverty is consistent with increasing inequality. A distinction needs to be made between (a) the percentage of population who are poor and (b) the absolute numbers of the poor. With growth it is conceivable that the percentage of the population in absolute poverty may decline; however, their absolute numbers may increase because of an increase in population.

Research and empirical evidence seem to suggest that if the growth rate stays high enough and is sustained for a sufficiently long period, it is most likely that the percentage of the population living in absolute poverty will decline. Whether the absolute number of people in poverty declines will depend on the interaction between population growth and economic growth. The dynamics of population growth and its interaction with economic growth are not fully known. There are a number of available empirical studies demonstrating that high growth has been associated with an increase in the absolute number of the poor. However, an interpretation as to why it is so is not easy; one encounters so many contributing factors that it is difficult, if not impossible, to ascribe it to a unique set of factors. Furthermore, there are controversies as to the nature of data and their correct interpretation. At the same time, however, very few countries in the last three decades or so have been able to maintain very high rates of growth and for sufficiently long enough time.

However, even if the growth rate is high and sustained over a long time—

long enough to make a dent on poverty—there are forces at work in a market economy that may delay the process of alleviating poverty. This may be due to interventions in the market mechanism and pricing process by governments or organized economic pressure groups. Attempts to force wages above what the supply and demand for labor would otherwise determine, or to push interest rates down below the opportunity-cost of capital, will reduce employment as well as encourage the substitution of capital for labor. High wages for those who are employed keep out of employment the rest of the labor force. Monopolies tend to restrict output and employment, and hence contribute to inequality and poverty. For the process of economic growth to prevent the emergence of forces tending to aggravate inequality, the free market mechanism must be allowed to function freely, and prices must reflect scarcity. But free market forces do not pre-empt the growth of monopolies, and market imperfections are not necessarily due to state intervention. To suggest that market forces must not be interfered with is to deny the existence of vested interests, as well as unequal opportunities arising out of the unequal ownership of assets. Those who advocate economic growth first and distribution later, aver that poverty can be alleviated through redistribution, only when the size of national wealth and the level of income have grown to high enough levels; otherwise, there will be a redistribution of poverty rather than an alleviation of poverty.

It is contended that premature attempts to redistribute income or assets or to undertake poverty-alleviating measures would divert resources away from growth, i.e., to consumption away from investment, and to current expenditures away from savings, and would in the long run hurt efforts to redress poverty. To have growth, savings and investment must be accelerated; in a private enterprise system, it is high private incomes and thus inequalities of personal income as well as high corporate profits that are needed to generate high savings. This implies a slow growth or stagnation in the income and the consumption of the poor. In a socialist system, it would imply a constraint on the consumption of all, including the poor, unless at the early stages of development, income and assets are redistributed so that growth builds itself upon an already egalitarian base of assets and income distribution.

The protagonists of growth first and distribution later are unable to demonstrate that those who secure the lion's share of the benefits of growth and major control of wealth and income—hence political and economic power—will volunteer to surrender their wealth, privileges and power for the benefit of the poor and that they willingly encourage or readily sponsor the adoption of poverty-alleviating or redistributive measures. Furthermore, in a context of rising expectations of the poor, a long period of waiting for the process of "trickle down" to work itself out is neither feasible nor politically acceptable. Sustained poverty is likely to destabilize a socio-political system through the aggravation of tensions and conflicts, and can endanger growth itself.

Social expenditures for the benefit of the poor or the provision of assets or income-earning opportunities for the poor can, in many ways and within limits, help the process of overall growth and capital accumulation as well. Not only do social expenditures improve their health and enhance their contribution to the national product, but investment or assets accumulated by the poor or for the poor also yield high returns that are no less than those obtained on investment by the rest. This is evidenced by the productivity of small farms or small-scale industries.

Small farmers respond to incentives, undertake innovations, save and invest; they generate returns on investment and produce output per hectare no less than the big farmers. Redistribution of land, therefore, helps achieve growth, equity and poverty alleviation. There are ways of taking advantage of the economies of scale in farming operations, to the extent that they exist, as for example, in the marketing and processing operations or in the use of specific capital equipments for irrigation or land development. The institutions, such as the cooperatives or mutual aid arrangements, help combine the advantages of scale with those of individual incentives. Under the conditions of heavy pressure of population on land, not enough land is available to provide all with viable farms; some form of joint or communal production arrangement, combined with expansion in nonagricultural employment, promote equity without sacrificing output.

Along with equity and growth, stability of income and consumption has become an equally important goal. Growth, equity and stability are interrelated. Important sources of instability in developing countries are instability in food and agricultural production caused by variations in weather or in the availability of purchased inputs, or by pests and diseases, etc. Instability in trade characterized by fluctuations in price and volume of trade in turn causes instability in consumption. Trade related instability affects income and employment in export sectors and, via fluctuations in availability of imports, affects the import-competing or import-dependent sectors.

Instability of employment and income has its worst effects on those who are already at a low level of income; at times of economic downturn, the consumption and welfare of the poor are affected most severely. Food shortages cause starvation among the poor; they do not have adequate purchasing power to offset the fall in income or in self-production of food, nor have they the means to hold stocks during periods of scarcity. Instability of prices and supplies adversely affects expectations of the future and slows down savings and investment. Stability in prices and supplies of food, adumbrated in the concept of food security, has come to be accepted as a very important objective of development policy.

People's participation, both as a means as well as an end of socioeconomic development, has been in the forefront of development thinking and dialogue in recent years. People's participation is independent of the objectives of equity and alleviation of poverty—the latter defined as the poor benefiting from the fruits of development. People's participation is intended to secure their involvement in the decision-making process, in the determination of development objectives, targets and policies, as well as in the formulation, implementation, monitoring and evaluation of programs and projects. Two prerequisites for people's participation are: (1) representative and democratic organizations of the poor; and (2) decentralization at the grassroots or the local level of the decision-making process in which people's organizations can effectively participate.

Both these prerequisites are linked with political processes, systems and structures of a country; basically, they relate to the way in which political power is organized, mobilized and exercised. Democratic, representative and self-governing institutions and people's organizations are not universally accepted in all societies. Administrative decentralization by itself does not ensure people's participation. The structure of power is often centralized; those who are in control of political power at the centre, be they politicians or bureaucrats, are afraid or loathe to give up or delegate authority or functions to lower levels at the grassroots level.

Many governments are reluctant to encourage people's organizations, lest they weaken the control of the governments. Participatory democracy is the very antithesis of centralized control. Issues of efficiency of centralized versus decentralized political processes continue to arouse controversy; they involve appropriate criteria for judging the efficiency of political organizations in the context of society's goals and objectives.

People's organizations that are not in tune with or do not work in conformity with the political power structures of the day can seldom play a developmental role without creating tensions and conflicts. How to generate a stable political system in which voluntary, freely functioning people's organizations can play a developmental role without creating "unmanageable tensions" is a challenge to the practitioners of development policy. Development without change is not conceivable. It is not without risks to the existing political and social order; successful development depends upon unleashing people's energies and motivating them towards the attainment of growth and equity in a participatory process. The appropriate balance between conformity and dissent, control and freedom, harmony and conflict is to be sought by each society in its own way; the process of adjustment and conciliation may change over time. Economic development is not an entirely economic process; it is to an important extent a political and social process.

Some Considerations on the Transfer of Technology: A Latin American Perspective

ENRIQUE IGLESIAS

The most important aspect of the issue of science and technology is probably that over 95% of all research and development activities in the world are carried out in the industrialized countries. This gives a preliminary—but very clear—idea of the importance of this issue to the developing world.

This enormous imbalance automatically leads to excessive dependency, which in turn creates many problems for the developing world, from a substantial loss of scarce resources (specialized personnel, through the brain drain, and foreign exchange), to the imposition or involuntary acquisition of foreign norms, structures and cultural values; the neglect of scientific and technological creativity in the developing countries; and a deepening of the economic and social duality of those countries.

That is why, even though it is true that science and technology are essential components of the national development process, it must be recognized and accepted that the achievement of their tremendous potential is increasingly dependent on cooperation and action at the international level.

Hence, the Vienna Program of Action on Science and Technology for Development adopted by the United Nations Conference on Science and Technology for Development, held at Vienna in August 1979, rests on two basic foundations: the strengthening of the science and technology capacities of the developing countries, and the restructuring of the existing pattern of international scientific and technological relations, which obviously includes the transfer of technology.

There is a close interrelation and interaction between the two issues. As regards the transfer of technology, for the acquiring country to make full use of a technology, it must expand its capacity to generate, control, absorb, adapt and apply the scientific and technological know-how. It must also create or expand institutions capable of advising the government or the industry concerned with regard to the quality or suitability of the technology and the manner and terms of its acquisition.

At the same time, in their effort to strengthen domestic scientific and technological capabilities, the countries will have to resort, to the extent and under the terms and modalities deemed appropriate, to cooperation with countries that are further advanced in the area concerned.

International cooperation for development in the area of science and technology must help the developing countries strengthen their creative and innovative capacities and thus encourage their own independent scientific and technological development. This means that fundamental changes must be made in the existing patterns of international relations in this area, to allow for a substantial expansion of international cooperation and thus increase the opportunities available to the developing countries for developing and strengthening their scientific and technological capabilities in accordance with each country's needs, its own situation and its vision of the future. Changes must also take place in the international technology transfer process, so that such transfers might be considerably increased and facilitated. This is the case particularly when transfers are made to the developing countries which must be given much greater access to the technologies they need, including advanced technology.

Having made these preliminary remarks and outlined the major features of the problem of the transfer of technology, we shall now review those issues that are most discussed in academic circles, as well as in the bodies most involved in direct action.

As is well known, the flow of technology to the developing countries, including the Latin American ones, takes place through the following channels: direct foreign investment; turnkey arrangements for the supply of machinery, equipment and plants, that is, embodied technology; the establishment of joint ventures; and the licensing of know-how, either patented or not.

In quantitative terms, the most important of these channels are those that provide for the supply of machinery and equipment and for direct foreign investment; however, the licensing of know-how, patented or unpatented, has in recent years become increasingly important, especially in absolute terms.

It has been estimated, on the basis of a sample considered representative of the developing countries, that royalty payments increased from $200–300 million in 1965 to over $1 billion in 1975, and to over $2 billion in 1980; this shows how important the market of the developing countries is to the industrialized countries. These figures indicate that the share of the developing countries in the worldwide exchange of technology rose from 8% in 1965 to 14% in 1980. This may be attributed, on the one hand, to general economic growth and, on the other, to the industrialization efforts made by the governments of those countries.

Another salient feature of the flow of technology to the developing countries of which Latin America is a good example is that these flows originate primarily with the transnational corporations. It is estimated that TNCs account for 60 to 70% of the technologies traded; this, in turn, represents 90% of the flow of technology to the developing world. A large part of this flow is from parent company to subsidiary, the latter being either totally or partially owned by the former.

Another distinctive feature of this flow is the increasingly important role played by governments in the transfer of technology. This trend is typical of, although not exclusive to, the developing countries. Even in the developed countries, the state has taken on an increasingly important role, either as a promoter or as a direct operator.

Suffice it to note, for example, the incentives provided during the years following the Second World War, by the United States Government and some European governments, not to mention the governments of the countries with centrally planned economies. This action was not limited to encouraging or providing incentives to scientific and technological development and its applications, but rather—especially in the case of the United States—it also consisted of the careful regulation of the terms on which technology was to be transferred internally, so as to prevent any possibility of monopolistic action. In Japan, as is well known, there was considerable control over the import of technology—although this did not, of course, prevent the Japanese from acquiring those technologies they considered necessary—which has only recently been somewhat relaxed.

After having explored, to different extents and in different ways, the route of import substitution, the developing countries, and specifically the Latin American ones, subsequently chose the route, which was to some degree complementary to the former one, of regulating the technology imports. Thus, several countries of the region, Argentina, Bolivia, Brazil, Colombia, Ecuador, Guatemala, Mexico, Peru and Venezuela, created administrative units responsible for recording and controlling such imports. This control was and is exercised on a case-by-case basis and tends, as is well known, to discourage or, as the case may be, prohibit transactions that, for various reasons, are considered to be contrary to the national interest (contracts with restrictive or leonine clauses, strictly nominal transfer of a technology, excessively high prices, etc.).

This action, aimed at obtaining more fair and just terms for the transfer of technology and ensuring that they are reflected in any contracts signed, constituted the central concern of that first phase, which might well be called a "defensive" one and which was strongly influenced by fiscal considerations and the need to save foreign exchange. A good illustration of this type of action—which was very useful and necessary, but not sufficient—is the case of the Andean Group whose first provisions aimed at controlling the transfer of technology (Decision 24) were basically defensive in nature, although from the beginning it was evident that the Andean Group's concern went beyond that simple aspect.

Nevertheless, and this is what is really important, there has recently been a tendency to reorient this regulation of technology imports in such a way that, without neglecting the aforementioned objectives (which might be termed of an immediate nature), other more complex goals are borne in mind that relate to more substantive aspects of the national development processes and of the adequate insertion into them of the technology variable.

This new orientation means that policies and ways must be designed to enable the countries to improve their capacity to make a suitable selection of technologies and to ensure that the acquisition, assimilation and development of the technologies chosen are compatible with and complementary to the growth of the country's scientific and technological capacity in the relevant production and service sectors. There is no question that the developing countries must be given as much access as possible to scientific and technological innovations, but it is essential that, in the process of selecting technologies, they consider their socioeconomic situation and, of course, their development policies and objectives. In this regard, it must not be forgotten that their undue dependency on foreign technology has not only inhibited, in many cases, the growth of national scientific and technological capabilities, but it has also led to the use of technologies that are not necessarily the most appropriate ones in light of the national objectives and conditions.

With regard to the unique policies and problems which arise in this area of the transfer of technology, special mention should be made of the cases when this transfer is made to the developing countries by small and medium-sized enterprises located in the developed countries, a matter which is the subject of a recent study by UNCTAD, *Organizational Forms of Transfer of Technology to Developing Countries by Small and Medium-sized Enterprises: A Case Study of Equity Joint Ventures and Technology Agreements in Latin America,* Publication No. TD/B/C.6/77. (New York: UNCTAD, 1977).

Reference has already been made to the "traditional" form of transfer—that made by the large transnational corporations that usually set up subsidiary companies or joint ventures in other countries, maintaining total or majority control over them and transferring to them, on terms favorable to their own interests, technologies regarding technical know-how and industrial processes.

As an alternative to this system, new forms of association have been created between small- and medium-sized enterprises in developed countries—employing fewer than 500 persons and representing almost 90% of the total number of enterprises—and enterprises in developing countries that are neither subsidiaries of nor dependent on the transnationals. This new experiment was initiated by European enterprises that have established links with Latin American enterprises, many of the former having a great technological density and a highly specialized production. The branches of industry involved are: clothing, paper, chemicals, rubber, plastics, metallurgy, machinery and construction. It is estimated that at present, small- and medium-sized enterprises represent from 15 to 30% of all subsidiaries of European enterprises operating in Brazil, Mexico, Peru and Venezuela. Provisional conclusions (not many cases have been studied) on this system confirm the impression that these arrangements can offer concrete advantages to both parties. This is particularly true in the case of agreements on technology, as these enterprises in the developed countries appear to be—as far as licensing is concerned—more willing to adjust to local conditions; because they are not monopolistic or oligopolistic, they also appear to be more willing to share their production know-how with partners in developing countries and thus "indigenize" their commercial operations in those countries. It may be said, therefore, that the advantages of these arrangements seem to lie in that which they lack the most: the monopolistic power to impose high prices, restrictive terms and global plans so as to get around or manipulate national policies.

The above considerations have certainly taken us beyond the scope of what is strictly referred to as the transfer of technology to other questions of equal importance, such as the selection of suitable technologies and the need for a given technology to be acquired on convenient terms that will allow for it to be quickly asssimilated, adapted and developed within the context of local conditions and resources. To this might be added also the requirement that it should not replace activities that can be carried out locally. It is well known that efforts have been made in the region to "unwrap" the "technological package", so that simple technologies that are available locally are not "slipped" in with the complete technologies that the region must necessarily acquire.

The national policies of the developing countries are increasingly reflecting this concern in that they are approaching the problem as a natural whole; that is, they are concerned not only with the tools for regulating the import of technologies, but also with strengthening their own capacity to select and acquire those technologies that best meet their needs.

This means that their capacity to evaluate technologies must be strengthened; the technologies must be broken down to allow the country to make a financial and technical evaluation of them, to determine the effect they will have on the local technology structure and, especially, to set up an information system with which users can make informed choices among the technologies potentially available to them.

This is the goal sought by a change introduced recently by one Latin American country in its system for controlling the transfer of technology. In so doing, this country has recognized the crucial role of information and has made available to those acquiring technologies a state advisory service for the selection of technologies. What is open to question is that, at the same time, this country dismantled—albeit partially—its apparatus for controlling the import of technologies.

The example of the Andean Group may also be mentioned in this connection. The aforementioned provisions on control over the import of technologies in the member countries have been supplemented with several measures that go beyond the provisions and deal with the question of selecting technologies. One of these is the recent creation of the Andean Technological Information Service, which is bound to become an essential tool in this technology selection effort and a very important part of the Andean approach to science and technology.

At the regional level, it is worth mentioning the Latin American Technological Information Network (RITLA), an initiative sponsored by SELA which is still in the formative stage. At the world level, there is the Technological Information Exchange System (TIES), an initiative supported by UNIDO, which is designed to facilitate a broad exchange of information among similar registries or agencies concerned with controlling the import of technology. Also at the world level, regional technology transfer centers in Bangalore (India) and Dakar (Senegal), have been established with the assistance of UNCTAD and UNIDO.

The preceding paragraphs clearly show the highly important role of international cooperation in the selection, acquisition and transfer of technology, an area which, by its very nature, is particularly suited to international cooperation. In this regard, the exchange of information is undoubtedly one of the key areas to be considered; it can be supplemented with another tool that, although difficult to manage, is potentially decisive—that is, joint negotiations by two or more developing countries, or subregional and regional groups, with suppliers from developed countries.

So much for our consideration of the problems and modalities of the reception of foreign technologies in the region. Let us look briefly at the aggregated, and even structural, implications of this flow for the economies of the region.

It is becoming more and more evident that this massive influx of foreign technology has unleashed in several of the recently industrializing countries a series of structural phenomena on which more thorough research is essential. The reception of foreign technology has been associated, *inter alia*, with: (1) the appearance of an incipient national technological capacity that is growing very rapidly and is taking shape through the adaptation of foreign technology to local conditions of use and through the generation of some national technological designs; (2) an increasing capacity for exporting manufactures of relatively sophisticated technology; and (3) the beginning of a trend towards the export of local capitals and technologies to other regions of the underdeveloped world.

These structural features—still in the embryonic stage but, in our view, undergoing a definite maturing process—go hand in hand with characteristics such as a rapid increase in the participation of foreign capital in manufacturing and in the economy in general; a growing concentration of the main sectors of production; a certain trend towards deterioration of the patterns of income distribution and, consequently, of the relative well-being of those sectors of the population that have not participated in technical progress, etc. In brief, it may be said that the technological modernization process has been associated with a complex socioeconomic evolution in the countries of the region, particularly in those of greater relative economic development.

As a counterpart to the costs of imitative and adaptive technological change mentioned above—the excessively high cost of the technologies transferred, restrictive clauses, repeated payments for the same purpose, and so on—it should be noted that the region's capacity to export final and semiprocessed industrial goods has increased, particularly in the larger countries. In the initial stages, these exports were confined almost entirely to the markets of bordering countries and, in general, to the LAFTA area; but they have subsequently expanded quite dynamically to the industrial countries.

Despite the aforementioned restrictions, costs and problems involved in the modalities of technological transfer in the countries concerned, their industrial progress has also gone hand in hand with a slow but clearly perceptible growth of their capacity for absorption and adaptation and, on a lesser scale, for technological creativity. Thus, in many cases, the manufactured goods exported are not mere replicas of the original designs conceived in countries of the developed world; this is also true of the manufacturing processes used in making such goods. The rapid growth of industrial productivity and its relationship with the increase in manufactured exports has been aided, to some extent, by domestic technological efforts.

Closely related to this growing export capacity is the "internationalization" within and beyond Latin America, of certain Argentine, Brazilian and Mexican enterprises. The study of this question has only recently begun. It is still an incipient phenomenon, and its further development will have a significant impact on the field of technology and foreign trade.

We therefore consider it worthwhile to present some quantitative information on this promising beginning of a transfer of technology that originates, this time, in the countries of the region and is exported to other countries within and outside Latin America.

According to more or less recent data (produced by the IDB/CEPAL Program of Research on Scientific and Technological Development in Latin America and by the Inter-American Development Bank—*Economic and Social Progress in Latin America: The External Sector* (Washington, D.C.: 1982)), on the countries of greater relative economic development in the region for which information is available, exports of non-embodied technology (licensing arrangements, consultancy and technical assistance contracts, industrial and civil projects and engineering services and sales of plants under turnkey arrangements) are estimated to have reached around \$410 to 420 million (1973–80) in Argentina, over \$3.1 billion (1966–80) in Brazil, and over \$800 million (1974–9) in Mexico.

Although the above figures may not be spectacular, they are significant enough to introduce a factor of optimism. Indeed, they show that despite all their well-known difficulties and limitations, the countries of the region were able, at least in certain activities and sectors, to complete the evolutionary sequence that began with import-substitution industrialization and the incorporation and adaptation of foreign technologies, followed by the export of relatively sophisticated manufactures and, finally has led to the export of technologies, a considerable share of which were exported outside the region.

4

TECHNOLOGICAL TRANSFORMATION OF THE WORLD

Transfer of Technology: Theory and Policy

TOM GANIATSOS*

Introduction

If we define technology broadly as the application of knowledge and ideas to the development of more efficient or reliable ways of producing goods and services, it is evident that all societies have understood its fundamental importance in improving man's material well-being. Yet, acceptance by economists of technological development as an explicit national objective in its own right and of the legitimacy of public intervention to accelerate the process of technological transformation in developing societies is a rather recent phenomenon.[1] Although a step ahead of academic economists, international organizations and agencies too have been slow to adopt technology questions into their agendas and programs of work. This paper gives an account of the change in perceptions that has occurred over the past two decades or so regarding the nature of technological advance as it relates particularly to developing economies and the significance of governmental, including intergovernmental, action to accelerate it. Section 2 reviews the recent evolution in the status of the concept of technology in the main branches of economic theory underlying development economics. Section 3 summarizes the main issues raised by the transfer of technology to developing countries. Section 4 gives a brief account of the role of public intervention to promote technological advance in developing countries at the national and international levels. The final section draws together the main conclusions.

One inescapable conclusion is that UNCTAD has raised the level of consciousness of the international community on technology's role in economic development. The guiding force behind all of UNCTAD's work on technology has been Surendra Patel. Perhaps because of its intangible nature, technology had not been at the forefront of Third World concerns at the first UNCTAD which focused primarily on commodities, trade in manufactures, and money and finance. From a peripheral activity incorporated into the agenda only at

*The author is a staff member of the United Nations Conference on Trade and Development. The views expressed here are personal and not necessarily those of the organization or its staff.

the Third Conference, technology can be said to have become a mainstay of UNCTAD not only because it is one of the few success stories in an otherwise dismal record of North-South encounters, but also for the recognition that the quality of the intellectual output has received in academic circles. Surendra Patel's success has been the result of a rare combination of political skill, an unwavering sense of commitment and the ability to draw the best out of his colleagues.[2]

Technology in Economic Theory[3]

The main object of economic development in the 1950s and 1960s, as perceived in the expanding body of specialized literature on the subject at that time, was economic growth through capital accumulation. The obstacles to growth were seen to arise less from technological backwardness *per se* as from two sorts of economic constraints—the savings-investment gap and the import-export gap.[4] This is not to say that technology and technological change were overlooked. Textbooks on economic development did treat three issues related to technology: entrepreneurship, the importance of ability to take risks, initiate new ventures and undertake production decisions; labor skills, the education and training of qualified manpower; and choice of techniques. Sen succeeded in demonstrating that much of the lively theoretical controversy surrounding the latter subject was essentially a debate about the relationship between the growth implications and welfare assumptions of capital- versus labor-intensive investment strategies.[5] Thus, while technology was implicit in the analysis of development problems, there was no systematic consideration either of the impact of technical progress on development or of how such change took place or could be helped along.

It is often noted that there has never been a theory of economic development but rather a collection of writings on development problems drawing from a variety of economic disciplines. That development economics in the 1950s and 1960s and the policies that followed from it did not give technology the concentrated attention it may have deserved, is at least in part attributable to the restricted role of technology in three of the main bodies of underlying (neoclassical) economic theory dealing, respectively, with production, economic growth and international trade.

The conventional theory of production assumed that the firm whose behaviour it was intended to describe was faced with a production function—typically of the Cobb-Douglas form—defined for a given technology. The process of obtaining new technological information was not at all considered, since the assumption was that all such new technology was freely and instantaneously available. Moreover, technology itself was conceived of as a single aggregate entity rather than of different elements (e.g., product specification, process know-how and organization etc.), some of which are created internally rather than purchased from outside suppliers.

The variety of neoclassical, Keynesian, Marxist and hybrid growth models and the literature on them is great and will not be reviewed here. It can be said generally that their main preoccupation is the analysis of the properties of steady-state (i.e., equilibrium) growth, defined as that rate at which all of the relevant variables such as investment, population, employment etc. grow at a constant rate.[6]

At the level of the economy, economists of the neoclassical school relied on the concept of an aggregate production function, again of the Cobb-Douglas variety, to depict the contribution of technological progress to increased productivity. An increase in output per man arising from a rise in the amount of machinery per man was interpreted as a move along a given production function, whereas an increase due to technological progress was interpreted as a shift in the production function. The modern Cambridge School rejected the notion of a production function because of the unacceptability, in its eyes, of the assumed tendency towards full employment equilibrium growth and the payment of factors at prices equal to their marginal productivities. But such rejection permitted less insight on the contribution of technology and other inputs to growth (however useful its theorizing may have been for shedding light on the important questions of the demand determinants of steady-state growth and distribution).

What *causes* the production function to shift—that is, what causes technological progress—was excluded from the basic neoclassical model: like manna from heaven, technical change just happened and was treated as being totally exogenous. Modifications of this assumption by Kaldor and Arrow related the origins of technological progress to present or accumulated past capital formation, as justified by the argument that learning takes place along with investment, hence contributing to knowledge.[7] Technological change was thus made endogenous.

Solow's empirical application of the neoclassical growth model to the measurement of productivity growth utilized data on the relative shares of capital and labor (equal by assumption to their respective marginal products) and the rates of growth of per capita output and capital to obtain the contribution of capital to the overall growth rate. His conclusion was that the residual 87.5% of the per capita growth of the United States during 1909–45 was attributable to technological progress. Further refinements were made subsequently by other economists and Solow himself to shrink the size of the residual and break down empirically the different forces that contribute to the overall growth rate. Although the logical aspects of growth theory were better understood as a result of the work of growth economists, it has been generally agreed that the behaviour of the residual factor in growth identified with technology advance is still not well understood. Consequently, the theory does not offer much practical guidance either as regards technological or other policies affecting growth.

The place of technology in the classical theory of international trade developed by Ricardo (and improved upon by Mill) was to explain the structure, direction and volume of trade between countries on the basis of labor's productivity. Mutually beneficial trade between countries arose, not because of the absolute technological superiority (higher productivity) of labor in one as compared with another country, but because of the relative technological superiority of labor in producing one product as compared with others in the same country.

In the standard, static approach to trade theory of Heckscher, Ohlin and Samuelson with two sectors and two factors, trade and comparative advantage are determined by factor proportions. Given identical technologies for any one product across countries and under the assumptions of constant returns to scale and immobility of factors of production internationally, the theory states that countries well-endowed in capital will export capital-intensive goods and countries well-endowed in labor will export labor-intensive goods and that equilibrium will be reached when factor prices are equalized. Thus, the theory does not allow for, nor is it intended to allow for, technology-related trade—either from differences between countries in the degree of technological advancement or from variations in the capacities of different countries to exploit economies of scale.

The classical trade theory has been "dynamized" in the modern literature to show how different types of Hicksian technical progress (neutral, capital-saving and labor-saving) may affect the growth of trade and movement in the terms of trade.[8] It has been noted that an expansion of output—whether from technical progress or increased factor availability—can under certain conditions (for example, if the country's exports are relatively capital-intensive) lead to a deterioration in the terms of trade of such magnitude as to more than offset the gains to the country in question from the greater output. Such an outcome has been called immiserizing growth.[9] But, once again, the treatment of technology as an exogenous variable adds little to understanding of the process of technological change.

The coming of age of technology as one of the focal points of the development dialogue may be said to owe its intellectual origins to two strands of economic theorizing. One consists of research in the neoclassical tradition devoted to the analysis of technological innovation as a source of trade; the other is the economics of foreign investment and transnational corporations. A third would be an extension of the work of development economists belonging to what may be loosely called the structuralist "dependencia" tradition.

The standard classical theory of trade based on factor proportions began to be subjected to rigorous testing as a result of the findings of Leontief that the exports of the United States, a capital abundant country, were less capital-intensive than its imports.[10] The work of Posner, Hufbauer, Vernon and others in the 1960s viewed technology much like a natural resource endowment whose fruits, while it is immobile, can only be acquired by importing the goods that it produces. Posner's work showed that one country could acquire a comparative advantage it previously did not possess by successfully launching an innovation. The resulting trade would gradually disappear as the innovation was discovered and imitated in other countries.[11] Hufbauer articulated the theory further and tested it with data on trade in synthetic materials, distinguishing between "technological gap" trade and "low wage" trade.[12] The latter stemmed from the transfer of technology to countries where its application was less costly than in the originating country. Left unanswered by the theory was the question of why the innovations first occurred in some rather than in other countries.[13] Vernon's theory of the product cycle provided explanations for the initial location of production in the innovating country and for its gradual transfer to other countries that could eventually switch from importing to exporting of the product.[14] It postulated that the life of each product passed through three stages: the new product, the maturing product and the standardized product—during which the locus of production moved via the transfer of technology, from the originating industrialized country in the first stage to the non-industrialized country in the third stage. Since the United States was the country with highest per capita income and labor costs, Vernon argued that high-income or household labor-saving products would be likely to be innovated there first.[15] The reasons for the initial location of production in the innovating country are as follows: first, the cost of production is not a critical issue because of the high degree of product differentiation and monopoly position initially; second, the development and commercialization of a new product is easier where it is possible to be assured of swift communication with customers and suppliers; and third, the wage cost as distinct from the wage rate differential is likely to be low or even favorable insofar as the embodiment of skills is high. The product will thus be produced in the country for which it is designed, and exported as part of technological gap trade to markets in other countries where there is also a demand. The expansion of output from the originating country to other industrialized countries—stage two—ensues in response to the expansion of markets in those countries, increasing ratio of transportation cost to product price, rising import barriers and growing standardization. In Vernon's third stage, the product has achieved a general penetration of world markets, it has been completely standardized—implying virtually no need for interaction with suppliers and customers and, production runs are long and routinized through assembly lines. At this point, Vernon argues, developing countries have the potential to produce and compete in the export of those goods, once the missing capital is found, as now routinized production methods permits greater use of unskilled labor which is in abundance, and as standardization of the product lowers the cost of communication between exporters and importers.

Empirical research has supported the technological gap-product cycle type of hypotheses, though Vernon himself has emphasized that much still remains to be explored before complete understanding of technology in international trade is secured.[16] This includes, for example, the work of Gruber *et al.*, Keesing and Hufbauer on the structure and behavior of United States manufactured exports[14] and the work analyzed by Mansfield and Romeo comparing the mean age of technologies transferred by United States-based transnational corporations to their overseas subsidiaries in developed and developing countries.[18]

Of crucial importance in the impact of technology is the manner in which the firm possessing a technological advance exploits its monopoly advantage. Broadly speaking, the firm has three possible options open to it: (1) it can locate its production facilities in one country and meet demand from the rest of the world through exports; (2) it can establish production facilities and meet demand from local production; or (3) it can forego the production option and market the technological advantage itself directly. The neotechnology theory of trade of Vernon and others addressed itself to the choice between these options bud did not consider it in great detail. The emergence in the 1950s and 1960s of the transnational corporation as a potent vehicle for the transformation of international economic and technological relationships naturally focused attention on the factors underlying the foreign investment decision in the resulting proliferation of theoretical treatises on direct foreign investment and, more generally, on the multinational corporation.

Traditionally, the theory of international capital flows viewed long-term movements as a reflection of the process of equating the marginal efficiency of capital in different countries. Such a theory was designed to explain portfolio investment which was in fact the principal form of long-term international investment prior to the Second World War. But, as first articulated by Hymer, direct investment occurs in order to exploit some kind of monopolistic or oligopolistic advantage, whereas portfolio investment occurs in response to opportunities to earn higher rates of return on capital arising from its greater scarcity abroad than at home.[19] Moreover, while the latter is best viewed as flowing from one economy to another, the former is seen as flowing from one sector in one country to the same sector in another. Caves and other writers—following from Hymer—have, therefore, based their explanations of the decision to invest abroad on oligopolistic interdependence and the possession of certain monopolistic advantages, to offset the disadvantages of crossing international boundaries.[20] The sources of advantage are not unlike those of established firms compared with potential entrants in an oligopolistically structured industry, as propounded by Bain in industrial organization theory.[21] They may include privileged access to financing, raw materials and other inputs; superior managerial skills; economies of scale; command over technology (including the technology required for producing differentiated products); access to markets; and other sources. The oligopolistic setting in which foreign investment occurs has been brought out by a number of studies concerning the behaviour of United States direct investment. Within manufacturing, the activities of US transnationals were found to be directed mainly towards "technology intensive industries and/or to those producing differentiated high income consumer goods and, within resource based industries, to capital and/or technology intensive industries supplying products to markets in which the investing firm has a favored access".[22] The consensus is that, among all of the sources of monopoly power that may impel a firm to initiate or build up its overseas production activities, the most strategic are technology and marketing, though opinion is divided on the relative importance of each of the two.

From a theoretical standpoint in a static world of perfect competition and free movement of capital, the incentive for technology transfer through establishment of overseas production facilities would not exist. The incentive would not exist because local firms under the idealized conditions would have access to all the advantages of any other firm in the world economy, such as access to the same technology and pool of managerial talent and the ability to borrow capital at home or abroad at the same rate of interest. To operate across national boundaries, the firm possessing monopolistic advantages must face the diffficult situation of running production facilities in an unfamiliar and uncertain environment which is a source of disadvantage *vis à vis* local firms; whereas it is possible to service the foreign market through licensing or exports. The first question, then, is why a firm with certain advantages chooses to exploit them by locating production abroad (either through foreign investment or licensing) rather than *via* exports from extension of its existing facilities? Given that production abroad is advantageous, the second question is why the firm in question does not sell or license its advantage rather than to establish a subsidiary?

With respect to the first question, the product cycle theory discussed above predicts internationalized production when a product has achieved market penetration and standardization of technology. The specific factors that enter into play include the degree to which the overseas market can be better serviced and local tastes and requirements better catered to through the location of production in the market; the role of tariffs and other trade barriers in restricting the access of exports; production cost differentials and locational factors, particularly in assembly-type operations; and, finally, oligopolistic reaction to other transnational firms already present in the market.

With respect to the second question, Lall and Streeten have put the considerations governing licensing versus direct investment very succinctly:

> In general, the more attractive and (politically) stable the market, the newer and more tightly controlled the technology involved, the larger and more internationally involved the firm in question, the more broadly based its sources of market power, and the lower the "absorptive" capacity of the potential licensee for the assets to be transferred, the more will direct investment be preferred over licensing. On the other hand, the more widely diffused the technology, the smaller and riskier the market, the more inexperienced, risk-adverse or nationally oriented the firm, the more specific its particular "advantage" and the more capable the potential licensee, the more will licensing be preferred over direct investment.[23]

The determinants of why technology is transferred tell us little about its intercountry welfare effects. One of the first attempts to analyze the effects of the direct foreign investment form of transfer was by MacDougall in a comparative-static model.[24] He showed how the gains from a marginal increment in foreign investment were divided between labor and domestic and foreign capital, and concluded that, under full employment and perfect competition, the host country gained mainly as a result of taxes on foreign profits, economies of scale and external economies. The latter included training of local labor which became available to the economy generally and productivity gains to (technological development of) local firms induced by emulation of the foreign subsidiary. With respect to capital inflows generally, the factor proportions approach to trade and growth has, as previously noted, analyzed welfare effects through the impact of such inflows on the terms of trade. In this context, Kemps's and Jones's considerations of the notion of an optimal tax on international movements of capital, analogous to the optimal tariff on international trade, suggested that restrictions on capital flows in both the importing and exporting countries could lead to an improvement of welfare. All of these exercises have concluded that the net effect on real output in the host country from foreign capital inflows is positive unless there are heavy losses due to monopoly power, high tariffs or other types of government intervention in the market, or negative externalities.[25] But the presumption was that any losses suffered on one score would be more than offset by gains on another—so that the overall effect from foreign capital inflows was favorable. Since direct investment was a principal carrier of technology, the conclusions regarding welfare effects of direct investment flows were also assumed to apply to effects of the technology accompanying these flows.

Transfer of Technology and Developing Countries

Starting in the early 1970s, however, specific attention began to be devoted to the factors explaining the impact of the transfer of technology *per se*, particularly in developing countries. This shift in emphasis was largely reflected in the studies and reports produced by UNCTAD. Animated by the writings of the Latin American structuralists and the "dependencia" school,[26] the creation of UNCTAD in 1964 had already brought into question the extent to which the poor countries of the periphery actually benefited from the existing pattern of trade with the industrialized countries of the center. If the distribution of the gains from trade in commodities was less than equal, could one not also challenge the presumption that the benefits from trade in technology were equitably divided? This led to another fundamental question based on the historical observation that the industrialization of "late-comer" countries of Continental Europe and Japan in the latter half of the 19th and early 20th centuries owed at least as much to the absorption of innovations from the countries that had industrialized earlier as it did to endogenously generated technical change.[27] What was the nature of the factors that limited the ability of today's developing countries to exploit the accumulated fund of technical knowledge needed to industrialize? Clearly, the answer was in large part bound up with historically determined economic and social patterns. But, as argued in the work of UNCTAD, part of the answer also had something to do with the process of technology transfer itself—that is, the nature of the market for technology in today's world and the position of developing countries in that market.[28]

Before considering the market for technology, it is useful to dwell briefly on technology as a concept. Technology is sometimes mistakenly identified as consisting simply of plants or machinery. But this represents only that part of technology that occurs in embodied form. Without the remainder—that is, disembodied knowledge of the processes underlying production activity—the hardware is useless. Both forms can be transferred to a technology-receiving firm, but the ability to make effective use of them—that is, technological capacity—cannot. It is this and related weaknesses that are at the heart of the external technological dependence of developing countries, a point to which we will return later.

Within industry, technological knowledge gives rise to some or all of the following technological activities:[29]

—Feasibility studies, market survey and other pre-investment services;
—Determination of the range of technologies and choice of technology;
—Engineering design and detailed engineering;
—Machinery installation and plant construction;
—Acquisition of industrial processes and product designs;
—Improvements to processes and product designs;
—Management and operation of production facilities;
—Training of technical and managerial personnel; and
—Marketing.

Moreover, the transfer of technology may take place through a number of different mechanisms, ranging from commercial transactions such as technology contracts, purchase of machinery and direct foreign investment, to informal mechanisms such as visits and meetings.

The large number of technological activities associated with industrial projects would imply in principle that the potential investor has a wide range of commercial sources for obtaining technology, including both local and foreign sources. In fact, the degree of packaging—that is, the variety of sources of technology used, depends on the desire and ability of the buyer to diversify his sources and the willingness of suppliers to supply the elements separately. Quite often, in developing countries especially, most of the elements are acquired from abroad as part of a single package that may include not only several elements of the technology itself (e.g. processing know-how, designs and management), but also capital and the brand name of the product. In sum, according to UNCTAD, technology is a heterogeneous concept that may take many forms; the mechanisms of its acquisition are also many and varied; and, although the elements of technology are nominally separable, it may be difficult in practice to acquire them separately from one another or independently of other inputs.

As mentioned previously, the commercial transfer of technology, whether by foreign direct investment, licensing or other means, takes place in order to exploit or defend a monopoly advantage. The monopolization of technology may arise simply because it has recently been developed and competing firms have no access to it yet (first stage of the product cycle); or it may be the result of ownership of a patent on the technology. The advantage may also arise from control over other components of the package such as scarce inputs, markets or trademarks. While the latter do not constitute technology, they are associated with its transfer and, linked with it, they increase the monopoly rents of the technology supplier by differentiating the final product in the minds of consumers.

The existence of structural imperfections in the market for technology has been supported by empirical evidence on the concentrated ownership of patents.[30] Geographically, this control of proprietary knowledge is concentrated in a few industrialized countries. Moreover, within those countries it is a relatively small number of transnational corporations that account for the bulk of expenditure on R & D.[31]

Imperfections in the behaviour of the market for technology are related to informational aspects and to its character as a public good. With respect to the former, it is evident that, as in all other markets, the prospective consumer of an item requires information to be able to make a rational decision. However, the peculiarity of technology as a good is that it is itself information, and what is required by the purchaser is information about information. The problem is how the potential purchaser is to obtain sufficient information about the technology to assure him that it is not being falsely represented, without destroying the whole basis for the transaction. In the words of Arrow, "its value for the purchase is not known until he has the information, but then he has in effect acquired it without cost".[32] The public goods character of technology derives from the fact that once it is developed in one enterprise its use by other enterprises does not preclude its continued use by the originating enterprise.[33] Hence for the enterprise that already has access to a particular technology, the incremental cost in its use or transfer is likely to be low—much lower than the average cost of its generation and transfer. The incremental cost of transfer is not zero, since there are costs of local training and of minor adaptation to scale, tastes and other local conditions. Transfer costs have been estimated by Teece, for a sample of 26 international transfers, as ranging from 2.25 to 59% of total investment outlays in a product, with a mean of 19.16%.[34] For the firm seeking to acquire the technology, the incremental cost of reproducing the same technology with its own technical capabilities may be several multiples of the transfer cost. The upper limit of the price of technology will depend not only on its cost of reproduction to the buyer, but also on the elasticity of the demand for its output. The equilibrium price will fall somewhere between this maximum and its transfer cost. If we think of such a market setting as being oligopolistic, therefore, the supply of technology and its price will not be the result of smoothly functioning, impersonal market forces, but rather the outcome of negotiations between buyers and sellers. The final returns and their distribution are a function of the relative power of the bargainers, and that of the supplier is typically the strongest.[32]

As suggested by UNCTAD in some of its ealiest work, developing countries as a group suffer from a particular weakness in the process of bargaining for the acquisition of technology.[36] In contrast to industrialized countries—where technology transfer proceeds in both directions—the flow of technology in developing countries is in an inward direction only. Thus, the average developing country firm has less experience to draw upon in negotiating at the bargaining table. It also means that any adverse welfare effects resulting from the limitations of the market mechanism are compounded against rather than cancel out in favor of the developing country. The concept which best characterizes the unequal position of developing countries in the international market for technology, in UNCTAD's view is dependence.[37] The technological dependence of one country on another—such as that of France on the United States—is not a cause for alarm if it makes greater prosperity possible through the division of labor. What distinguishes the technological dependence of developing countries is its one-sided or asymmetrical nature. There are several aspects of this asymmetry between developing and developed countries which deserve examination, including: commodity patterns—the types of consumer goods consumed in developing countries reflect the influence of those consumed by the rich in industrialized countries;[38] means of production—sharply contrasting abilities to make capital goods; technical knowledge—lack of knowledge in developing countries that was generally available in developed countries and lack of knowledge in both groups of countries that was particularly suitable to developing countries; skills—disparities especially in those skills acquired through learning by doing; finance—the shortage of finance for investment constrains decision-making; control—decisions affecting developing countries were taken in developed countries; and initiative—incapacity to challenge the received technology even when technical ability to do so existed.

The concept of technological dependence provides the backdrop for interpreting the findings emerging from the considerable amount of the early empirical work carried out by UNCTAD and by others concerning technology transfer to developing countries.[39] First, although a significant proportion of contractual agreements for the transfer of technology in developing countries related to the modern sector of industry, a high proportion of contracts was still centered in traditional industries, suggesting a possible scope for reducing dependence on imported technology. Second, the relative hold of majority-owned direct investment as compared with other organizational forms of technology transfer (joint ventures, licensing, direct machinery purchase) appeared to be greater according as to whether the technology was more advanced or rapidly changing and difficult to commercialize. Third, technology suppliers exercised their bargaining power to impose restrictions or limiting conditions on sales of technology which added to the costs of transfer for the importing countries and reduced the benefits in the form of local skill formation and potential stimulation of technological development. These included, for example, tied purchase of inputs of all kinds, export restrictions, limitations on competing imports and on the diffusion of the imported technology, and imposition of guarantees on governments concerning tax and tariff treatment and foreign exchange policies. Fourth, an estimate of the direct foreign exchange costs of technology transferred to developing countries—covering only licensing of patents and trademarks and supply of consultancy services—indicated that by the end of the 1960s the costs may have reached as high as $1,500 million or about 0.5% of their GDP and two and one-half times their domestic expenditures for R & D. If account were taken of other, more difficult to estimate, forms of tacit (over-invoicing of imported intermediate inputs and under-invoicing of exports) and embodied (outlays for machinery and dividend outflows) foreign exchange payments for technology, the total amount would be much larger. Hence, the sums spent by developing countries on technology activities were sufficiently large that their efficient utilization was a matter of significance. Fifth, the operation of the international industrial property system governing patents and trademarks and legitimized by the Paris Convention of 1883 was not sensitive to conditions in developing countries. The great bulk of patents in developing countries were, according to UNCTAD, in foreign hands, and consequently had no connection with local inventive activity.[40]

Over 95% of these patents were never worked and hence offered no stimulus to local production and innovation. Instead, they served as an impediment to domestic production while enhancing monopoly advantages in the import markets of these countries. Trademarks had no technological content but, like technology, they were considered intangible assets and were often associated with licensing of patents. Of approximately four million trademarks in force throughout the world in 1974, one million were granted in developing countries and one half of these were owned by foreigners and most of them for items of consumption for upper-income groups.[41] Although the substantial advertising expenditures for promoting foreign trademarks were borne by consumers in the developing countries, the monopoly rents they generated accrued mainly to their owners abroad and added to the foreign exchange cost described above of the transfer of technology to developing countries.

Policy Intervention

The above points cast doubt on the efficiency of market structures in optimizing the spread of technical progress in the less developed part of the world and provide a strong argument for public intervention. Some time in the late 1960s and early 1970s governments began to take a more self-conscious role in the field of technology. There was and is, however, a fundamental conflict between the commercial interest of individual firms seeking to maximize profits and the national interest of promoting technological autonomy, As emphasized by Baranson, the state's adherence to national goals was inevitably at the expense of profits of established enterprises for whom technological dependence on a foreign patent or trademark was a form of risk aversion aimed at reinforcing the effectiveness of their exploitation of a protected domestic market.[42] One by one, therefore, various countries set up institutional mechanisms to govern the area of technology in the national interest. In some instances this antedated, in others it was in response to the initiative of UNCTAD—the first major United Nations forum to take up the subject in a comprehensive fashion in its Third Conference in Santiago in the spring of 1972.[43] Broadly speaking, the creation of institutional mechanisms has consisted of (1) laws, regulations and administrative controls over the conditions and terms governing the commercial importation of technology, and (2) parallel institutions and measures for strengthening local technological capabilities.

The control of technology imports has included such functions as: screening of patent and trademark licensing, direct foreign investment and other forms of technology transfer in accordance with national development priorities; monitoring of the size and duration of payments obligations; assuring the inclusion of basic obligations of the technology supplier; control over restrictive clauses; and regulation of the registration of patents and trademarks to prevent abuses adversely affecting economic and technological development.

Beginning most notably with the Andean Pact countries in 1970 and Mexico in 1972, some 25 developing country governments—mainly in Latin America and Asia and to a more limited extent in Africa—have overhauled their legislation, introduced new administrative rules and other new measures during the 1970s to regulate the conditions under which imported technology is commercialized in their economies.[44] This has included not only closer surveillance of licensing of patents and trademarks, but also changes in foreign investment laws and industrial property legislation. While the activities of the UNCTAD secretariat are presented under three main sections, emphasis is placed on an integrated approach to all of the activities. In fact, several studies, as well as operational activities, are carried out by joint teams comprising staff members from various subprograms. In particular, research and analysis carried out under several subprograms have gradually built up a significant stock of knowledge of issues related to development and transfer of technology, which is fully utilized by the Advisory Service on Transfer of Technology in its technical cooperation activities. At the same time, the activities of the Advisory Service have made important contributions to research and analysis. This linkage and feedback help to orientate the research work better to the needs of recipient countries, leading to improved technical co-operation delivery, which in turn makes for improved research work.

Based in large part on the analytical and empirical work of UNCTAD, developing countries have also pushed for reform of international law through the negotiation of an International Code of Conduct on the Transfer of Technology and the revision of the Paris Convention for the Protection of Industrial Property. As a sequel to resolution 39 (III), the idea of a Code of Conduct as such first surfaced in a proposal by developing countries in the UNCTAD Intergovernmental Group on Transfer of Technology in early 1973,[45] and by 1974 formed a part of the General Assembly's Program of Action on the Establishment of a New International Economic Order. It took four more years of UNCTAD negotiations by specially created intergovernmental expert groups before the basic structure and content of the Code was agreed upon and a composite text (based on drafts submitted by developing, developed market economy and socialist countries) submitted to the UN Conference on an International Code of Conduct on Transfer of Technology. The draft text of the Code embodies several major objectives sought by developing countries. Of greatest importance, the Code reinforces the right of technology-receiving countries to regulate the transfer of technology. It also condemns certain restrictive business practices, establishes obligations of the parties—chiefly technology suppliers; sets forth modalities for international cooperation and affirms special treatment for developing countries.[46]

Although negotiations over the current revision of the Paris Convention for the Protection of Industrial Property have been taking place under the aegis of the World Intellectual Property Organization (WIPO), it is unlikely the WIPO would have embarked on such a revision were it not for the gentle pressure that has built up in the aftermath of Conference resolution 39 (III) (which, as mentioned, had called for a new study of the patent system), and as a result of the conclusions of governmental expert group meetings based on the findings contained in the joint study of UNCTAD, WIPO and the UN Department of Economic and Social Affairs and in subsequent UNCTAD reports.[47] By 1975, revision of the patent system was included by the Seventh Special Session of the UN General Assembly as one of the priority international measures in its resolution on Development and International Cooperation (3362 (S-VII)). The main purpose of the revision, as seen by its developing country proponents, is to give greater recognition to the technological and economic needs of developing countries, as distinct from the more than adequately recognized rights of patent-holders, through reinforcing the right of developing countries to take national action against abuses of the patent privilege, including especially the lack of use or insufficient use of patented inventions and the presence of restrictive business practices in licensing arrangements.[48]

A very interesting interaction has taken place between the establishment of a legal framework at the national level and the international effort to negotiate a

Code of Conduct on the Transfer of Technology. On the one hand, the draft Code reflects many provisions of national legislation already enacted by a number of governments. On the other hand, the exposure of governments from other countries to the issues raised by the Code negotiations has influenced the establishment of national control measures in those countries. A similar, though less pronounced interaction has marked the revision of the Paris Convention.

It would, nevertheless, be a mistake to jump to the conclusion that existing international law already gives developing countries ample scope to legislate on technology transfer and that the Code of Transfer of Technology and the current revision of the Paris Convention are, therefore, unnecessary. While it is true that such scope exists, it is not unlimited. One of the most notable limitations on national action is the principle of national treatment that is built into both the patent system and the network of bilateral foreign investment treaties that several large industrialized country governments have promoted over the years between themselves and other countries, especially developing countries. "National treatment" gives the foreign investor or patent holder a status equal to that of national citizens. Developing countries that are signatories of either the Paris Convention or one of the above bilateral treaties and that attempt to promote indigenous technology by taking measures favoring their national firms over foreign firms (or over local subsidiaries of the latter) would be in violation of this principle. Moreover, existing "international law" relating to transfer of technology does not correspond to a simple, well-defined corpus of legal principles, but to an admixture of legal traditions based on investment, patent, antitrust and conflict of laws and principles.[49] Thus, what legal principles apply in any given situation is typically a matter of interpretation. Classical international law as such offers little protection against economic and other forms of pressure by technology supplying firms and their governments on individual developing countries seeking to initiate regulation of technology transfer. The influence of rich industrialized countries over international organizations such as WIPO is also a factor. With respect to the Paris Convention, the Model Law on Patents and Trademarks developed by WIPO and disseminated to developing country governments has been based on a narrow interpretation of the Paris Convention and the industrial property system that falls significantly short of the full potential permitted to national regulatory authorities under the Convention.

It follows from what has been said thus far that a restructuring of the international legal framework for transfer of technology, through agreement by North and South on the same rules and goals, is a necessary concomitant of national action by the South. Whatever may be the long-range benefits to the developing countries from this international restructuring—in terms of strengthening technological capabilities—its most important immediate significance is in legitimizing the right of all states to regulate the transfer of technology in their national interest. Closely related to the legitimizing function of international agreement on rules is their pedagogic value for those countries requiring a demonstration of the kinds of regulatory issues to be addressed in the formulation of technology transfer policies.[50] Finally, the resulting universality of norms and standards serves the useful function of mitigating the natural tendency of countries to compete with one another in offering the greatest exemptions and most permissive treatment to potential technology suppliers.

The preceding discussion has centred on one of the two broad types of public intervention in the field of technology—namely, establishment of standards for the control of transfer of technology. The other type of intervention, consisting of measures for strengthening local technological capabilities, is closely related to the first. It involves the establishment of institutions and integrated policies for progressively acquiring the ability to search for and select the various elements of technology (see Section 3 on the transfer of technology) from among alternative sources and to negotiate favorable terms; reinforcement of the capacity to absorb and adapt technology; and creation or strengthening of the scientific and technological infrastructure consisting of design and engineering firms, research institutes and laboratories, and capital goods industries that can design equipment suitable to the evolving circumstances and needs of the country.[51]

Whereas primary emphasis in the early 1970s was on legal and regulatory control mechanisms, some countries in recent years have also turned to affirmative action of the kind described above.[52] Japan is the outstanding precursor of this dual approach,[53] though it does not necessarily represent the unique model for all latecomers to industrialization. The increasing emphasis on affirmative action is also reflected in the published work of UNCTAD which has gradually evolved from general guidelines on technology policies and planning to a strategy for technological transformation and analysis of individual sectors and specific issues such as pharmaceuticals, food processing, capital goods, energy and international skill exchange.[54]

Because of the scarcity of information on the implementation of technology policies, much of what has been written on the subject of technological development consists of working hypotheses and must be treated with a certain degree of caution.[55] Nevertheless, a number of propositions emerge from the limited research that has already been done in UNCTAD and elsewhere.

As stated earlier in this paper, technological knowledge can be transferred, but the ability to use it cannot; it must be mastered through indigenous effort. The justification for the indigenous effort involved in the confrontation with and solution of the range of problems that arise when at least some components of technology are purchased separately rather than passively acquired in packaged form (as in direct foreign investment or turnkey projects) is that it creates the experience that is a requisite for learning. Increased local effort will not necessarily reduce the cost of acquiring technology in the short-run. In fact, it is likely to be less expensive in an industry to rely on a single supplier, with proven expertise, for all of the elements of technology.

There is hence a trade-off between the elevated short-run cost of doing some things oneself and the long-run increase in acquired technical skills and engineering and other know-how. The economic rationale for developing the skills and know-how locally is that they will (1) permit imported technology to be used with greater efficiency, and (2) strengthen capacity to make improvements on imported technology and to create efficient productive techniques not generally available from overseas sources that are appropriate to local conditions. As stated by Lall[56], the promotion of local technological effort is justifiable only when the discounted value of R & D expenditures, higher production costs, lower output quantity and quality, etc. over a period (sufficiently long for the intended results of the effort to be realized) are lower than the benefits yielded by higher rates of growth, more appropriate techniques and products, greater exports, more use of local materials and several forms of externalities.

The pursuit of industrialization policies has a particular focus when technological development is also a goal. First, within the consumer goods sector, the manufacture of standardized goods with a relatively stable technology affords developing countries a greater comparative advantage than, and is to be preferred over, the manufacture of differentiated goods characterized internationally by a constant stream of product innovations. Second, in so far as there is replacement of selected imports, such replacement extends beyond consumer goods to certain of the at least simpler or more standardized types of capital goods and machinery, because of the strategic importance of the latter in generating technological diffusion and raising the productivity of investment throughout the economy.[57] Third, both countries and industries within countries differ in the extent to which they can dispense with turnkey projects and the classic reliance on direct foreign investment as channels for the importation of technology. Much depends on the supply of skilled manpower and the capacity to carry out R & D. Moreover, economies of scale may constrain the pattern of industries that can be created and, therefore, limit the range of exploitable technological opportunities in the domestic market, particularly in smaller countries. Alternatively, a strong case may exist in more advanced developing countries, especially where scale economies are a limitation, for exploiting such opportunities through the selective promotion of infant export industries; though the technological learning possibilities are obviously more restricted if the industries are made up of foreign-owned subsidiaries rather than nationally-owned enterprises.

State intervention to promote industrial R & D may involve the funding of research in government laboratories or the granting of tax exemptions and other incentives to enterprises for in-house R & D. The scattered information available suggests, however, that funding of centralized laboratories and university research facilities has not always generated scientific work that is related to the needs of the productive system and that the industrial innovations that are generated lead to relatively few commercial applications. Research and development appears to be more effective in promoting technological progress when it takes place in research facilities linked directly to manufacturing enterprises or within the enterprises themselves.

Conclusions

Something close to a revolution has occurred since the 1950s and 1960s in the perceptions of academic economists and decision-makers on technology's role in economic development. In part, this has been fed by an improved understanding of the quantitative significance and endogenous nature of technological change as a factor in economic growth in industrialized societies and as a determinant of the composition and direction of international trade. Of equal importance in changing these perceptions has been an increase in knowledge of the behavior and welfare effects of transnational corporations in the exploitation of their technological advantages.

Under the leadership of Surendra Patel, UNCTAD's Technology Division has been in the forefront of the United Nations system in bringing the issue of technology to prominence in the North-South dialogue. Within UNCTAD the launching of negotiations on the Code of Conduct on Transfer of Technology remains a major initiative with lasting repercussions in redefining what constitutes the legitimate interest of developing countries in this domain. UNCTAD has also become the forum for policy debate on another issue of major concern to developing countries—the brain drain or "reverse transfer of technology". (I have not been able to go into this subect in the limited confines of the present paper). Where the specialized, technical nature of the subject has not permitted intergovernmental negotiations in UNCTAD, the pioneering work of Surendra Patel's Technology Division has been instrumental in goading the relevant specialized agencies into action, as in the case of the revision of the Paris Convention on Industrial Property by the World Intellectual Property Organization. It has also been the catalyst behind the proposed code on pharmaceutical industry practices presently being discussed at the international level by the Group of 77 and nongovernmental organizations, another topic I have not covered here due to space limitations.

At the national level, governments of both industrialized and nonindustrialized countries have become aware of the imperfections in the market for technical know-how and of the substantial externalities to be appropriated by society from more rapid and widespread diffusion of such know-how. The issue is no longer *whether* to intervene with the "free play" of market forces in making decisions relating to something as crucial as science and technology, but rather the form that such intervention will take. Although there is disagreement on the relative effectiveness of the various policy approaches—ranging from indicative guidelines and fiscal and other indirect measures, to detailed, *dirigiste* technology planning—no government today can afford to be without some kind of science and technology strategy. In August 1979, this awareness was formalized in the adoption by the international community of the Vienna Program of Action on Science and Technology for Development. Despite important areas of disagreement between North and South, both accepted the basic principle of the need to implement deliberate policies for (1) strengthening scientific and technological capacities of developing countries at national, regional and international levels; (2) restructuring of the existing pattern of scientific and technological relations; and (3) strengthening the role of the United Nations system in science and technology.[58] Moreover, an organization with impeccably conservative credentials as those of the World Bank has been carrying out an ambitious program of research on "technological mastery".[59]

Looking back over the recent past, it now seems that the studies produced by UNCTAD in its early work on technology and by the small number of researchers writing about the subject at that time may have been unduly pessimistic about the prospects for technological advance in that part of the world characterized as being both poor and technologically dependent. For a small group of higher income developing countries with contrasting technology policy regimes, there have been a number of indications of technological development during the 1970s. Many of the countries that nationalized their extractive sectors a decade or more ago have demonstrated by now that they have absorbed and mastered the technology required to manage their nationalized enterprises. The newly industrialized countries have succeeded in expanding exports of manufactured items utilizing standardized and well-established technology and have also begun to move into capital goods exports and technology transfer to other developing countries. Of course, to the extent that some of this trade has taken place between affiliates of transnational corporations, it is not all indicative of indigenous technological progress. It also remains to be seen how these countries will be able to respond to the latest wave of automation that is striking traditional industries. For the majority of developing countries, Surendra Patel himself has argued, on the basis of a striking array of figures on the change in composition of output and trade and other selected indicators of development, that the cliché-ridden image of extreme technological backwardness of the 1950s is no longer applicable today and that there has been an impressive strengthening of these countries' technological foundation.[60]

A daunting task for future research will be to analyze technological change in individual countries in some detail and relate these country experiences to the contrasting policy environments in which they occurred. Research is also needed to explore further the interrelationships between technology transfer and technological innovation, as well as between the latter and industrialization generally. Meanwhile, reflecting on the experience of previous late-industrializing countries in the 19th and early-20th centuries , Surendra Patel maintains that the present world economic crisis may well afford an opportunity for a large part of the Third World to accelerate its technological transformation if only it will take destiny in its own hands.

Notes

1. Two recent collections of articles on the transfer of technology testify to the present intellectual acceptance. They are: R. Hawkins and A. Prasad (Eds.), *Technology Transfer and Economic Development*, Research in International Business and Finance, Vol. 2, (London: JAI Press Inc., 1981); and A. Heston and H. Pack, Special Eds., *Technology Transfer: New Issues, New Analysis, The Annals of the American Academy of Political and Social Science* (November 1981).
2. It should also be noted that he has from the outset been able to rally support from an able and loyal professional staff. The veterans of longest standing include Rolf Andreasson, Daniel Chudnovsky, Rumman Faruqi (who has since left UNCTAD), Gloria Koch, Masafumi Nagao, Peter O'Brien (who has since left UNCTAD), Jorge D'Oliveira e Sousa, Pedro Roffe, Ulrike von Wartensleben and myself. It is not possible to list here the large number of consultants whose contributions to UNCTAD's technology program must also be acknowledged. Three persons who should be singled out both for the duration and size of their inputs are Arthur Ewing, A. K. Sen and K. K. Subrahmanian.
3. Neither this nor the next section have any pretense of being a survey of the literature on technology and development. Rather, the attempt has been in the limited space available to review some of the main trends in the underlying economic theory. I have, for example, omitted a discussion of writings on the impact of technology on employment, consumption patterns and environment.
4. H. B. Chenery and A. M. Strout, "Foreign assistance and economic development", *American Economic Review* (September 1966); and R. I. McKinnon, "Foreign exchange constraints in economic development", *Economic Journal* (June 1964).
5. A. K. Sen, *Choice of Technique* (Oxford: Clarendon Press, 1960).
6. The literature on growth, relating in one way or another to the role of technology, has been exhaustively surveyed by F. H. Hahn and R. C. O. Mathews, "The theory of economic growth", *Economic Journal* (December 1964); C. Kennedy and A. P. Thirwall, "Technical progress: A survey", *Economic Journal* (March 1972); A. K. Sen, *Introduction to Growth Economics* (Harmondsworth: Penguin Books, 1970); and R. R. Nelson, "Research on productivity growth and productivity differences: Dead ends and new departures", *The Journal of Economic Literature*.
7. N. Kaldor and J. A. Mirrlees, "A new model of economic growth", *Review of Economic Studies* (June 1962); and K. J. Arrow, "The economic implications of learning by Doing", *Review of Economic Studies* (June 1962).
8. See H. G. Johnson, "Economic development and international trade", in his *Money, Trade and Economic Growth* (Cambridge, Mass.: Harvard University Press, 1962).
9. J. Bhagwati, "Immiserizing growth: A geometrical note", *Review of Economic Studies* (June 1958).
10. W. W. Leontief, "Factor proportions and the structure of American trade", *Review of Economics and Statistics* (November 1956).
11. M. V. Posner, "International trade and technical change", *Oxford Economic Papers* (October 1961).
12. G. C. Hufbauer, *Synthetic Materials and the Theory of International Trade* (London: Duckworth, 1965).
13. H. G. Johnson, *Technology and Economic Interdependence* (London: Trade Policy Research Centre and Macmillan, 1975).
14. R. Vernon, "International investment and international trade in the product cycle", *Quarterly Journal of Economics* (May 1966).
15. In subsequent versions this assumption was relaxed to allow for first innovation in other high-income countries.
16. A review of this research and of attempts to extend his model is contained in R. Vernon, "The product cycle hypothesis in a new international environment", *Oxford Bulletin of Economics and Statistics* (November 1979). Also relevant is R. Vernon, (Ed.), *The Technology Factor in International Trade* (New York: National Bureau of Economic Research and Columbia University Press, 1970).
17. W. Gruber, D. Mehta and R. Vernon, "The R and D factor in international trade and international investment of US industries", *Journal of Political Economy* (February 1967); and G. C. Hufbauer, *op. cit.*

18. E. Mansfield and A. Romeo, "Technology transfer to overseas subsidiaries by US-Based Firms", Working Paper, Department of Economics, University of Pennsylvania (1979) cited in D. Teece, "Technology transfer and R & D activities of multinational firms: Some theory and evidence", in R. Hawkins and A. Prasad, *op. cit.*
19. S. H. Hymer, *The International Operation of Nations Firms: A Study of Direct Foreign Investment* (Cambridge, Mass.: MIT Press, 1976), (originally presented as doctoral thesis in 1960).
20. R. E. Caves, "International corporations: The industrial economics of foreign investment", *Economica* (February 1971); and "Industrial economics of foreign investment: The case of the international corporation", *Journal of World Trade Law* (May/June 1971).
21. J. S. Bain, *Barriers to New Competition,* (Cambridge, Mass.: Harvard University Press, 1956).
22. J. H. Dunning, "Explaining changing patterns of international production: In defense of the eclectic theory", in S. Lall, guest editor, Special Issue on the Multinational Corporation, *Oxford Bulletin of Economics and Statistics* (November 1979).
23. S. Lall and P. Streeten, *Foreign Investment, Transnationals and Developing Countries* (London: Macmillan, 1977).
24. G. D. A. MacDougall, "The benefits and costs of private investment from abroad: A theoretical approach", *Bulletin of the Oxford University Institute of Statistics* (August 1960).
25. " . . . if the foreign firm simply replaced imports by domestic output, charging the same price and paying the going wages for domestic labor, and remitting the interest on its capital and the rent on its superior knowledge as profits, the host country would gain nothing. In fact, if the foreign investment were attracted by a tariff or by fiscal subsidies, the host country might well lose". H. G. Johnson (1975), *op. cit.*
26. For example, R. Prebisch, "The economic development of Latin America and its principal problems", *Economic Bulletin for Latin America* (February 1962); and C. Furtado, *Development and Underdevelopment* (Berkeley: University of California Press, 1964).
27. S. Patel, "Transfer of technology and developing countries", *Foreign Trade Review* (January-March 1972). See also D. Landes, *The Unbound Prometheus: Technological Change and Industrial Development in Western Europe from 1750 to the Present* (Cambridge University Press, 1969).
28. *The Channels and Mechanisms for the Transfer of Technology from Developed to Developing Countries,* prepared for UNCTAD by C. Cooper with the collaboration of F. Sercovitch (TD/B/AC. 11/5) (April 1971): and UNCTAD, *Guidelines for the Study of the Transfer of Technology to Developing Countries,* UN Sales Publication, No. E.72.II.D.19.
29. *Ibid.*
30 UNCTAD, *The Role of the Patent System in the Transfer of Technology to Developing Countries,* UN Sales Publication, N. E.75.II.D6.
31. The evidence is summarized in UNCTAD, "Technological Dependence: Its Nature, Consequences and Policy Implications", (TD/190) in *Proceedings of the United Nations Conference on Trade and Development, Basic Documents* (Fourth Session, Nairobi, 5–31 May 1976), Vol. III, UN Sales Publication No. E.76.II.D.12.
32. K. J. Arrow, *Essays in the Theory of Risk Bearing* (University of Chicago Press, 1971).
33. K. J. Arrow, "Economic welfare and the allocation of resources for invention", in *The Rate and Direction of Inventive Activity: Economic and Social Factors*—A Report of the National Bureau of Economic Research, (Princeton University Press, 1962).
34. D. Teece, "Technology transfer by multinational firms: The resource cost of international technology transfer", *Economic Journal* (June 1977).
35. C. Vaitsos, *Intercountry Income Distribution and Transnational Enterprises* (Oxford: Clarendon Press, 1974).
36. UNCTAD, "Transfer of Technology", (TD/106) in *Proceedings of the United Nations Conference on Trade and Development, Financing and Invisibles* (Third Session, Santiago, April–May 1972), Vol. **III**, UN Sales Publication No. E.73.II.D.6.
37. UNCTAD "Technological Dependence . . . ", *op. cit.*
38. The notion that internationally transferred "consumption technology" may be inappropriate for the conditions in less-developed countries was first expounded by F. Stewart in "Choice of techniques in developing countries", *The Journal of Development Studies* (October 1972), and is further articulated in her *Technology and Underdevelopment* (London: Macmillan, (Second Edition), 1978).
39. There was a questionnaire survey based on both developed and developing countries; country case studies, including Spain, Hungary, Chile, Ethiopia, Sri Lanka, Japan, Iraq and others;

and several related reports on restrictive business practices prepared in connection with UNCTAD's program of work on manufactures. Some of the work outside of UNCTAD included K. K. Subramaniam, *Import of Capital and Technology: A Study of Foreign Collaborations in Indian Industry*, (New Delhi: People's Publishing House, 1972); C. Vaitsos, *op. cit.;* and J. M. Katz, *Importación de Tecnología, Aprendizaje Local e Industrialización Dependiente: Documento de Trabajo* (Buenos Aires: Instituto Torquato di Tella, January 1972). These early findings are brought together in UNCTAD, *Major Issues Arising from the Transfer of Technology to Developing Countries*, UN Sales Publication, No. E.75.II.D.2.

40. UNCTAD, *The Role of the Patent System in the Transfer of Technology to Developing Countries*, UN Sales Publication, No. E.75.II.D.6.
41. UNCTAD, *The Role of Trademarks in Developing Countries*, UN Sales Publication, No. E.79. II.D.5.
42. J. Baranson, *North-South Technology Transfer: Financing and Institution Building* (Mt. Airy, Maryland: Lomond Publications, 1981).
43. An intergovernmental Group on Transfer of Technology, which first met in 1971, paved the way for consideration of technology at the Third Conference. Conference resolution 39 (III) Transfer of Technology *inter alia* invited developing countries to: (1) "establish institutions, if they do not already have them, for the specific purpose of dealing with the whole range of complex questions connected with the transfer of technology from developed countries"; (2) "take specific measures they deem necessary to promote an accelerated transfer of adequate technology to them under fair and reasonable terms and conditions"; (3) "apply the provisions of the International Development Strategy concerning allocation of resources for research and development"; (4) "develop an efficient infrastructure geared to the specific socio-economic needs of each country as a solid basis for the adoption and/or adaptation of imported technology, the creation of national technology and the application thereof and strengthening of its domestic, scientific and technological capacities"; and (5) among themselves at the regional and interregional level to consider various forms of technological cooperation. Most significantly, it also laid the basis for the International Code of Conduct on Transfer of Technology by resolving to request the Secretary General of UNCTAD and the Director General of the World Intellectual Property Organization to "study the possible bases for new international legislation regularizing the transfer from developed to developing countries of patented and non-patented technology". UNCTAD *Proceedings* (Third Conference), *op. cit.* Vol. **I.** *Report and Annexes.*
44. UNCTAD, *Commom Approaches to Laws and Regulations on the Transfer and Acquisition of Technology*, TD/B/C.6/91 (October 1982).
45. UNCTAD, "Report of the Intergovernmental Group on Transfer of Technology on its Second Session", *Official Records of the Trade and Development Board, Thirteenth Session*, TD/B/424, Annex II (New York: United Nations, 1974).
46. UNCTAD, *Draft International Code of Conduct on the Transfer of Technology as of 10 April 1981*, TD/ CODE TOT/33 (May 1981).
47. Resolution 3 (I) of the Committee on Transfer of Technology, for example, sets out the main considerations that should guide the revision of the Paris Convention on the basis of the agreed conclusions and recommendations of the group of governmental experts convened by the UNCTAD Secretary General in September 1975 to study the role of the patent system in the transfer of technology to developing countries. UNCTAD, *Official Records of the Trade and Development Board*, Seventh Special Session, Supplement No. 4, TD/B/593, Annex I.
48. See UNCTAD, *Technology: Restructuring the Legal and Juridical Environment. Issues under Negotiation*, TD/237 in Proceedings of the United Nations Conference on Trade and Development, fifth session (Manila: May/June 1979), United Nations publication, Sales No. E.79.II.D.16.
49. A. Fatouros, "International controls of technology transfer", in T. Sagafi-Nejad, R. Moxon and H. Perlmutter (Eds.) *Controlling International Technology Transfer—Issues, Perspectives and Policy Implications* (New York: Pergamon Press, 1981); and D. Miller and J. Davidow, "Antitrust at the United Nations: A tale of two codes", *Stanford Journal of International Law* (Summer 1982).
50. I have heard of at least three recent cases of developing country governments that have used the draft Code of Conduct in negotiating contractual arrangements with technology suppliers.
51. For a detailed consideration of the nature and purposes of technology policies and planning, see UNCTAD, *Handbook on the Acquisition of Technology by Developing Countries*, UN Sales Publication,

No. E.78.II.D.15, and *Planning the Technological Transformation of Developing Countries*, UN Sales Publication, No. E.81.II.D.10.

52. Much less information is available on actual government policies of this kind than on laws and regulations. However, see J. Baranson, *op. cit.* and R. Driscoll and H. Wallender III, "Control and incentives for technology transfer: A multinational perspective", in Sagafi-Nejad, Moxon and Perlmutter, *op. cit.*, for evidence of a shift in policies.
53. See UNCTAD, *Case Studies in the Transfer of Technology: Policies for Transfer and Development of Technology in Pre-War Japan*, TD/B/C.6/26 (April 1978); and T. Ozawa, "Technology transfer and Japanese economic growth in the postwar period", in Hawkins and Prasad, *op. cit.*
54. Some selected recent references include: UNCTAD, *A Strategy for the Technological Transformation of Developing Countries*, TD/277 (January 1983); *Guidelines on Technology Issues in the Pharmaceutical Sector in the Developing Countries*, UN Sales Publication, No. E.82.II.D.15; *Outline of a Policy Framework for the Transfer, Application and Development of Technology in the Food Processing Sector*, TD/B/C.6/AC.6/3 (April 1982); *The Capital Goods and Industrial Machinery Sector in Developing Countries*, TD/B/C.6/AC.7/2 (May 1982); *The Energy Sector in Developing Countries: Issues in the Transfer, Application and Development of Technology*, TD/B.C.6/AC.9/2 (August 1982); *The Feasibility of Measuring International Flows of Human Resources*, TD/B/C.6/AC.8/2 (May 1982); and *Towards an Integrated Approach to International Skill Exchange: Proposals for Policy and Action on Reverse Transfer of Technology*, TD/B/AC.35/2 (June 1983).
55. The research findings of a vast programme of empirical studies on the economics of technological change at the firm level in the six largest Latin American countries (IDB/ECLA/IDRC/UNDP Research Programme on Science and Technology in Latin America) are in the process of being written up. Some preliminary conclusions have been presented by the director of the programme, J. Katz, in his paper entitled "Domestic Technological Innovation and Dynamic Comparative Advantage: Some Reflexions on a Comparative Case-Study Programme" which he delivered at the Seventh World Congress of the International Economic Association at Madrid, Spain, September 1983. A research programme on "technological mastery" is underway at the World Bank. See L. Westphal and C. Dahlman, "The Acquisition of Technological Mastery in Industry", mimeo (Washington D.C.: The World Bank, 1981). See also F. Stewart, "International technology transfer: Issues and policy options", *World Bank Staff Working Paper*, No. 344 (1979); and S. Lall, *Developing Countries as Exporters of Technology: A First Look at the Indian Experience* (London; Macmillan, 1982).
56. S. Lall, *op. cit.* (1982).
57. N. Rosenberg, *Perspectives on Technology* (Cambridge University Press, 1976); and UNCTAD, *The Capital Goods and Industrial Machinery Sector*, TD/B/C.6/AC.7/2.
58. *Report of the United Nations Conference on Science and Technology for Development* (Vienna: 20–31 August 1979), United Nations Publications, Sales No. E.79.I.21.
59. See reference in note 55.

60 S. Patel, "The age of the Third World", *Third World Quarterly* (January 1983).

Science, Society and Endogenous Development

GEORGE SKOROV

The second half of the 20th century has shown quite clearly that science has had an immense impact on all countries of the world, developed and developing. The achievements of the latter in health care and education, the use of natural resources and the development of agriculture, industry, transport and communications would have been impossible without the application of modern science and technology. It is the scientific and technological revolution that has created unprecedented possibilities for less developed countries to skip many intermediate steps in the development of the productive forces in a, historically speaking, relatively short term. But the grasping of these opportunities lags behind expectations. The gap between the developed and the developing countries continues to widen. The scientific and technological revolution, while solving some problems, is creating others, often even more complex. This happens in all countries, but more often in developing ones.

Who would have thought, for example, that a lowering of infant mortality, lengthening of the expectation of life, a rise in general education, the "green revolution", and the use of highly productive modern technology would bring developing countries such new problems as the "demographic explosion", a rapid disintegration of the old social structures and the acceleration of a new class stratification, the flight to the cities and the exacerbation of the employment problem, which are so much more difficult, and perhaps more costly, to cope with now than only a few decades ago? Why have these new phenomena, problems and contradictions arisen? What did we leave out of account? Where did we go wrong? Is it possible to eliminate the negative consequences of the modern scientific and technological revolution?

There is no simple answer to these questions, but analysis of the ways and forms in which the scientific revolution unfolds and manifests itself in developing countries may help to clarify the subject.[1] Progress in dealing with some of the problems of national development is primarily the result of their own efforts, but it is also true that it is based as a rule on the application of foreign scientific discoveries and technical innovations. The first steps can be made in this way in any field and a partial solution of one problem or another achieved, but the overall development problem cannot be dealt with by relying on the achievements of others. What is needed is a rapid development of national science and massive applied research, so that the results of the R & D are brought to the stage of technological innovation. Most developing countries are unable to do this.

The use of science for development purposes and technical re-equipping of the economy call for the appropriate social climate. Sweeping social reforms are therefore a most important precondition for the successful solution of this task. Wherever these conditions are not met, the development of science yields no tangible results, or almost none, while the use of imported technology often leads to negative social consequences.

As a rule, science and technology in developing countries is of an enclave character. Innovations affect some industries only, mainly exporters, and leave the bulk of the economy untouched. In many cases, low priority problems remote from the population's needs are tackled by means of science. This deformed, limited development leads to bottlenecks and other imbalances among differing industries and sectors of the economy and to marked contrasts between the standard of living of the elite and of the bulk of the population.

Because those matters have appeared *ad nauseum* in the specialist literature and analyzed in many United Nations documents,[2] there is no point going into them here, save to note that the causes holding back the use of science and technology in the interests of development include certain aspects of the development of science itself, the domestic socioeconomic conditions in developing countries, and the international environment. Only, it would seem, by acting on all these factors simultaneously, can we expect a greater contribution of science to developing countries' problems.

Any appreciation of the advances and drawbacks in forming a national science capability in the developing countries cannot simply be a statement of the facts; it calls for a differentiated approach to the various groups of countries as well as certain criteria. An attempt to assess the science capacity of the developing world as a whole will inevitably distort the real picture because of the immense diversity in the science potential of each country. The nations of Asia, Africa and Latin America, moreover, started building their scientific capability at different times, so that to evaluate success and failure in this field we have also to allow for the time scale and the starting point. The criteria for Mexico and Brazil, for example, will be quite different from those for Chad, Upper Volta or Nepal. While weak and unorganized links between science laboratories and industry can be adjudged a drawback in the former, the founding of even one or two such elementary research units in the latter is an undoubted advance.

The main achievement of the developing countries in this field is, perhaps, the growing consciousness of the importance of creating their own national science capacity; and this consciousness is the basis of their attempts to foster the interaction of its different components. Thus, there is now hardly a single country that does not have at least a minimum science capability. In it is the growth of a national system of education, especially at the highest level, which has played an important part. In the early 1980s, the total number of students in the universities and higher schools of developing countries was over 18 million, or almost 20% more than in all the Western industrial countries, taken together in the mid-1960s.[3] In some countries considerable efforts have been made to bring the content of education and teaching methods close to the needs of national development. A tendency has emerged in post-primary education to add practical and technical disciplines to the curriculum and further diversify it. Vocational and technical training has been expanded. Attention to out-of-school education, to the founding of adult education centers, and to the use of radio and television for instructional purposes, has markedly increased. In some countries, steps have been taken toward introducting or expanding teaching in the mother tongue.

As a result of these positive developments, there has been a marked increase in the numbers of scientists, engineers and other professional and technical cadres, and parallel with this growth, an increase in the number of laboratories and the beginnings of a number of R & D services.

Appropriations to develop science have also increased; there are examples of the use in developing countries of certain forms and methods of financing science employed in the industrially advanced countries of the West. In Peru, for example, they are trying out the financing of science through taxes on industrial output.

The recognition that developing countries have given to the idea of formulating a national scientific and technological policy can also be considered a major achievement. Efforts to centralize the management of science, to plan research objectives on a national scale, and to devise national problem-oriented programs are being actively encouraged. Individual countries have already gathered some experience in formulating concrete development plans for science and technology.

We must also note that certain developing countries have advanced to foremost positions in the basic research. For all the contradictory nature of this process, we cannot shut our eyes to the positive significance of this achievement.

Another point deserving a positive evaluation is that developing countries are not shutting themselves in simply on their own problems, but are striving to take part in the solution of matters occupying the world's best brains. An over-enthusiastic concern for international problems, however, is often detrimental to the tackling of national tasks; and achievement can turn into a drawback.

The establishment of a network of industrial consultancy services that gather scientific information, provide advice on planning and design work, carry out geological, geophysical, meteorological, and other surveys, and help raise the qualifications of specialists, can be counted an asset in certain developing countries.

An important achievement, and, at the same time, a means for future development of the national science capacity, is consolidation of the government's coordinating activity as part of a national development strategy on which scientific and technological policy is based and financed. It is to this purposeful activity of their governments that developing countries owe their recent progress in expanding and consolidating their R & D capability.

Together with the unquestionable achievements in building a science capability, developing countries have also come up against many unresolved problems and difficulties; and if these are not eliminated in the years to come, the efficiency of using science for development will remain comparatively low, and may even deteriorate.

One of these problems stems from the small scale and inadequate quality of their science capability. In many countries, it is a matter of a score or so of small, poorly equipped, understaffed laboratories. Though they often do useful work, they are unable to carry out major research.

Although the consolidated figures do not accurately reflect the position in the individual countries, they nevertheless say quite a lot. The number of scientists, engineers and technicians per 10,000 of the economically active population in developing countries in 1975 was only a fortieth of the numbers in industrially developed countries.[4] The expenditure of developed countries on R & D in the mid-1970s was 43 times that of the developing countries,[5] which have scarcely reached the \$5 billion recommended in the Second United Nations Development Decade.[6] More than half of all this outlay, moreover, has been made by seven countries: India, Indonesia, Argentina, Brazil, Mexico, Iran and Venezuela. It is not surprising that the share of all developing countries in the patents issued in the world in 1980 was only 9% and in the production of technically sophisticated goods 3%.[7]

To come back to the matter of scientific cadres, the low quality of their training is an acute problem, apart from their lack of numbers. Scientists, engineers and technicians are the key to a science capability. If they lag behind present-day qualifications, it is hard to expect a country with such personnel to be able to ensure the requisite standard of research and development. Furthermore, apart from India and a few other countries where there are centers of advanced science, the problem of post-graduate training and refresher courses for young scientific workers is far from solved in the developing world. Even in the most advanced developing countries, the training of high-level manpower cannot be considered fully solved. It is no accident that, with this background of a general lack of technical expertise, there is the paradoxical situation of "overproduction" of engineers and scientific workers. In India, for example, unemployment among graduate engineers was around 17% in the mid-1970s; and that at a time when there was an acute shortage of highly qualified technical experts in many branches of the economy.

While having to overcome immense difficulties in training scientific, engineering, technical and medical workers, developing countries are losing a large number of qualified people, who leave in search of work in the West, or are remaining abroad after their training in colleges and universities. More and more often this is being referred to in United Nations documents as a "reverse transfer of technology", given the positive economic effect it has in consolidating the R & D capability of developed capitalist countries. The UNCTAD Secretariat has estimated that, from the beginning of the 1960s to 1975, 420,000 professional and technical workers migrated from developing countries to the USA, Canada, Great Britain and Western Europe.[8] While there is no estimate yet for more recent years, it is the opinion of several experts that the inflow of skilled people into these countries has not declined significantly, which means that the total gross migration of skilled personnel to the West would amount to between 600,000 and 700,000 people in the early 1980s.

A drain of high-level manpower on such a scale has a negative effect on the science capacity of developing countries; it leads to a waste of reserves of trained personnel and is threatening successful fulfilment of social and economic programs. The direct economic loss of the "brain drain" is enormous. It has been put at $46 billion for the period 1961–72, or around $3.8 billion a year,[9] more than the annual expenditure on R & D.

A factor seriously holding back the development of research is the weak technological equipment of university laboratories and research institutes in developing countries: the lack of elementary laboratory and other scientific equipment and the absence of domestic production thereof. The majority of developing countries are therefore completely dependent on imports of such equipment from industrially developed nations.

So far, there has been a grave discrepancy in most African, Asian and Latin American countries between national development needs and the orientation of research. A common defect still found is the irrelevance of the research themes of many scientific institutions to the basic development needs of their countries. The efforts of many scientists are concentrated on lines of research corresponding primarily to the priorities of industrially developed countries or to the development of basic science. The economic effect of such an orientation ultimately benefits the international community, but not the developing country concerned, the source of the discovery. Frequently about half the appropriations of newly independent countries are earmarked for R & D in such fields as atomic energy, space research, defence, and telecommunications, to the detriment of work on health protection and the struggle against tropical diseases; the development of agriculture; the use of alternative sources of energy; the wider use of local resources; and industrial research.

The deformation of the development of science in these countries is made worse by the operations of transnational corporations (TNCs), which employ the most valuable, skilled national cadres in their own interests. As a rule, the subsidiaries of the TNCs were accustomed to having at their disposal research services whose job it was to adapt production processes to local conditions or to control the quality of raw materials and the end-product. Now the research work of TNCs is being organized more and more along the lines of international specialization, with each laboratory doing fundamental or applied research in the context of the corporations' general programs. Even the indirect link between the work of the laboratories and local needs is then ended. This policy can be qualified as a "brain drain" right on the territory of developing countries.

The worst shortcoming in the building of a science capability in developing countries preventing it from acquiring a genuine R & D capacity is perhaps the almost total absence of adequate experimental and prototype design facilities. This is a principal cause of the exceptionally low returns on investment in science development, since the results of research cannot be taken to the stage of technological innovations without such a base. Regardless of the scale of research inputs and development of science, if this lack is not overcome in the near future, the efforts will remain ineffective.

Efficient use of a national science capability is still a problem in developing countries because of the lack of an institutional machinery linking the separate elements of the R & D and production capacity together in a single, functioning whole. Something has already been done in this direction through the development of a science and technology policy and the adoption in some cases of plans for science and technology development. But the problem of implementing declared policy and adopted plans, i.e., of reaching planned objectives, is still far from solved, even in the advanced countries of the developing world, not to mention the less developed ones. Yet it is one of the most important issues in development and will remain on the agenda for years to come.

Finally, we must mention yet another factor responsible for the present not wholly satisfactory state of the R & D capability of developing countries: the transfer of technology. For the greater part of the postwar period, this was considered in many developing countries, and in some is still so considered, as the main, and almost the only, means of overcoming technical backwardness and raising the economy to a new technological level. The experience of the three and a half postwar decades, however, has quite clearly shown that, although the import of foreign technology does help to raise the technological level of production, it does not in itself in any way lessen technological and financial dependence on foreign countries; rather the contrary, it increases dependence, albeit in other forms.

It suffices to recall that payments for the import of foreign technology have risen, according to UNCTAD estimates, from \$1.5 billion at the end of the 1960s to \$9 billion at the end of the 1970s (constituting 6% of the total investment in fixed capital), and allowing for various overpayments of \$30 to \$35 billion.[10] According to the available data, the cost of imports of technology may come to between \$80 billion and \$140 billion a year by the year 2000.[11]

The absurdity of such a policy is evident, because it has not led, and will not lead, to the ending of technological dependence. It will also not change the place of developing countries in the international division of labor by the year 2000, contrary to their often-expressed aspirations; this is why their resolve to set up and consolidate their own R & D capability hardened in the United Nations system of specialized agencies in the second half of the 1970s; hence, too, the corresponding tasks for the development and orientation of a national science, extension of the experimental and design base, and the strengthening of production capacity.

Dissatisfaction with the result of development based on maintenance of the former colonial periphery's dependence on the industrial centers of capitalism[12] has inspired a search for new models and theories of development. One of the most popular of these is the concept of endogeneous development, according to which the development of any country should be governed by internal and not external impulses and interests, and should proceed without any pressure whatsoever from outside.

Since the main sources of development are located inside a country, the development process should be based mainly or even wholly on the use of internal resources. For that purpose, society's resources have to be redirected to meeting the population's "basic needs" and preventing an elitist type of development. Such are the general outlines of this concept.

It is spearheaded against developing countries' growing dependence on industrially developed market economies, which is a consequence of their involvement in an unequal system of international division of labor. At the same time, it rejects the theory of stages of growth and the model of "catching up growth" built into this system.

The strategy of endogenous development also envisages a conscious respect for national traditions and cultural values that must be organically built into the new structure.[13] A special place is allotted in this concept to the so-called "appropriate technology", the purpose of which is to utilize local resources and labor to the maximum. Although this term cannot be considered adequate, it draws attention to the problem of developing an original technology meeting the special needs of developing countries.

It would be proper here to make the reservation that endogenous development is still only a concept and not a reality, just like the new international economic order, with which it is closely associated. If we take it that the closest practical embodiment of this concept is self-reliant development, then distinct components of endogenous development can obviously be found in such countries as Algeria, India, Peru, Tanzania, Zambia, Ethiopia. These components were characteristic of the development of the USSR from a backward agrarian economy to an industrially and technologically advanced one. It can be said, therefore, with some justification, that the concept of endogenous development in its spirit, if not the letter, is close to the socialist model of development.[14]

In any case, the concept stands in need of further theoretical deepening, as well as the elaborating of its concrete aspects. In particular, such issues as the role of science and technology in the transformation of established social structures, the relation of world and national cultures, the delimitation of progressive and archaic traditions, the filling of old forms, hallowed by ages, with a new content, need to be more closely defined.

A central place among the practical issues in the strategy of endogenous development belongs to the use of national resources. It is generally accepted that these are limited, that their utilization at present leaves much to be desired, and that a considerable part of them are flowing abroad. At the same time, the prospect of broad, massive international financial aid, given the unabated arms race, remains more than problematical. From that angle, the strategy of endogenous development not only relies on internal resources, but also implies careful planning of their use on both the macroeconomic and the microeconomic levels (which also fully applies to the resources for developing science). The concentration of resources and efforts on priority problems that will spin off benefits for other fields is particulary important here, as well as coping with these problems through the efforts of national scientists and technologists.

Endogenous development entails the reinforcement of the manpower engaged in R & D activities, wider use of the traditional technological base, and systematic modernization of the country's productive system. Endogenous development also requires substantial strengthening of the experimental infrastructure so as to ensure the use of research results. An important element of such development is a careful selection of foreign technologies and the limiting of imports mainly to those items that are essential for transformation of the country's production base.

How this strategy will be implemented in practice the future will show. It seems to be important, however, to stress that it would be wrong to construe the idea of endogenous development and self-reliance as a new variant of autarky, a rejection of the international division of labor, and developing countries' self-isolation from the outside world. That extremist view would be unrealistic and harmful. It seems essential that the developing countries continue to use international trade and technology transfers, and, more generally, world scientific and technical experience, as effective instruments of restructuring their economies in accordance with their own real needs, and not in order to enrich TNCs and serve the needs of industrially developed countries. Reducing dependence is not equivalent to curtailing links; it simply means in the long run, putting these links on to a more equal basis.

Such a variant of development is quite feasible. It has its difficulties, of course, and drawbacks, but its undoubted advantage is the unleashing of local initiative, an increase in the use of local resources, and opportunity for development in the interests of the population, assertion of national dignity, and in the long run, the achievement of economic self-reliance.

Because the scientific, technical and economic potential of most developing countries is not unlimited, and implementing many development schemes calls for ever bigger resources, the attempts to pool the efforts of separate countries to tackle common problems has become increasingly significant in recent years; self-reliance is being seen more and more as collective self-reliance. This trend is clearly visible in the development of regional cooperation in the R & D field.

The interest of developing countries in regional cooperation stems from the common tasks they face in applying science and technology. This cooperation is taking various forms. The most common is the exchange of scientific and technical information on a bilateral or multilateral basis. A large part of this exchange takes place at various international conferences and symposia—from meetings of experts, to conferences of ministers responsible for science policy. In addition, it takes the form of an exchange of publications between various national and regional scientific centers in Asia, Africa and Latin America. And, of late, there has been active exchange of technological information within the organizations of countries exporting petroleum, mercury, bauxite, iron ore, copper, coffee, sugar and other products.

Another widely practiced form of regional scientific and technical cooperation is the organization of short-term refresher courses for experts in various trades. Such courses are regularly conducted by UNESCO, UNIDO, FAO, WHO, UNCTAD and other United Nations institutions in all regions.

Another form of cooperation is the joint elaboration and implementation of large-scale development projects. Their distinguishing feature is joint research and development and the use of modern technology. The schemes include, in particular, comprehensive exploitation of the basins of big rivers (e.g., the Mekong in Asia, the Niger in Senegal, and Volta river basin in West Africa); the building of transcontinental highways (Trans-Sahara, Asiatic); joint mapping and geological prospecting work; and joint measures to combat mass epidemics and dangerous infectious diseases (schistosomiasis, trypanosomiasis, etc.).

International and regional centers are playing an increasing part in promoting cooperation in science and technology. The best known are the international agricultural research centers in Mexico, the Philippines and Nigeria; the regional research and teaching institutes in Bangkok, Beirut, Dakar and Santiago; the center for the transfer and development of technology in Asia (Bangalore); and the similar, recently founded centers in Africa and Latin America. The matter of setting up a Latin American network of technological information (RITLA) is now on the agenda.

Regional cooperation in training engineers, technicians and scientific cadres is expanding, and now includes the following: exchange of syllabuses; seconding of lecturers and students for practical work; aid to less developed countries from more developed ones (e.g., from Mexico to Venezuela, Ecuador and Senegal); study in the universities of neighboring developing countries; and the founding of special teaching establishments for foreign students.

A promising form of cooperation is the founding of joint industrial undertakings (especially in industries producing means of production or working for export) and the rendering of technical assistance in the equipping of industrial enterprises by advanced developing countries (India, Brazil, Mexico and Argentina). The biggest projects of this kind are the steelworks being built jointly by Brazil (51% of the capital) and Mexico (49%), the Mexican-Jamaican aluminum works, the Venezuelan-Colombian petrochemical and transport undertakings. The building of Argentine-Brazilian steelworks, the output of which would be exported in part to Afro-Asian countries, is under discussion. Cooperation in the building and operation of the infrastructure projects (railways, power lines, communications, airways) is gradually growing. Agriculture presents a broad potential area of cooperation in which an intensive exchange of experience has already long been underway through the efforts of the FAO.

The plans of action (world and regional) for the application of science and technology to development prepared at the beginning of the 1970s performed an informative function of no little importance. For all their shortcomings, these documents undoubtedly helped in the drafting of national science and technology plans, especially in emphasizing the importance of regional cooperation in this area. A logical continuation of this activity was the World Plan of Action for technological cooperation among developing countries adopted in Buenos Aires in 1978.[15] An important source of information for the delegates at the Buenos Aires conference was the catalogue of services in the field of technological cooperation among developing countries published by UNIDO in 1977, in which 900 organizations in 67 countries were listed that provide this kind of cooperation for other developing countries. The list named teaching establishments and institutions for refresher courses, research centers, pilot plants, and centers providing instructors or information on specific matters.

Though regional scientific and technological cooperation has been enriched in the 1970s and 1980s by new ideas, agreements and machinery, it would be wrong to say that it has reached a level that would enable developing countries to cope successfully with the problems facing them. The organization of a really broad and effective collaboration in this area is still, apparently, a matter of the future.

To say what needs to be done in the immediate future, it is advisable to take into consideration which of the planned but unfulfilled or partially carried out objectives still remain pressing: the new task arising of late; the experience of the more advanced countries' scientific and technical development; and the limited resources that developing countries can really count on in the foreseeable future.

In light of these considerations, the following would seem appropriate as a means of improving the contribution of science to national development problems:

—Maximum efforts should be made to consolidate national R & D capability, especially by improving the training and rational employment of science cadres and raising the technical equipping of laboratories, research centers, and universities. In reality, a science capability does not exist in and of itself, but is a component of a country's R & D and industrial capacity. It is, therefore, particularly important to provide for the building of an adequate experimental and design infrastructure along with the consolidation of the science infrastructure, as well as a production base which together form the key conditions for industrial application of the scientific results obtained, and generation and diffusion of technological innovations. The quest for institutional forms that link the constituent elements of the R & D capability into a single, coordinated, efficient mechanism is particularly important; without it, a rise in the effectiveness of developing countries' use of their science potential is not easily envisaged.

—The level of national science and technology policy has to be raised and efforts made to bring it to the stage of specific plans of science and technology development, and (which is especially important) to ensure the conversion of these plans into reality. This task is closely linked with that of improving the efficiency of planning in general, since science and technology plans derive in principle from the overall economic development plans and are a constituent of them. Realization of this task will call for a further strengthening of the role of national governments, consolidation of the public sector of the economy, consistent renewal of the social structures of society, and elimination of all the social obstacles in the way of development. Spread of scientific knowledge among the population and an increase in the science component in school education would have a positive influence on this plane.

—The orientation of science to obtain end results and deal with the most pressing task of economic development needs to be strengthened. The list of these tasks should correspond to a country's actual conditions, and the number of them should be commensurate with its possibilities, so as to avoid dissipation of efforts. From that follows the significance of the right choice of priorities in the development of science and of raising the role of applied research so as to yield beneficial economic effect in the near future. This task also includes broader use of the knowledge amassed in the national culture concerned, which would help preserve national distinctiveness and originality and consolidate and further develop each country's cultural heritage.

—Scientific and technical cooperation between developing countries on regional and subregional levels needs to be established and extended in the context of a policy of collective self-reliance. On that plane, the following could play a big role:

(a) the creation of a modern scientific and technical information service meeting the needs of several neighboring countries and its eventual linking with the UNISIST system;

(b) the holding of regular consultations between the national bodies responsible for the management of R & D activities with a view to exchanging research plans, and subsequently to coordinating research efforts to cope with the most important common problems;

(c) a substantial strengthening of regional centers for the development of science and transfer of technology as an important link between the development of science and technique in industrially advanced and developing countries. In this area, the most complicated task is the use of technological innovations on a collective basis, since it is connected with heavy national investments of capital, patent and license legislation, the sharing of profits and losses, etc. These difficulties are not insuperable given goodwill by the countries involved.

The working out of measures linked wth the use of science and technology within the context of the programs of the New International Economic Order will need to be continued, including a code of conduct for transfer of technology, a code governing the operations of transnational corporations, legal regulation of exploitation of the wealth of the world's oceans, etc. At the same time, it is necessary to allow fully for the lessons of the international economic negotiations of the 1970s. The main one was that the restructuring of existing international relations is a lengthy, complicated business, and that progress toward a New International Economic Order will obviously be slow and thorny. Many developing countries have concluded from this that they must, while not relaxing their fight to restructure international economic relations, rely more on their own abilities and forces, actively mobilizing internal resources.

A major issue of the 1980s, on which the fate of our planet hangs, is where the development of international relations is heading—to a relaxation of tension that will provide a chance to tackle urgent global problems, or once more to Cold War, which is fraught with the danger of a world thermonuclear confrontation. The solution of all other matters will hang on that, including the use of science and technology for development purposes. The outcome of the 9th and 10th Special Sessions and 37th regular session of the United Nations General Assembly, like other recent international negotiations on development issues, convincingly demonstrated the negative effect of the worsening international political situation on the tackling of pressing world problems. It is quite evident that if the international situation remains as tense as at present and the arms race continues unabated, the outlook for a restructuring of international economic relations will remain extremely problematical. The utter insanity of a continuation of the arms race is obvious. Not a single one of the urgent problems now facing the world can be solved from a position of strength, which is why a return to detente and arms control, and subsequent disarmament, is a vital business for everyone who aspires to a more just international economic order and a more efficient use of science in the interests of development.

Notes

1. See, in particular, G. E. Skorov (Ed.), *Science, Technology and Economic Growth in Developing Countries.* Translated by Jenifer Warren (Oxford: Pergamon Press, 1978).
2. See, for instance, the Report and other papers of the United Nations Conference on Science and Technology for Development (Vienna: 20–31 August 1979).
3. UNESCO, *Statistical Yearbook* (1972, 1980, 1982). Estimates for 1982 based on Table 2.2, p.II–11 in 1982 Yearbook.
4. UNCTAD, *Technology: Development Aspects of the Reverse Transfer of Technology* (Manila: UNCTAD V, May 1979), p. 29.
5. *Ibid.*
6. *Science and Technology for Development. Proposals for the Second United Nations Development Decade* (New York: United Nations, 1970), pp. 14, 43.
7. Derived from WIPO, *Industrial Property Statistics* (1980). UNCTAD, *Technology: Development Aspects of the Reverse Transfer of Technology, op. cit.*, p. 29.
8. UNCTAD, *ibid*, pp. 2–3.
9. *Ibid.*
10. UNCTAD, *Major Issues Arising from the Transfer of Technology to Developing Countries* (22 April 1974), p. 43; UNCTAD, *Specific Matters Arising from the Resolutions, Recommendations and Other Decisions Adopted by the Conference at its Fifth Session.* Requiring attention or action by the Board at its 20th session (27 February 1980), p. 2.
11. UNCTAD, *Major Issues Arising from the Transfer of Technology to Developing Countries, op. cit.*, p. 41.
12. Dr. Osvaldo Sunkel and a number of other Latin American economists (usually referred to as the "Prebisch School") have provided a detailed analysis of *dependencia* (a development mode based on dependence).
13. Interpretations of endogenous development can be found in a number of UNESCO publications: for instance, the secretariat document, *Concept of Development Centered on Man* (Paris: UNESCO, 20 April 1979); *Medium-term Plan* (1977–82) Doc. 19 C/4 (Paris: UNESCO, 1977), Chaps. III and IV; I. Alioshina, *UN Contribution to the Elaboration of New Theoretical Development Concepts,* Doc. SS-80/CONF.621/6 (Paris: UNESCO, 1980), Chap. II.
14. UNCTAD, *USSR's Experience in Building Up Technological Capacity,* study prepared by Prof. G. E.Skorov at the request of the UNCTAD secretariat (Geneva: UNCTAD, 1980).
15. "The Buenos Aires Plan of Action for Promoting and Implementing Technical Co-operation Among Developing Countries" (12 September 1978).

ANNEX I

Bibliography of Surendra Patel's Personal and Principal Official Publications, 1945 to 1983*

I. *Books:*

1. *Agricultural Labourers in Modern India and Pakistan* (Current Book House) Bombay, 1952; Russian translation with an introduction by G. G. Kotorsk: published under the title Sel'skokhozyaistvinniye Rabochiye v Indii i Pakistanye (Moscow, 1955).
2. *Peasant Uprisings in India (1880–1900)*, (Bombay 1953); also reprinted in Desai, A. R. (ed.), *Peasant struggles in India*, Oxford University Press, 1979.
3. *Essays on Economic Transition* (Asia Publishing House), Bombay, 1965.
4. *The India We Want: Its Economic Transition* (Manaktals), Bombay, 1966.
5. *Editor, Pharmaceuticals and Health in the Third World*, (Pergamon Press, October 1983); also published as a special issue of the journal *World Development*, Vol. **11**, No. 3, March 1983.

II. *Journals edited:*

1. *Guest editor, World Development*, "Special issue on patents", Vol. **3**, No. 9, September 1974.
2. *Guest editor, World Development*, Special issue on "Trademarks in developing countries", Vol. **7**, No. 7, July 1979.

III. *Research papers, articles, etc.:*

1. "Effect of War-time Prices on Agricultural Indebtedness" in the *Indian Journal of Economics*, April 1945.
2. "Is Agriculture in Bhal Region a Paying Occupation?—An Economic Survey", in the *Journal of the Gujarat Research Society*, Bombay University, January 1947.
3. "British Economic Thought and the Treatment of India as a Colony", in the *Indian Journal of Economics*, April 1947.
4. "Marxism and Recent Economic Thought", in *Science & Society*, Vol. **XI**, No. 1, New York, 1947.
5. "Agricultural Labourers in Modern India and Pakistan", in the *Indian Journal of Economics*, July 1952.
6. "Trends in Unit Yields of Foodgrains: 1936–1950", in the *Indian Journal of Economics*, July 1953.
7. "Distribution of the National Income of India: 1950–1951", in the *Indian Economic Review*, February 1956.
8. "Growth of Income and Investment in India and China: 1952 and 1960", in the *Indian Economic Review*, February 1956.
9. "Long-term Trends in Output and Income in India: 1896–1960", in the *Indian Economic Journal*, January 1958.
10. "Investment and Efficiency in the Indian Railways During the Second Five-Year Plan", in the *Indian Economic Review*, February 1958.

*This list does *not* include any of the publications which were prepared for and published by the United Nations system during the author's association with the organization from 1950 to 1984.

11. "Savings, Investment and Economic Growth: A Dynamic Approach", in the *Indian Economic Review*, August 1959.
12. "Export Prospects and Economic Growth: India", in the *Economic Journal*, September 1959.
13. "Planning in India and China: Its Relevance for a Theory of Economic Growth", in the *Economic Weekly*, Annual Number, January 1960.
14. "Economic Perspectives for the Decades to Come", *Perspective*, January 1961.
15. "On the Eve of the Third Plan", *Economic Weekly*, Annual Number, 4 February 1961.
16. "Rates of Industrial Growth in the Last Century", in the *Economic Development and Cultural Change*. Essays presented to Professor S. S. Kuznets (University of Chicago) April 1961.
17. "Observations upon Comments on Export Prospects and Economic Growth: India", *Economic Journal*, June 1961.
18. "Export Strategy for the Next Decade", *Economic Weekly*, Annual Number, July 1961.
19. "Stage of Economic Growth", in *Perspective*, December 1961.
20. "Economic Outlook for our Children", in *Indian Journal of Labour Economics* (Lucknow University), April 1962.
21. "Economic Consequence of Disarmament", in the *Bulletin of the Atomic Scientists*, (Chicago) October 1962.
22. "Données générales pour le premier plan quinquennal de la République du Congo (1965–1969): "Rapport Patel" in Ministère du Plan et du Développement Industriel, République du Congo, *Travaux de la Conférence Nationale des Ministres Provinciaux du Plan* (Leopoldville, February 1964); pp. 33–146.
23. "Esquisse de la transition économique en Afrique", in the *Etudes Congolaises* (Bruxelles), February 1964.
24. "What is holding up agricultural growth?", in the *Economic Weekly, Annual Number*, February 1964.
25. "Economic Distance between Nations: Its Origin, Measurement and Outlook" in the *Economic Journal*, March 1964.
26. "Main Features of Economic Growth over the Last Century", in the *Indian Economic Journal*, March 1964.
27. Rejoinder to "A comment on S. J. Patel's Analysis of Exports", in *Indian Economic Journal*, March 1964.
28. "India's Economic Future", in the *Economic Weekly*, Annual Number, July 1964.
29. "Economic Transition in Africa", in the *Journal of Modern African Studies*, Cambridge University Press, Vol. **2**, No. 3) 1964.
30. "An Outline of Economic Prospects in East Africa", East African Institute of Social and Cultural Affairs, *Problems of Economic Development in East Africa* (EAISCA), Nairobi 1965.
31. "Main Lessons of Economic Growth over the Last Century", in *Mercurio*, No. 1 (Rome 1965); also *Mainstream* (New Delhi), Nehru Number 1964.
32. "Prospects of Industrial Development in Africa" (with A. F. Ewing), in the *Foundation Man and Africa* (Churchill Ltd., London 1965).
33. "Educational Distance between Nations: Its Origin and Outlook", in the *Indian Economic Journal* (Bombay), July–September 1965.
34. "Planning for Children and Youth in Asia—An Economic Framework" ECAFE Document Bangkok, September 1965.
35. "An Outline of Development Possibilities in Africa", in V. B. Singh, (ed.) *Essays on Africa*.
36. Review note on 3 books on Indian Planning in the *Economic Journal* (December 1964).
37. Review of M. Kaser's *Comecon* in the *Economic Weekly* (Bombay, 2 October 1965).
38. Review of B. K. Madan's *Aspects of Economic Development and Policy*, in *Kyklos* (Basle), Fasc. 3, 1965.
39. Reply to Kothari's note on the "Economic Distance", in the *Economic Journal*, September 1965.
40. "Reflections on the Development Problems in Cambodia"—*ECAFE document*, Bangkok, 22 November 1965.
41. Review of H. Rao's Agricultural Production Functions, in the *Indian Journal of Labour Economics*, July–October 1965.
42. "Asia in the Changing World Economy", in the *Journal of Development Studies* (London), Vol. **3**, No. 2, January 1967.
43. Review of Sicat's (ed.) *Philippine Economy in the 1960's*, in *Kyklos*, Vol. **XIX**, 1966, Fasc. 3.
44. "World Economy in Transition, 1850 to 2060", in Feinstein (ed.), *Socialism, Capitalism and Economic Growth*, Essays presented to Maurice Dobb (Cambridge University Press, 1967).

45. "Prospects of Trade between the Developing and the Socialist Countries", in *Foreign Trade Review* (New Delhi), Vol. **II**, No. 1, April–June 1967.
46. "A Note on the Incremental Capital Output Ratio and Rates of Economic Growth in the Developing Countries", in *Kyklos*, Vol. **XXI**, 1968, Fasc. I.
47. Rejoinder to Boserup's "Warning against Optimistic ICOR Statistics", in *Kyklos*, Vol. **XXII**, 1969, Fasc. 4.
48. Review of M. D. Chaudhry's *Regional Income Accounting in an Underdeveloped Economy—A Case Study of India*, in *Kyklos*, Fasc. 1, 1968.
49. "Depressed Exporters: A Special Case", in *Foreign Trade Review*, April–June 1969.
50. Review of Z. Y. Hershlag's *Turkey: The Challenge of Growth*, in *Kyklos*, Fasc. 1, 1970.
51. "Goals of Planning in the Seventies—A Note", in *Economic and Political Weekly*, Vol. **VI**, Nos. 30–32. Special Number, July 1971.
52. "Depressed Exporters: The Hard Core of the Development Problem", in *Economia Internazionale*, Genoa, August–November 1971.
53. "Some Implications of the Structural Change in Exports of Developing Countries", in *Foreign Trade Review*, Annual Number, January–March 1971.
51. Review of Mr. Tanzer's *The Political Economy of International Oil and the Underdeveloped Countries*, in *Kyklos*, Fasc. 1, 1971.
55. "Transfer of Technology and the Second Development Decade", in *Economic Times*, Bombay, 1972.
56. "Transfer of Technology and Developing Countries", in *Foreign Trade Review*, Annual Number, January–March 1973; also in *Technos*, April–June 1972; also in *Foro Internacional*, Vol. **XXX**, No. 1, Mexico, 1972.
57. "UNCTAD III and the Transfer of Technology to Developing Countries", in *Foreign Trade Review*, UNCTAD–III Special Number, October–December 1972.
58. "Transfer of Technology and Third UNCTAD", in *Journal of World Trade Law*, Vol. **7**, No. 2, March–April 1973.
59. "The Cost of Technological Dependence", in *Ceres*, FAO, Rome, Vol. **6**, No. 2, March–April 1973.
60. "Note on Project Report on Pulp Mill Complex in Guyana" on project submitted by the Parsons and Whittmore-Lyddon Organization, 18 June 1974 (restricted).
61. "Code of Conduct on Transfer of Technology", in *Foreign Trade Review*, Vol. **IX**, No. 2, July–September 1974.
62. "Notes on the Cerro Colorado Copper Project, Panama", 11 November 1974 (Restricted).
63. "The Patent System and the Third World", in *World Development*, Vol. **2**, No, 9, September 1974.
64. "The Technological Dependence of Developing Countries", in the *Journal of Modern African Studies*, Vol. **12**, No. 1, 1974.
65. "Collective Self-Reliance of Developing Countries", in Cairncross, Sir Alec and Puri, M., (eds.) *Employment, Income Distribution and Development Strategy, Essays in honor of H. W. Singer*, (McMillan, 1976) also in the *Journal of Modern African Studies*, Vol. **13**, No. 4, 1975; in European Association of Development Research and Training Institutes, General Conference 1975, Linz, Austria; and in *Revue Tiers-Monde*, T. XVII, No. 65, January–March 1976.
66. Contribution to the Colloquium of the Université of Nice, published in Jean Touscoz, *Transferts de technologie, Sociétés transnationales et Nouvel ordre international*, 1978.
67. "Energy policies and collective self-reliance of the Third World", in *Resources Policy*, June 1978.
68. "Towards the technological transformation of the Third World", in *Impact of Science on Society*, UNESCO, Vol. **28**, No. 4, October–December 1978; also in *Internationale Entwicklung*, Vienna.
69. "United Nations Conference on an International Code of Conduct on Transfer of Technology", in *Guardian, Third World Review*, September 1978.
70. "Trademarks and the Third World", in *World Development*, Vol. **7**, 1979.
71. "Technology for the Poor: A Note on the Technology Code", in *The Economist*, 18–24 November 1978.
72. "Third World: Historical Antecedent", in *Mainstream*, October 1979.
73. "Planned Development in India, Review of major changes 1950–1975", in *Mainstream*, Republic Day Number 1980.
74. "Planned Development in India—II, Reflections on Choices: 1980 to 2000", in *Mainstream*, 16 February 1980.
75. Technological transformation of the Third World, Baljit Singh Commemoration Volume, 11 December 1980 (being published).

76. "The technological transformation of the Third World—Main Issues", in *Man and Development*, Vol. **II**, No. 4, December 1980.
77. "International Code of Conduct on Transfer of Technology: UNCTAD negotiations", *Scientific American*, 29 January 1982.
78. *Desarrollo planificado en la India:* I. Revisión de los cambios importantes ocurridos entre 1950 y 1975. II. Reflexiones en torno a posibles opciones: 1980–2000, El Colegio de México, 1981.
79. "The technological transformation of the Third World", *Economic and Political Weekly*, 27 February 1982.
80. "L'Age du Tiers Monde est Proche", in *Le Temps Stratégique*, (Genève), No. 2, Automne, 1982.
81. "The New International Economic Order and Technological Transformation of the Third World" in *The New International Economic Order: Commercial, Technological and Cultural Aspects*; Hague Academy of International Law, Martinus Nijhoff, The Hague/Boston/London, 1981.
82. "The Age of the Third World", *Third World Quarterly*, January 1983, Vol. **5**, No. 1; also in *Mainstream*, New Delhi, 5 March 1983 and in Spanish in *Comercio Exterior* (El Colegio de México) June, 1983.
83. "Reflections on the economic crisis and the Third World", *Mainstream*, 5 November 1983.
84. "Technology strategy for the Third World", *Mainstream*, November 12, 1983.
85. "India's regression in the world economy", Essays in honour of Dr. I. G. Patel (Oxford University Press, New Delhi, 1985); a revised version published as a Discussion Paper by the Institute of Development Studies, University of Sussex, July 1985.
86. "Third World initiatives on pharmaceuticals: a documentation for the record" in *Health, Society and Politics*, Institute of Development Studies Bulletin, (University of Sussex); Vol. **14**, No. 4; October 1983.
87. Editor's introduction to *Pharmaceuticals and Health in the Third World* (Pergamon Press, 1983).
88. Foreword to Chudnovsky, D. and Nagao, M., *Capital Goods Production in the Third World: An Economic Study of Technology Acquisition* (Frances Pinter, London 1983).
89. "China's Strategy for its Technological Transformation", *The Economic and Political Weekly*, Bombay, March 3, 1984.
90. "Economic Crisis and Transition from Capitalism" paper presented to the ICSSR Seminar on Marx, Schumpeter and Keynes (New Delhi, January 1984); published as a Discussion Paper of the Institute of Development Studies, University of Sussex, 1985; for an earlier version see *Mainstream*, New Delhi, 24 March and 31 March 1984.
91. "Towards a New Technological Order", *Mainstream*, New Delhi, March 3, 1984.
92. "A Technology Strategy for the Third World" in R. Lalkaka and Wu Mingyu (eds.), *Managing Science Policy and Technology Acquisition* (UNFSSTD, New York and Dublin 1984); reprinted in *Mainstream*, New Delhi, 12 November 1984.
93. "India in the World Economy", *Mainstream*, Republic Day Number, New Delhi, 26 January 1985.
94. "A swift transformation of the educational profile in the Third World", *Journal of Higher Education*, New Delhi, Spring 1985.
95. "Educational 'Miracle' in the Third World" (being published).

ANNEX II

Governmental Meetings in UNCTAD on Technology Issues, 1971 to 1984

A. *Intergovernmental Group on the Transfer of Technology*

1. First session (14–21 June 1971).
2. Second session (29 January to 9 February 1973).
3. Third session (15–26 July 1974).

B. *Committee on Transfer of Technology*

4. First session (24 November to 5 December 1975).
5. Second session (4 to 15 December 1978).
6. Third session (17 to 28 November 1980).
7. Fourth session (29 November to 10 December 1982).
8. First special session (13–22 February 1984).

C. *International Code of Conduct on the Transfer of Technology*

9. Intergovernmental Group of Experts on a Code of Conduct on Transfer of Technology (5–16 May 1975).
10. Intergovernmental Group of Experts on a Code of Conduct on Transfer of Technology, resumed session (24 November to 3 December 1975).
11. First session of the Intergovernmental Group of Experts on an International Code of Conduct on Transfer of Technology (8 to 19 November 1976).
12. Second session of the Intergovernmental Group of Experts on an International Code of Conduct on the Transfer of Technology (25 March to 5 April 1977).
13. Third session on the Intergovernmental Group of Experts on an International Code of Conduct on the Transfer of Technology (26 to 29 July 1977).
14. Fourth session of the Intergovernmental Group of Experts on an International Code of Conduct on the Transfer of Technology (31 October to 11 November 1977).
15. Fifth session on the Intergovernmental Group of Experts on an International Code of Conduct on the Transfer of Technology (13 to 24 February 1978).
16. Sixth session on the Intergovernmental Group of Experts on an International Code of Conduct on the Transfer of Technology (26 June to 7 July 1978).
17. First session of the United Nations Conference on an International Code of Conduct on the Transfer of Technology (16 October to 11 November 1978).
18. Resumed session of the United Nations Conference on an International Code of Conduct on the Transfer of Technology (26 February to 9 March 1979).
19. Preparatory Meeting for the second session of the United Nations Conference on an International Code of Conduct on the Transfer of Technology (24 to 28 September 1979).
20. Second session of the United Nations Conference on an International Code of Conduct on the Transfer of Technology (29 October to 16 November 1979).
21. Third session of the United Nations Conference on an International Code of Conduct on the Transfer of Technology (21 April to 6 May 1980).

22. Fourth session of the United Nations Conference on an International Code of Conduct on the Transfer of Technology (23 March to 10 April 1981).
23. Fifth session of the United Nations Conference on an International Code of Conduct on the Transfer of Technology (17 October to 4 November 1983).

D. *The Role of the Industrial Property System in the Transfer of Technology to Developing Countries.*

24. Group of Governmental Experts on the Role of the Patent System in the Transfer of Technology (1 to 12 September 1975).
25. Group of Governmental Experts on the Role of the Industrial Property System in the Transfer of Technology (6 to 14 October 1977).
26. Group of Governmental Experts on the Economic, Commercial and Developmental Aspects of Industrial Property in the Transfer of Technology to Developing Countries (1 to 10 February 1982).

E. *Sectors and Areas of Critical Importance*

27. Meeting of Governmental Experts on the Transfer, Application and Development of Technology in the Food Processing Sector (1–10 June 1982).
28. Meeting of Governmental Experts on the Transfer, Application and Development of Technology in the Capital Goods and Industrial Machinery Sector (7–16 July 1982).
29. Meeting of Governmental Experts on the Transfer, Application and Development of Technology in the Energy Sector (25 October to 2 November 1982).

F. *Reverse Transfer of Technology*

30. Group of Governmental Experts on Reverse Transfer of Technology (27 February to 7 March 1978).
31. Intergovernmental Group of Experts on the Feasibility of Measuring Human Resource Flows (30 August to 6 September 1982).
32. Meeting of Governmental Experts on the Reverse Transfer of Technology (28 August to 7 September 1983).

G. *Regional/Sub-regional/Inter-regional Seminars, Workshops and Training Programs*

33. First UNCTAD/SIDA training course on Transfer of Technology (UNCTAD/TT/1). Nairobi, 2 to 25 April 1975).
34. Informal Group of Experts on the Establishment of Centres for the Transfer and Development of Technology, 8 to 12 December 1975.
35. Second UNCTAD/SIDA training course on Transfer of Technology (UNCTAD/TT/AS/1) Sri Lanka, 30 September to 2 October 1976; followed by a study tour conducted in India, 3 to 9 October 1976.
36. Third UNCTAD/SIDA training course on Transfer and Development of Technology (UNCTAD/TT/AS/6). San José, Costa Rica, 3 to 16 April 1978, followed by a study tour in Caracas, Venezuela, from 17 to 21 April 1978.
37. Fourth UNCTAD/SIDA training course on Transfer and Development of Technology (UNCTAD/TT/AS/10). Dakar, Senegal, from 1 to 14 October 1979, followed by a study tour in Algiers, Algeria, from 15 to 21 October 1979.
38. Workshop on Technology Policies and Planning. Dar-es-Salaam, United Republic of Tanzania, from 13 to 24 October 1980.
39. Workshop on Trade and Technology Policies in the Pharmaceutical Sector in the Caribbean Region (UNCTAD/TT/41/Rev.1). Georgetown, Guyana, 18 to 29 August 1980.
40. Workshop on Technology Policies and Planning for Technological Transformation (UNCTAD/TT/42 Rev.1). Dar-es-Salaam, United Republic of Tanzania, 13 to 24 October 1980.
41. Asian Regional Workshop on Technology Policies and Planning for Technological Transformation (UNCTAD/TT/50). Bangalore, India, 15 to 26 June 1981.
42. Workshop on legal aspects of the Transfer and Development of Technology (UNCTAD/PSS/TCS/40). Kuwait, 19 to 23 September 1981.
43. Workshop on Trade and Technology Policies in the Pharmaceutical Sector (UNCTAD/TT/48). Ivory Coast, 12 to 23 October 1981.

44. UNCTAD/PRODEC Seminar on Technology Transfer and Contracting Projects, Helsinki, Finland, 10 to 21 October 1983.
45. UNCTAD/PRODEC Seminar on Planning and Administration of Project Imports, Helsinki, Finland, 7 to 30 May 1984.

H. *In-house and On-the-job Programs*

46. Twenty-seven different groups/programs on technology policy issues attended by 167 government officials from 55 developing countries. Each group between 3 and 8 participants. Geneva; two to three weeks each.

I. *In-field Programs at National, Sub-regional/Regional Levels*

47. Training program on Transfer and Development of Technology. Lima, Peru, from 16 to 27 November 1981; Bogota, Colombia, from 30 November to 4 December 1981.
48. Training program in the Pharmaceutical Sector. Caribbean Community Secretariat, October 1981.
49. Training on Techno-economic Evaluation of Projects and Technology Agreements. Calcutta, India, 16 March to 8 April 1982. (In co-operation with M. N. Dastur and Co., Consulting Engineers).
50. Training program on Transfer and Development of Technology. New Delhi, India, from 29 March to 16 April 1982.
51. Training Program on Trade and Technology Policies in the Pharmaceutical Sector, Tunis, 4 to 30 October 1983.
52. Workshop on the Pharmaceutical Industry in Ethiopia, Addis Ababa, 9 to 11 March 1983.
53. National Seminar on Science and Technology Policies, Sierra Leone, 9 to 17 April 1984.
54. National Seminar on Strategy for Technological Transformation, Ethiopia, 14 to 18 May 1984.

ANNEX III

Resolutions, Decisions and Agreed Conclusions on UNCTAD's Work in Technology, 1968 to 1984

A. *Adopted by the United Nations General Assembly, 1971 to 1983*

1. Resolution 2821 (XXVI) 16 December 1971	Transfer of technology, including know-how and patents.
2. Resolution 3202 (S–VI) 1 May 1974	Program of Action on the Establishment of a New International Economic Order.
3. Resolution 3362 (S–VII) 16 September 1975	Development and international economic co-operation.
4. Resolution 3507 (XXX) 15 December 1975	Institutional arrangements in the field of the transfer of technology.
5. Resolution 31/183 21 December 1976	Establishment of a network for the exchange of technological information.
6. Resolution 31/184 21 December 1976	United Nations Conference on Science and Technology for Development.
7. Resolution 32/178 19 December 1977	Network for the exchange of technological information and the industrial and technological information bank.
8. Resolution 32/188 19 December 1977	United Nations Conference on an International Code of Conduct on the Transfer of Technology.
9. Resolution 32/192 19 December 1977	Reverse transfer of technology.
10. Resolution 33/151 20 December 1978	Reverse transfer of technology.
11. Resolution 33/157 20 December 1978	United Nations Conference on an International Code of Conduct on the Transfer of Technology.
12. Resolution 34/195 19 December 1979	United Nations Conference on an International Code of Conduct on the Transfer of Technology.
13. Resolution 34/200 19 December 1979	Development aspects of the reverse transfer of technology.
14. Resolution 35/60 5 December 1980	International Code of Conduct on the Transfer of Technology.
15. Resolution 35/62 5 December 1980	Reverse transfer of technology.
16. Resolution 36/140 16 December 1981	United Nations Conference on an International Code of Conduct on the Transfer of Technology.
17. Resolution 36/141 16 December 1981	Reverse transfer of technology.
18. Resolution 36/145 16 December 1981	United Nations Conference on Trade and Development.
19. Resolution 37/251 21 December 1982	Development of the energy resources of developing countries.
20. Resolution 37/207 20 December 1982	Development aspects of the reverse transfer of technology.

21. Resolution 37/210 20 December 1982	International Code of Conduct on the Transfer of Technology
22. Resolution 38/151 19 December 1983	Development of the energy resources of developing countries.
23. Resolution 38/153 19 December 1983	International Code on Conduct on the Transfer of Technology.
24. Resolution 38/154 19 December 1983	Development aspects of the reverse transfer of technology.

B. *Adopted by the High-level Committee on the Review of Technical Co-operation among Developing Countries, 1981 to 1983*

25. Resolution 2/4 8 June 1981	Methodologies for promoting co-operation among developing countries in scientific and technological research.
26. Decision 3/6 6 June 1983	Co-operative exchange of skills among developing countries through technical co-operation among developing countries: study by the secretariat of the United Nations Conference on Trade and Development.

C. *Adopted by Sessions of the Conference, 1972 to 1983*

27. Resolution 39 (III) 16 May 1972	Transfer of technology.
28. Resolution 87 (IV) 30 May 1976	Strengthening the technological capacity of developing countries.
29. Resolution 88 (IV) 30 May 1976	Industrial property.
30. Resolution 89 (IV) 30 May 1976	International Code of Conduct on the Transfer of Technology.
31. Resolution 101 (V) 30 May 1979	Contribution of UNCTAD to economic, commercial and development aspects of the Industrial Property System in the context of its ongoing revision.
32. Resolution 102 (V) 30 May 1979	Development aspects of the reverse transfer of technology.
33. Resolution 112 (V) 3 June 1979	Strengthening the technological capacity of developing countries including accelerating their technological transformation.
34. Decision 113 (V) 3 June 1979	International Code of Conduct on the transfer of technology.
35. Resolution 143 (VI) 2 July 1983	Towards the technological transformation of developing countries.

D. *Adopted by the Trade and Development Board, 1968 to 1984*

36. Resolution 48 (VII) 21 September 1968	Transfer of Technology, including know-how and patents.
37. Resolution 62 (IX) 12 September 1969	Transfer of Technology, including know-how and patents.
38. Resolution 74 (X) 18 September 1970	Transfer of Technology, including know-how and patents.
39. Resolution 104 (XIII) 8 September 1973	Transfer of Technology and code of conduct in the field of transfer of technology to developing countries.
40. Decision 117 (XIV) 13 September 1974	Terms of reference of the Committee on Transfer of Technology.
41. Decision 192 (XIX) 20 October 1979	Strengthening the technological capacity of developing countries including accelerating their technological transformation.
42. Decision 193 (XIX) 20 October 1979	Development aspects of the reverse transfer of technology.

43. Decision 211 (XX) 25 March 1980	Strengthening the technological capacity of developing countries including accelerating their technological transformation.
44. Decision 212 (XX) 25 March 1980	Development aspects of the reverse transfer of technology.
45. Resolution 219 (XXI) 27 September 1980	Development aspects of the reverse transfer of technology.
46. Resolution 227 (XXII) 20 March 1981	Development aspects of the reverse transfer of technology.
47. Resolution 230 (XXII) 20 March 1981	Transfer, application and development of technology: issues in individual sectors and other areas of critical importance to developing countries.
48. Resolution 240 (XXIII) 9 October 1981	Legislation and regulations on technology transfer.
49. Resolution 241 (XXIII) 9 October 1981	Transfer application and development of technology: issues in individual sectors and other areas of critical importance to developing countries.
50. Resolution 242 (XXIII) 9 October 1980	Technical co-operation among developing countries.
51. Decision 268 (XXVI) 28 April 1983	Interregional linkages towards the implementation strategies for the technological transformation of developing countries.
52. Resolution 279 (XXVII) 20 October 1983	Technical co-operation among developing countries: co-operative exchange of skills among developing countries.
53. Decision . . . 6 April 1983	Technical co-operation among developing countries: co-operative exchange of skills among developing countries.
54. Decision . . . 6 April 1983	Towards the technological transformation of developing countries in the pharmaceutical sector.

E. *Adopted by the Intergovernmental Group on the Transfer of Technology, 1971 to 1974*

55. Resolution 1 (II) February 1973	Methodology for studying the transfer of technology, major issues arising from the transfer of technology; progress in implementing conference resolutions 39 (III).
56. Resolution 2 (III) 26 July 1974	The role of the patent system in the transfer of technology to the developing countries.
57. Resolution 3 (III) 26 July 1974	The possibility and feasibility of an international code of conduct in the field of transfer of technology.

F. *Adopted by the Committee on Transfer of Technology, 1975–1984*

58. Decision 1 (I) 4 December 1975	Issues relating to an International Code of Conduct in the field of transfer of technology.
59. Resolution 2 (I) 4 December 1975	Promotions of the technological capability of developing countries.
60. Resolution 3 (I) 5 December 1975	The role of the industrial property system in the transfer of technology to developing countries.
61. Decision 4 (II) 6 December 1978	Role of the industrial property system in the transfer of technology to developing countries.
62. Decision 5 (II) 15 December 1978	UNCTAD's participation in the United Nations Conference on Science and Technology for Development.
63. Resolution 6 (II) 15 December 1978	Strengthening the technological capacity of developing countries and accelerating their technological transformation.
64. Resolution 7 (II) 15 December 1978	Development aspects of reverse transfer of technology.
65. Resolution 8 (III) 28 November 1980	Formulation of a strategy for the technological transformation of developing countries.
66. Resolution 9 (III) 28 November 1980	The role of small and medium-sized enterprises in the international transfer of technology.

67. Decision 10 (III) 28 November 1980	Environmental aspects of the transfer and development of technology.
68. Resolution 11 (III) 28 November 1980	Review of progress in implementing conference resolutions 87 (IV) and 112 (V).
69. Decision 12 (III) 28 November 1980	Report on the UNCTAD Work Programme in the development and transfer of technology.
70. Resolution 13 (III) 28 November 1980	Development aspects of the reverse transfer of technology.
71. Resolution 14 (III) 28 November 1980	Economic, commercial and developmental aspects of the industrial property system in the context of its ongoing revision.
72. Resolution 15 (III) 28 November 1980	Follow-up to the United Nations Conference on Science and Technology for Development and to the establishment of the Intergovernmental Committee on Science and Technology for Development.
73. Resolution 16 (III) 28 November 1980	Technical and operational assistance by the Advisory Service on Transfer of Technology.
74. Resolution 17 (IV) 10 December 1982	Technical and operational assistance by the Advisory Service on Transfer of Technology.
75. Resolution 18 (IV) 10 December 1982	Issues in individual sectors of critical importance to developing countries. Food processing, capital goods and industrial machinery, and energy.
76. Resolution 19 (IV) 10 December 1982	The role of small and medium-sized enterprises in the international transfer of technology.
77. Resolution 20 (IV) 10 December 1982	Restructuring the legal environment for technology: national laws and regulations.
78. Resolution 21 (IV) 10 December 1982	Economic, commercial and development aspects of the industrial property system in the transfer of technology to developing countries.
79. Decision 22 (IV) 10 December 1982	Follow-up to the United Nations Conference on Science and Technology for Development.
80. Decision 23 (IV) 10 December 1982	Review of progress in implementation conference resolutions 87 (IV) and 112 (V).
81. Resolution 24 (IV) 10 December 1982	Work program of UNCTAD in the development and transfer of technology.
82. Resolution 25 (IV) 10 December 1982	A strategy for the technological transformation of developing countries.
83. Resolution 26 (S–I) 22 February 1984	Further elaboration and implementation of a strategy for the technological transformation of developing countries.

G. *Adopted by governmental and intergovernmental groups of experts on technology issues, 1975 to 1983*

84. Conclusions and recommendations of the group of governmental experts on the role of the patent system in the transfer of technology (1 to 12 September 1975).
85. Conclusions and recommendations of the group of governmental experts on the role of the industrial property system in the transfer of technology (6 to 14 October 1977).
86. Decision 1 (the revision of the Paris Convention for the protection of industrial property) of the group of governmental experts on the role of the industrial property system in the transfer of technology (6 to 14 October 1977).
87. Decision 2 (the impact of trademarks on the development process of developing countries and the review of the Paris Convention for the protection of industrial property) of the group of governmental experts on the role of the industrial property system in the transfer of technology (6 to 14 October 1977).
88. Conclusions and recommendations of the group of governmental experts on the economic, commercial and developmental aspects of industrial property in the transfer of technology to developing countries (1 to 10 February 1982).
89. Conclusions and recommendations of the meeting of governmental experts on the transfer, application and development of technology in the food processing sector (1 to 10 June 1982).
90. Conclusions and recommendations of the meeting of governmental experts on the transfer,

application and development of technology in the capital goods and industrial machinery sector (7 to 16 July 1982).

91. Conclusions and recommendations of the meeting of governmental experts on the transfer, application and development of technology in the energy sector (25 October to 2 November 1982).
92. Conclusions and recommendations of the group of governmental experts on the reverse transfer of technology (27 February to 7 March 1978).
93. Conclusions and recommendations of the meeting of governmental experts on the reverse transfer of technology (29 August to 7 September 1983).

ANNEX IV

A Bibliography of UNCTAD's Major Studies and Reports on the Subject of Technology*

I. *General studies*

1.	Transfer of technology, including know-how and patents. Elements of a program of work for UNCTAD, July 1970.	TD/B/310 and Corr. 1
2.	The channels and mechanisms for the transfer of technology from developed to developing countries. Study by Charles Cooper with the collaboration of Francisco Sercovitch, April 1971.	TD/B/AC.11/5
3.	Transfer of technology: Report by the UNCTAD secretariat, November 1971–January 1972.	TD/106 and Corr. 1
4.	Guidelines for the study of the transfer of technology to developing countries. A study by the UNCTAD secretariat (United Nations publication, Sales No. E.72.II.D.19).	TD/B/AC.11/9
5.	Major issues arising from the transfer of technology to developing countries. Report by the UNCTAD secretariat (United Nations publication, Sales No. E.75.II.D.2).	TD/B/AC.11/10/Rev.2
6.	Technological dependence: Its nature, consequences and policy implications. Report by the UNCTAD secretariat, February 1976.	TD/190 and Corr. 1
7.	Towards the technological transformation of the developing countries. Report by the UNCTAD secretariat, March 1979.	TD/238 and Corr.1
8.	A strategy for the technological transformation of developing countries. Report by the UNCTAD secretariat, January 1983.	TD/277

II. *International Code of Conduct on the Transfer of Technology*

9.	The possibility and feasibility of an international code of conduct on transfer of technology. A study by the UNCTAD secretariat, June 1974.	TD/B/AC.11/22 and Corr.1
10.	An international code of conduct on transfer of technology. Report by the UNCTAD secretariat (United Nations publication, Sales No. E.75.II.D.15).	TD/B/C.6/AC.1/2/Supp.1/Rev.1
11.	Report of the Intergovernmental Group of Experts on a Code of Conduct on Transfer of Technology, Geneva, 5–16 May 1975, May 1975.	TD/B/C.6/1

*Only the most important works are selected for listing here. For more details, see "Select bibliography of documents on transfer and development of technology" (TD/B/C.6/INF.2/Rev.3), 15 November 1982.

12.	Negotiation of an international code of conduct on the transfer of technology. Present status of negotiations and issues outstanding. Note by the UNCTAD secretariat, August 1983.	TD/CODE TOT/37
13.	Draft international code of conduct on the transfer of technology as of 4 November 1983. Note by the UNCTAD secretariat, December 1983.	TD/CODE TOT/41

III. *Economic, commercial and developmental aspects of industrial property in the transfer of technology to developing countries*

14.	The role of the patent system in the transfer of technology to developing countries. Conclusions of experts from developing countries, November 1975.	TD/B/C.6/12
15.	The role of the patent system in the transfer of technology to developing countries. Report prepared jointly by the United Nations Department of Economic and Social Affairs, the UNCTAD secretariat and the International Bureau of the World Intellectual Property Organization (United Nations publication, Sales No. E.75.II.D.6).	TD/B/AC.11/19/Rev.1
16.	The international patent system: The revision of the Paris Convention for the Protection of Industrial Property. Report by the UNCTAD secretariat, June 1977.	TD/B/C.6/AC.3/2
17.	The role of trade marks in developing countries. Report by the UNCTAD secretariat (United Nations publication, Sales No. E.79.II.D.5).	TD/B/C.6/AC.3/3/Rev.1
18.	The role of trade marks in the promotion of exports from developing countries. Report by the UNCTAD secretariat, October 1981.	TD/B/C.6/AC.5/2 and Corr.1
19.	Review of recent trends in patents in developing countries. Report by the UNCTAD secretariat, November 1981.	TD/B/C.6/AC.5/3 and Corr.1
20.	Trade marks and generic names of pharmaceuticals and consumer protection. Report by the UNCTAD secretariat, December 1981.	TD/B/C.6/AC.5/4

IV. *Laws and regulations*

21.	The implementation of transfer of technology regulations: A preliminary analysis of the experience of Latin America, India and the Philippines. Study prepared by Lilia R. Bautista at the request of the UNCTAD secretariat, October 1980.	TD/B/C.6/55 and Corr.1
22.	Control of restrictive practices in transfer of technology transactions: Selected principal regulations, policy guidelines and case law at the national and regional levels. Report by the UNCTAD secretariat (United Nations publication, Sales No. E.82.II.D.8).	TD/B/C.6/72
23.	Common approaches to laws and regulations on the transfer and acquisition of technology. Study by the UNCTAD secretariat, October 1982.	TD/B/C.6/91

V. *Technology policy, planning and institutional infrastructure*

24.	Action to strengthen the technological capacity of developing countries: Policies and institutions. Report by the UNCTAD secretariat, February 1976.	TD/190/Supp.1
25.	Handbook on the acquisition of technology by developing countries. Prepared by the UNCTAD secretariat (United Nations publication, Sales No. E.78.II.D.15).	UNCTAD/TT/AS/5 and Corr.1

26. Planning the technological transformation of developing countries. Study by the UNCTAD secretariat (United Nations publication, Sales No. E.81.II.D.10).	TD/B/C.6/50

VI. *Sectors and areas*

27. The capital goods and industrial machinery sector in developing countries: Issues in the transfer and development of technology. Study by the UNCTAD secretariat, May 1982.	TD/B/C.6/AC.7/2
28. The impact of electronics technology on the capital goods and industrial machinery sector: implications for developing countries. Study by the UNCTAD secretariat, May 1982.	TD/B/C.6/AC.7/3
29. Energy supplies for developing countries: issues in transfer and development of technology. Study by the UNCTAD secretariat, October 1978 (United Nations publication, Sales No. E.80.II.D.3).	TD/B/C.6/31/Rev.1
30. The food processing sector in developing countries: some recent trends in the transfer and development of technology. Report by the UNCTAD secretariat, October 1980.	TD/B/C.6/66
31. Major issues in transfer of technology to developing countries: A case study of the pharmaceutical industry (prepared by Dr. S. Lall in co-operation with the UNCTAD secretariat), October 1975.	TD/B/C.6/4
32. Technology policies and planning for the pharmaceutical sector in the developing countries. A report by the UNCTAD secretariat in co-operation with Dr. Nitya Anand, Director, Central Drug Research Institute, Lucknow, India, October 1980.	TD/B/C.6/56
33. Guidelines on technology issues in the pharmaceutical sector in the developing countries. Report by the UNCTAD secretariat (United Nations publication, Sales Bo. E.82.II.D.15).	UNCTAD/TT/49
34. Electronics in developing countries: issues in transfer and development of technology. Study by the UNCTAD secretariat in co-operation with Mr. Ashok Parthasarathi, Secretary, Electronics Commission, Government of India, in his personal capacity, October 1978.	TD/B/C.6/34
35. The role of small and medium-sized enterprises in the international transfer of technology: issues for research. Study prepared at the request of the UNCTAD secretariat by Mr. Eduardo White, with the collaboration of Mr. Silvio Feldman, November 1980.	TD/B/C,6/64
36. Transfer of technology: its implications for development and environment. Study by the UNCTAD secretariat prepared with the substantive and financial support of the United Nations Environment Programme (United Nations publication, Sales No. E.78.II.D.10).	TD/B/C.6/22

VII. *Reverse transfer of technology (brain drain)*

37. The reverse transfer of technology: Economic effects of the outlfow of trained personnel from developing countries. Study by the UNCTAD secretariat (United Nations publication, Sales No. E.75.II.D.1).	TD/B/AC.11/25/Rev.1
38. Co-operative exchange of skills among developing countries: Policies for collective self-reliance. Study by the UNCTAD secretariat (United Nations publication, Sales No. E.79.II.D.1).	TD/B/C.6/AC.4/8/Rev.1

39. The reverse transfer of technology: A survey of its main features, causes and policy implications. Study by the UNCTAD secretariat (United Nations publication, Sales No. E.79.II.D.10). TD/B/C.6/47
40. Co-operative exchange of skills among developing countries. A case for technical co-operation among developing countries. Note by the UNCTAD secretariat, May 1981. TCDC/2/INF/3
41. The feasibility of measuring international flows of human resources. Study by the UNCTAD secretariat, May 1982. TD/B/C.6/AC.8/2
42. Towards an integrated approach to international skill exchange: proposals for policy and action on reverse transfer of technology. Study by the UNCTAD secretariat, June 1983. TD/B/AC.35/2
43. Co-operative exchange of skills: institutional and policy issues. Study by the UNCTAD secretariat, February 1983. TD/B/943

VIII. *Technical and operational assistance*

44. Technical and operational assistance by the Advisory Service on Transfer of Technology. Report by the UNCTAD secretariat, October 1982. TD/B/C.6/93

IX. *Country studies and mission reports*

45. Brazil	(Study on capital goods) TD/B/C.6/AC.7/6 (May 1982)
46. Burundi	(Mission report) UNCTAD/TT/38 (October 1980)
47. Chile	(Study on transfer of technology issues) TD/B/AC.11/20 (May 1974)
48. Costa Rica	(Study on pharmaceuticals) INCTAD/TT/37 (May 1982)
49. Cuba	(Study on health and education technology) TD/B/C.6/46 (August 1979)
50. Cuba	(Study on pharmaceuticals) UNCTAD/TT/33 (December 1980)
51. Egypt	(Mission report) UNCTAD/TT/AS/7 (January 1980)
52. Ethiopia	(Study on transfer of technology issues) TD/B/AC.11/21 (June 1974)
53. Ethiopia	(Mission report) UNCTAD/TT/AS/4 (January 1979)
54. Ethiopia	(Study on pharmaceuticals) UNCTAD/TT/58 (1984)
55. Guyana	(Study on food processing) UNCTAD/TT/47 (November 1981)
56. Greece	(Study on transfer pricing) TD/B/C.6/32 (August 1978)
57. Hungary	(Study on transfer of technology issues) TD/B/AC.11/18 (May 1974)
58. India	(Study on pharmaceuticals) TD/B/C.6/20 (October 1977)
59. India	(Study on reverse transfer of technology) TD/B/C.6/AC.4/6 (January 1978)
60. India	(Study on iron and steel) TD/B/C.6/27 (June 1978)
61. India	(Study on petrochemicals) UNCTAD/TT/54 (July 1983)

62. India	(Study on capital goods) UNCTAD/TT/55 (September 1983)
63. Iran	(Study on resettlement programs) TD/B/C.6/83 (October 1982)
64. Iraq	(Mission report) UNCTAD/TT/AS/2 (June 1978)
65. Jamaica	(Study on technology transfer in tourism) TD/B/C.6/49 (November 1979)
66. Japan	(Study on transfer of technology issues) TD/B/C.6/26 (April 1978)
67. Mexico	(Study on food processing) TD/B/C.6/75 (March 1982)
68. Nepal	(Study on pharmaceuticals) UNCTAD/TT/34 (October 1980)
69. Parkistan	(Study on reverse transfer of technology) TD/B/C.6/AC.4/3 (January 1978)
70. Peru	(Study on food processing) TD/B/C.6/AC.6/4 (April 1982)
71. Philippines	(Study on reverse transfer of technology) TD/B/C.6/AC.4/5 (January 1978)
72. Philippines	(Study on pharmaceuticals) UNCTAD/TT/36 (October 1980)
73. Philippines	(Study on technology regulations) UNCTAD/TT/32 (October 1980)
74. Rwanda	(Mission report) UNCTAD/TT/51 (May 1980)
75. Senegal	(Study on food processing) TD/B/C.6/AC.6/6 (April 1982)
76. Senegal	(Study on modern and traditional medicine) TD/B/C.6/86 (October 1982)
77. Spain	(Study on transfer of technology issues) TD/B/AC.11/17 (April 1974)
78. Somalia	(Mission report) UNCTAD/TT/39 (October 1980)
79. Sri Lanka	(Study on transfer of technology issues) TD/B/C.6/6 (October 1975)
80. Sri Lanka	(Mission report) UNCTAD/TT/5 (November 1976)
81. Sri Lanka	(Study on pharmaceuticals) TD/B/C.6/21 (June 1977)
82. Sri Lanka	(Study on reverse transfer of technology) TD/B/C.6/AC.4/4 (January 1978)
83. Tanzania	(Study on Pharmaceuticals) UNCTAD/TT/35 (October 1980)
84. Tanzania	(Mission report) UNCTAD/TT/52 (May 1982)
85. Tanzania	(Study on capital goods) TD/B/C.6/AC.7/4 (May 1982)
86. Thailand	(Study on food processing) TD/B/C.6/AC.6/5 (April 1982)
87. Thailand	(Study on capital goods) TD/B/C.6/AC.7/5 (May 1982)
88. Tunisia	(Study on capital goods) UNCTAD/TT/53 (November 1982)
89. USSR	(Study on technology issues) TD/B/C.6/52 (September 1980)
90. USSR	(Study on planning of science and technology) UNCTAD/TT/56 (October 1983)
91. Venezuela	(Study on pharmaceuticals) UNCTAD/TT/25 (March 1982)

List of Contributors

Krishna Ahooja Patel
Editor, *Women at Work*, International Labour Office, Geneva, Switzerland

Samir Amin
Director of UNITAR 'Futures of Africa' project, Dakar, Senegal

Jagdish Bhagwati
Arthur Lehman Professor of Economics, and Director, International Economics Research Center, Columbia University, New York

Nikhil Chakravarthy
Editor, *Mainstream*, New Delhi, India

Frederick F. Clairmonte
Commodities Division, UNCTAD, Geneva, Switzerland

Gamani Corea
Former Secretary-General, UNCTAD

Anne Gordon Drabek
Managing Editor, *World Development*, Washington D.C.

Muchkund Dubey
Peramanent Representative of India to the United Nations and Other International Organizations, Geneva, Switzerland

Arthur F. Ewing
Formerly with the UN Economic Commission for Europe, the Economic Commission for Africa and UNCTAD

Johan Galtung
Visiting Professor, World Politics of Peace and War, Dept. of Politics, Princeton University, USA

List of Contributors

Tom Ganiatsos
Technology Division, UNCTAD, Geneva, Switzerland

Reginald Herbold Green
Fellow, Institute of Development Studies at the University of Sussex, Brighton, UK

Godfrey Gunatilleke
Director, Marga Institute, Colombo, Sri Lanka

Enrique Iglesias
Former Executive Secretary, UN Economic Commission for Latin America; presently, Foreign Minister, Uruguay

Nurul Islam
Assistant Director General, Economic and Social Policy, Food and Agriculture Organization, Rome

Vijay L. Kelkar
Economic Adviser, Ministry of Commerce, Government of India, New Delhi

Marc Nerfin
President, International Foundation for Development Alternatives, Nyon, Switzerland

Raul Prebisch
Former Secretary General, UNCTAD; presently Adviser to Argentine Government

Amartya K. Sen
Drummond Professor of Political Economy, Oxford University, UK

George Skorov
Institute of the USA and Canada, Academy of Sciences of the USSR, Moscow

Alice Thorner
Centre d'Etudes de l'Inde et de l'Asie du Sud, Ecole des Hautes Etudes en Sciences Sociales, Paris

Index

Accountability, associations 49, 50
Achievement, international comparisons 67
Advertising industry, TNCs 152
Advisory Service on Transfer of Technology 241
Advocacy activities 49
Afghanistan 79
Africa
  1980s, crises 183
  development coordination 187
  development strategy 159, 163
  external imbalance/import strangulation 175
  food supply 179
  income distribution and basic service access 181
  legitimacy of states 184
  participation 182
  Southern African liberation 186
Afro-Asian group 101
Agricultural development, India 128
Agriculture, GDP 167
Algeria 165
  income distribution 181
Andean Group, Transfer of technology 221
Andean Technological Information Service 223
Annexationist movement 144, 146
Appropriate technology 259
Asian development, cultural element 195
  diversity 196
  history 198, 199
  life styles 204
  new patterns of change 202
  religio-cultural values 209
  values 204
Asset-ownership 23
Associations, third system 47
Autarky 161, 260
Autocentered development 159

Balance-of-payments crises, Africa 175
Banks, US, transnationalization 140
Bargaining, North-South 97
Being and becoming, Asia 205

Bombay Plan 125
Brain drain 257, 283
Brazil 68
British American Tobacco Company 144

Canada 162
Capabilities, living standards 65
Capital accumulation 5, 7
Capital flows 114, 116, 234
Capitalism 3
  Africa 189
China
  growth rate 69
  poverty 22
  rise of 37
  world image 41
China option 22
Choice of Techniques Problem 13
Citizens and associations 47
Civil rights movement 56
CMEA trade 119
Code of Conduct on the Transfer of Technology 242, 271
COMECON 161
Collectivism, Orient 38
Commodity agreements 32
Commodity index 64
Commodity markets 147
Commodity prices 34
Common Fund 32
Common Market 113
Communications 57
Community Projects, India 127
Competition 4
  TNCs 144
Conglomerate power 146
Constant tastes 64
Consumers Association of Penang 50
Cooperation in science and technology 261
Credit restrictions 6
Credit, trade 118
Crisis 29
  Africa 183
Cross-subsidization 146
Cultural element in Asian development 195

Dentsu 153
Developing countries
  energy consumption 83
  harmonizing of interests 94
  role of 30
  transfer of technology 236
Development
  Asia
    cultural element 195
    values framework 207
  endogenous
    Africa 159
    science and society 253
  India 123
  since 1950s 213
  technology 229
  World Bank 5
Development coordination, Africa 187
Development economics 11
Development reactivation 31
Direct systems support 33
Disguised unemployment 14
Domestic savings 13, 19

Economic distance and living standards 63
Economic theory 12
  technology 230
Education
  developing countries 255
  international comparisons 68
EEC Commission 55
Egypt 165
Employment, India 129
Endogenous development 259
  Africa 159
Energy crisis 76
Energy demand, developing countries 83
Energy and international security 75
Energy sources 77
Entrepreneurial activity 38
Ethiopia 179
Ethnicity 189
European Economic Community 55, 112
Expansionary policies, centers 6
Expert processing zones 130
Exporting Promoting strategy 17
Export-led growth 111
  Africa 177
Export-pessimism 16
Exports and industrial growth 111
Exports, postwar growths 107
Extended Fund Facility 99

Factor mobility 110, 115, 116
Far Eastern Four, Exporting Promoting strategy 17

Financial supermarket 149
Fiscal reform, US 7
Food cropping agriculture 168
Food supply, Africa 179
Foreign aid 14, 19
Free market forces 215
Free trade zones 147
French third estate 53
Front Line States 186, 188
Ful-Tex Euro Services Texaco 152

Game theory, North-South bargains 97
Gandhi, I. 135
Gandhi, M. 124, 127
GATT 106, 110
Generalized System of Preferences 34
Global corporations 143
GNP
  China 69
  India 69
  international comparisons 68
Green Revolution 21
Group 77, 85, 94, 99, 101
Growth
  autocentered 167, 168
  China and India 69, 70
  inequality 213
  recovery following recession 31
  technology 231
  world trade in manufactures 105
Growth models 13
Growth rates 18

Haberler Committee 87
Harrod-Domar model 13
Health expenditures 23
Hechscher-Ohlin-Samuelson 110
Hidden resources 14
Hirschman, A. 14
Hindu 210
History and development, Asia 199
Human rights 209

Ideology, Asia 201, 211
IMF Compensatory Financing Facility 99
Import demand 29
Import strangulation, Africa 176
Import substitution 8, 13, 18
Income distribution 15, 21
  Africa 181
Income inequality and growth 214
Income taxes 6
India
  administrative machinery since independence 133

basic capabilities 68, 69
development of the idea of development 123
Green Revolution 21
Group 77, 94
technological capability 197
what's missing? 136
Industrial consultancy services 255
Industrial growth - exports 111
Industrialization 3
Africa 165
Inflation
TNCs 154
US 6
Insurance, TNCs 152
Interdependence 30
Interest rates 6
international trade 118
Intermediate goods, export growths 107
International Development Association 99
International development strategy 87, 89
International Labor Organization 55
International monetary system 33
International security and energy 75
International technology transfer 220
International trading system 34
International Workingmen's Association 56
Internationalization 139
Latin America 225
Intra-firm trade 116
Investment 12, 215
transnationals 116
Investment Allocation Problem 13
Iran 81

Japan
capital accumulation 7
import of technology 221
rise of 37
world image 41
Ziabatsus 144
Japanese-Chinese cooperation 39

Knowing and knowledge, Asia 208
Kreinin, M. 112

Land reform, India 128
Latin America 3
transfer of technology 220
Latin America Technological Information Network 224
Legitimacy of states 184
Lewis, W. A. 110
Liberation, Africa 186
Licencing 235
Living standards and economic distance 63

Mahalanobis Plan frame 136
Malawi 178
Manufactures
growth in world trade 105
periphery 4
TNC subsidiaries 147
Market forces 215
Mergers 146, 151
Mexico 68
Middle East energy resources 78
Military rule, Africa 183
Minimum incomes 15
Modernization, Asia 195, 199
anthropocentric and materialistic character 208
economic rationality 207
positivistic structure of knowledge 208

Namibia 186, 187
Nationalism, Africa 163
National science capacity 254
Nehru, J. 126
Neocolonialism 189
Networking 56
New International Economic Order 37, 101, 264
Nigeria 165
Nigeria-Benin 188
Nongovernmental organizations 47, 55
North-South negotiations 85
bargaining 97
format 101
formulation of country positions 92
formulation of regional positions 93
Group 77 94, 99, 101
North-West decline 40
Nurkse, R. 14

Occident and orient 43
Oil access and national security 76
Oil crisis 75, 76
Oil demand by developing countries 83
OPEC 77, 97
Open-market policy 8
Optimal Savings Problem 13
Opulence 64

Paris Convention for the Protection of Industrial Property 242, 243
Partial orderings 65
Participation 216
Africa 183
Patel, S. bibliography of publications 1945-1983 267
Patents 240, 241

Peace organizations 49
Periphery, role of 3, 30
'Plan' 52
Political economy of power, TNCs 143
Portfolio investment 234
Positive-sum game negotiations 97
Poverty 14, 215
  China 22
  India 124
Power of the people 47
Price incentives 16
Primary commodity trade 147
Primary education 23
Primitive Problem of Accumulation 13
Product cycle 233
Production function 231
Productivity fall at centers 5
Productivity, Japan 40
Progressive taxing of expenditures 7
Protection of Industrial Property 242
Pull up strategy 15

Quality of life 71

Real income 64
Recession 1980-1983 29
Redistribution 22, 23
Regional positions 93
Religio-cultural value system and development 209, 210
Research and development, TNCs 245
Resource flows 117
Reverse transfer of technology 257, 272, 276, 283
Rosenstein-Rodan, P. 12
Royalty payments 220
Rural productivity 167

Scarce resources 166
Science capability in developing countries 254
Scientific revolution and development 253
Sears Roebuck 150
Sectoral mix of workforce, Asia 202
Security and energy access 75
Self-reliance 260
Senegal 179
Services
  Africa 182
  TNCs 149
Shipping, transport of manufactured goods 118
Socialism and development, Africa 189
Social structure and development 253
Socioeconomic organization, Asia 203
South Africa, dependence on 188
Southern African Development Coordination 188
Southern African liberation 186
South-East triangle, rise of 37
South-South negotiations 100
Soviet Union
  energy planning 79
  ties with Japan 43
Spain 162
Special Drawing Rights 5
Special Fund, UN 99
Sri Lanka 68
Stability 216
Stagflation 29
Standard of living 65
Steel industry, Africa 166
Supplementary Financing Facility 99
Sweden, capital accumulation 7

Tanzania
  food supply 180
  import reduction 177
  income distribution 181
Tariff/non-tariff wall 42
Tariff reductions 106, 110, 111, 118
Technological gap - product cycle 233
Technological Information Exchange System 224
Technological Revolution, India 137
Technology
  concepts 229, 237
  economic theory 230
  markets for 238
  transfer to developing countries 236
  UNCTADs major studies and reports 281
Technology transfer
  Africa 190
  code of conduct 242, 271
  cost 238
  developing countries 236
  Latin America 219
  research and development 258
  theory and policy 229
  UNCTAD 271, 276
Texaco 152
Textiles, TNCs 148
Third system 47
'Third World' 51
Tinbergen, J. 87
Tokyo-Beijing axis 39
Trade
  centre-periphery 4
  engine of growth 87
  international system 34
Trade credit 118
Trade embargo 42

Trade flows, EEC 113
Trade growth in manufactured goods 105
Trade and technology 232
Trade theory 232
Trademarks 240, 241
Tradition and development 199
Transfer of technology
  Africa 190
  code of conduct 242, 271
  costs 238
  developing countries 236
  governmental meetings in UNCTAD 1971-1984 271
  Latin America 190
  research and development capability of developing countries 258
  reverse 257, 272, 276, 283
  theory and policy 229
  UNCTAD decisions 276
  UNCTAD reports 281
Transfer pricing 117
Transnational corporations 139
  Africa 190
  factor mobility 116
  insurance 152
  Japan 144
  major sectors 146
  manufactures 147
  political economy of power 143
  research and development in developing countries 257
  services 149
  transfer of technology 220, 234
Transnational integral conglomerates 149
Transnationalization 139
Transnational service conglomerates 149
Transportation costs 106, 118
Trickle down 30, 215

UNCTAD
  governmental meetings on technological issues 1971-1984 271
  Group 77 101
  major studies and reports on technology 281
  regional and subregional integration, Latin America 96
  resolutions, decisions and agreed conclusions on work in technology 1968-1984 275
  technology and development 229
  technology transfer 236
United Nations 54
Urbanization, India 129
US
  energy security 81
  fiscal reform 7
Utility, opulence 64

Value added 108, 114
Value systems, Asia 195, 204
Vernon, R. 233
Vienna Program of Action on Science and Technology for Development 219

Welfare economics 63
WHO/UNICEF 55
Women's liberation organizations 50
World Bank 5, 33
World economy, Africa 191
World Intellectual Property Organization 242
World trade growth 105

Ziabatsu 144